S0-CGT-851

ARCTIC
AND TROPICAL ARBOVIRUSES

Academic Press Rapid Manuscript Reproduction

Proceedings of the Second International Symposium on Arctic Arboviruses Held at Mont Gabriel, Canada, May 26–28, 1977

ARCTIC AND TROPICAL ARBOVIRUSES

edited by

EDOUARD KURSTAK

Comparative Virology Research Group
Faculty of Medicine
Université de Montréal
Montreal, Quebec, Canada

ACADEMIC PRESS New York San Francisco London 1979
A Subsidiary of Harcourt Brace Jovanovich, Publishers

ACADEMIC PRESS, INC.
111 Fifth Avenue, New York, New York 10003

United Kingdom Edition published by
ACADEMIC PRESS, INC. (LONDON) LTD.
24/28 Oval Road, London NW1 7DX

Library of Congress Cataloging in Publication Data

International Symposium on Arctic Arboviruses, 2d, Mont Gabriel, Québec, 1977.
Arctic and tropical arboviruses.

Organized . . . by the International Committee on Arctic Arboviruses and the Université de Montréal, under the aegis of the World Health Organization.''
1. Arthropod-borne viruses—Congresses. 2. Arbovirus diseases—Arctic regions—Congresses. 3. Arbovirus diseases—Tropics—Congresses. I. Kurstak, Edouard. II. International Committee on Arctic Arboviruses. III. Université de Montréal. IV. Title.
[DNLM: 1. Arctic regions—Congresses. 2. Tropical climate—Congresses. 3. Arboviruses—Congresses. 4. Arbovirus infections—Congresses. W3 IN916AF 2d 1977a / QW168.5.A7 I61 1977a]
QR201.A72I67 1977 576'.6484 78-11054
ISBN 0-12-429765-X

PRINTED IN THE UNITED STATES OF AMERICA

79 80 81 82 9 8 7 6 5 4 3 2 1

Contents

List of Contributors

Numbers in parentheses indicate the pages on which authors' contributions begin.

R. Ackermann (173), Department of Virology, Neurology Clinic, University of Cologne, 5 Köln 41, Federal Republic of Germany

V. A. Aristova (21), D. I. Ivanovsky Institute of Virology, Academy of Medical Sciences, Moscow, USSR

Harvey Artsob (39), National Arbovirus Reference Service, Department of Medical Microbiology, Faculty of Medicine, University of Toronto, Ontario, Canada

David M. Asher (179), Laboratory of Central Nervous System Studies, National Institute of Neurological and Communicative Disorders and Stroke, National Institutes of Health, Bethesda, Maryland 20014

M. Balducci (101), Institute of Hygiene, University of Rome, Rome, Italy

L. K. Berezina (21), D. I. Ivanovsky Institute of Virology, Academy of Medical Sciences, Moscow, USSR

U. K. M. Bhat (263), National Institute of Allergy and Infectious Diseases, National Institutes of Health, Rocky Mountain Laboratory, Hamilton, Montana 59840

James P. Bruen (157), University of California, School of Public Health, Berkeley, California 94720

M. Brummer-Korvenkontio (197), Department of Virology, University of Helsinki, Helsinki, Finland

Jordi Casals (173, 303), Department of Epidemiology and Public Health, Yale University School of Medicine, New Haven Connecticut 06510

S. P. Chunikhin (297), Institute of Poliomyelitis and Virus Encephalitides, USSR Academy of Medical Sciences, Moscow, USSR

Carleton M. Clifford (83), National Institute of Allergy and Infectious Diseases, National Institutes of Health, Rocky Mountain Laboratory, Hamilton, Montana 59840

J. Cory (211), National Institute of Allergy and Infectious Diseases, National Institutes of Health, Rocky Mountain Laboratory, Hamilton, Montana 59840

V. Danielová (173), Institute of Parasitology, Czechoslovak Academy of Sciences, Prague, Czechoslovakia

T. I. Dzhivanyan (297), Institute of Poliomyelitis and Virus Encephalitides, USSR Academy of Medical Sciences, Moscow, USSR

K. B. Fomina (21), D. I. Ivanovsky Institute of Virology, Academy of Medical Sciences, Moscow, USSR

Y. P. Gofman (21), D. I. Ivanovsky Institute of Virology, Academy of Medical Sciences, Moscow, USSR

V. L. Gromashevski (21), D. I. Ivanovsky Institute of Virology, Academy of Medical Sciences, Moscow, USSR

James L. Hardy (157), University of California, School of Public Health, Berkeley, California 94720

Keith Harrap (277), Department of Microbiology, College of Medicine and Dentistry of New Jersey, Rutgers Medical School, Piscataway, New Jersey 08854

T. Hovi (197), Department of Virology, University of Helsinki, Helsinki, Finland

N. V. Khutoretskaya, (21), D. I. Ivanovsky Institute of Virology, Academy of Medical Sciences, Moscow, USSR

S. M. Klimenko (21), D. I. Ivanovsky Institute of Virology. Academy of Medical Sciences, Moscow, USSR

N. G. Kondrashina (21), D. I. Ivanovsky Institute of Virology, Academy of Medical Sciences, Moscow, USSR

Edouard Kurstak (1), Comparative Virology Research Group, Faculty of Medicine, Department of Microbiology and Immunology, Université de Montréal, Montreal, Quebec, Canada

J. Ländevirta (197), Central Hospital of Savonlinna and the University of Helsinki, Helsinki, Finland

C. J. Leake (245), Department of Entomology, London School of Hygiene and Tropical Medicine, London WCIE 7HT, England

M. C. Lopes (101), Institute of Hygiene, University of Rome, Rome, Italy

D. K. Lvov (21), D. I. Ivanovsky Institute of Virology, Academy of Medical Sciences, Moscow, USSR

D. M. McLean (7), University of British Columbia, Department of Medical Microbiology, Vancouver, B.C., Canada

A. S. Novokhatski (21), D. I. Ivanovsky Institute of Virology, Academy of Medical Sciences, Moscow, USSR

N. Oker-Blom (197), Department of Virology, University of Helsinki, Helsinki, Finland

K. Penttinen (197), Department of Virology, University of Helsinki, Helsinki, Finland

Ralf F. Pettersson (231), Massachusetts Institute of Technology, Center for Cancer Research, 77 Massachusetts Avenue, Cambridge, Massachusetts 02139

S. B. Presser (157), University of California, School of Public Health, Berkeley, California 94720

Mary Pudney (245), Department of Entomology, London School of Hygiene and Tropical Medicine, London WCIE 7HT, England

William C. Reeves (157), University of California, School of Public Health, Berkeley, California 94720

E. Rehse (173), Department of Epidemiology and Public Health, Arbovirus Research Unit, School of Medicine, Yale University, New Haven Connecticut 06510

R. Rehse-Küpper (173), Department of Virology, Neurology Clinic, University of Cologne, 5 Köln 41, Federal Republic of Germany

P. Saikku (197), Department of Virology, University of Helsinki, Helsinki, Finland

B. G. Sarkisyan (21), D. I. Ivanovsky Institute of Virology, Academy of Medical Sciences, Moscow, USSR

Nava Sarver (277), Department of Microbiology, College of Medicine and Dentistry of New Jersey, Rutgers Medical School, Piscataway, New Jersey 08854

A. A. Sazonov (21), D. I. Ivanovsky Institute of Virology, Academy of Medical Sciences, Moscow, USSR

T. M. Skvortsova (21), D. I. Ivanovsky Institute of Virology, Academy of Medical Sciences, Moscow, USSR

Leslie Spence (39), National Arbovirus Reference Service, Department of Medical Microbiology, Faculty of Medicine, University of Toronto, Ontario, Canada

Victor Stollar (277), Department of Microbiology, College of Medicine and Dentistry of New Jersey, Rutgers Medical School, Piscataway, New Jersey, 08854

L. A. Thomas (211), National Institute of Allergy and Infectious Diseases, National Institutes of Health, Rocky Mountain Laboratory, Hamilton, Montana 598

Virginia Thomas (277), Department of Microbiology, College of Medicine and Dentistry of New Jersey, Rutgers Medical School, Piscataway, New Jersey 08854

Wayne H. Thompson (139), Department of Preventive Medicine, University of Wisconsin, Center for Health Sciences, 465 Henry Hall, Madison, Wisconsin 53706

T. Traavik (67), Institute of Medical Biology, University of Tromsø, N-9001 Tromsø, Norway

A. Vaheri (197), Department of Virology, University of Helsinki, Helsinki, Finland

M. G. R. Varma (245), Department of Entomology, London School of Hygiene and Tropical Medicine, London WCIE 7HT, England

P. Verani (101), Institute of Hygiene, University of Rome, Rome, Italy

C.-H. von Bonsdorff (197), Department of Virology, University of Helsinki, Helsinki, Finland

John P. Woodall (123), San Juan Laboratories, Bureau of Laboratories, Center for Disease Control, Public Health Service, U.S. Department of Health, Education, and Welfare, GPO Box 4532, San Juan, Puerto Rico 00936

C. E. Yunker (211, 263), National Institute of Allergy and Infectious Diseases, National Institutes of Health, Rocky Mountain Laboratory, Hamilton, Montana 59840

V. M. Zhdanov (21), D. I. Ivanovsky Institute of Virology, Academy of Medical Sciences, Moscow, USSR

Preface

In tropical and subtropical regions arboviruses cause severe and often fatal diseases. For a long time, these viruses have gained much attention and have been the subject of intense studies. The constant presence of arboviruses in these regions is directly related to the hot and humid climate that favors the rapid multiplication of arthropods, the vectors of these viruses. These biological vectors, namely, mosquitoes and ticks, also exist in the northern polar regions where, in spite of the cold temperatures for several months of the year, the short, hot, and humid summers make possible the proliferation of arthropods and their intense activity. Recently, it was also observed that in the northern regions arboviruses can be transmitted to man and animals by arthropods.

Studies on arctic arboviruses were necessitated by the increasing influx of man in the subarctic and arctic regions through hunting and fishing expeditions and as a result of the exploration for natural resources, particularly in Canada, the Soviet Union, and the United States (Alaska). In addition to the research projects on arboviruses in progress at present in several northern countries, the role of these northern territories in the epidemiology of the influenza viruses and in the dissemination of the rabies virus must be elucidated. It has been suggested that migratory birds could play a role in the transmission of influenza virus from the north to the south and that the arctic fox could be the vector for the rabies virus.

Taking these facts into consideration, the International Committee on Arctic Arboviruses (ICAA), which was founded in 1976, decided to expand its activity and change its name to the International Committee on Polar Viruses (ICPV). This new name justifies the desire of the membership to extend its scientific activities in the northern and southern polar regions. This new orientation was approved by the participants at the Second International Symposium on Arctic Arboviruses, held on May 26–28, 1977 at Mont Gabriel (Quebec), Canada.

This second symposium on arctic arboviruses was presided over by Professors E. Kurstak and D. M. McLean, and was a followup of the Helsinki symposium organized in 1975 by Professor N. Oker-Blom and his collaborators at the University of Helsinki. The second symposium was organized at Mont Gabriel by the International Committee on Arctic Arboviruses and the Université de Montréal under the aegis of the World Health Organization.

At this symposium, entitled "Comparison between Arboviruses from Arctic and Tropical Regions," it was decided that the ICPV would organize symposia at three-year intervals and edit the communications for publication. The general secretariat of the ICPV was established at the Université de Montréal with the consent of the University authorities and the members of ICPV. Canada, Finland, the United States, the Soviet Union, and the World Health Organization are the elected members of ICPV. Council members are Argentina, Australia, Norway, and New Zealand. The elected members of the ICPV are Professors E. Kurstak and D. M. McLean (Canada), N. Oker-Blom and M. Brummer-Korvenkontio (Finland), D. K. Lvov and B. F. Semenov (Soviet Union), J. Casals and C. E. Yunker (United States), and P. Brès (World Health Organization).

The principal communications presented at the symposium at Mont Gabriel were subsequently prepared for this volume. Its title "Arctic and Tropical Arboviruses" was chosen to designate the geographical scope of the epidemiological problem of arbovirus infections in man and animals. Several chapters are devoted to the recent investigations on arboviruses in the northern regions and on their vectors, mosquitoes, and ticks, as well as to the detection in the north of arboviruses originally isolated in the south. This bipolar distribution of arboviruses could be the result of the transport of arbovirus-infected ticks by migratory birds. The migration of birds from the arctic toward the antarctic and vice versa, passing through the intermediary regions, creates an ecological–epidemiological problem of great interest. A few chapters included in this book describe diseases the etiological agent of which could possibly be an arbovirus, such as in the case of nephropatia epidemica, rapid diagnostic techniques for the detection of arboviruses, and *in vitro* culture methods for arboviruses using arthropod cells.

With this book we aim to stimulate the interest of researchers not only in arboviruses but also in other viruses which may be a source of infection in polar regions. It is our hope that this volume will provide a useful tool for all concerned with viral diseases: virologists, epidemiologists, and ecologists.

The development of arctic and subarctic regions should not proceed without detailed studies on the medical and veterinary problems these viruses could cause. Those involved in the development of polar regions should provide the financial support for research projects on these viruses, their vectors, and the possible infections they may induce.

I wish to express my sincere gratitude to the contributors for the effort and care with which they prepared their chapters; to the members of the Committee and to the authorities of the Université de Montréal, in particular the Vice-Rector of Research, Mr. Maurice L'Abbé and the Dean of the Faculty of Medicine, Dr. Pierre Bois, for their aid in the organization of the symposium. We also thank the International Union of Biological Societies for their financial help and the staff of Academic Press for their part in the production of this volume.

Edouard Kurstak

A few participants at the Second International Symposium on Arctic Arboviruses (Mont Gabriel, Canada).

Seated from left to right: M. Brummer-Korvenkontio, D. K. Lvov, N. Oker-Blom, D. C. Gajdusek, E. Kurstak, J. Casals, D. M. McLean, C. E. Yunker, B. F. Semenov.

CHAPTER 1

CONSIDERATIONS ON ARBOVIRUS INFECTIONS IN NORTHERN REGIONS

Edouard Kurstak

I. INTRODUCTION

Arboviruses in arctic and subarctic regions as well as other viruses infecting animals and man have received little attention until now. The recent development of northern regions in several countries creates the need for urgent research on this type of viruses and the diseases they cause.

The problem of arboviruses of northern regions was raised during a symposium on arboviruses in Helsinki in 1975. The following year an International Committee on Arctic Arboviruses (ICAA) was created, including Canada, Finland, United States, USSR, and the World Health Organization (WHO). The University of Montreal houses the general secretariat of this committee. In order to promote research on these viruses, the ICAA organized the Second International Symposium on Arctic Arboviruses at Mont Gabriel, Canada, May 26-28, 1977. The theme of this symposium was the comparison between arboviruses from arctic and tropical regions (Kurstak, 1977). At the second symposium, it was decided to change the name of the ICAA to the International Committee on Polar Viruses (ICPV).

ISBN 0-12-429765-X

The objectives of the ICPV are the coordination of investigations on the ecology, epidemiology, molecular biology, diagnosis, and control of mosquito-borne and tick-borne viruses that may induce encephalitis or other febrile illnesses affecting man and animals in circumpolar zones. Other viral diseases in these zones, namely, influenza are also considered. The transmission of influenza viruses by wild birds from northern to southern regions is of importance and should be investigated urgently as an ecological and epidemiological problem. Another problem of importance in northern zones is caused by arctic foxes as reservoir of rabies virus. The ICPV will take the necessary steps to establish a bank of information on these viruses.

II. ISOLATION OF ARBOVIRUSES

In the Canadian arctic and subarctic regions (north of 60°N) California encephalitis virus (CEV), snowshoe hare subtype, was first isolated from *Aedes canadensis* mosquitoes (McLean *et al.*, 1970). Subsequently, CEV has been repeatedly isolated (McLean *et al.*, 1970; McLean, 1978) from *Aedes communis* and three other *Aedes* in the north boreal forest zones (66°N, 138°W), and the arctic tidewater region of Inuvik (69°N, 135°W). Two of these species are *Aedes hexodontus* and *Aedes punctor*. It appears that small mammals, particularly snowshoe hares, *Lepus americanus*, arctic ground squirrels, *Citellus undulatus*, and red squirrels, *Tamiasciurus hudsonicus*, are reservoirs of CEV. The natural cycle of infection by this bunyavirus during successive summers involving *Aedes* mosquitoes as vectors and small mammals as reservoirs occurs from the boreal forests through the open woodlands to the tundra regions. The situation is similar in Alaska where the snowshoe hare subtype of CEV was repeatedly isolated (McLean, 1978), as well as Northway virus isolated in Alaska (Calisher *et al.*, 1974) and in Canada (McLean *et al.*, 1977; Artsob and Spence, 1978). In addition, the isolation in the North Bay region of Canada of Powassan virus from *Ixodes marxi* ticks and high level of Powassan antibody in red squirrels suggested the extension of the distribution of this virus from southern to northern regions (Rossier *et al.*, 1974; McLean and Bryce-Larke, 1963).

Since 1969, in the northern zones of the European and of Far Eastern USSR, 241 strains of Tyuleniy virus (Togaviridae, Flaviviruses) and three Uukuniemi group viruses (Bunyaviridae) were isolated from *Ixodes putus* and *Ixodes signatus* ticks, the main parasites of seabirds (Lvov, 1978). It appears that *I. putus* ticks are the major vectors of arboviruses. As many as

236 virus strains were isolated from *I. putus* (Zaliv Terpeniya virus-85, Tyuleniy virus-53, Okhotskiy virus-47, Sakhalin virus-43, and Paramushir virus-8).

Tick-borne Uukuniemi and Humhinge viruses and mosquito-borne Inkoo (CE group) virus are prevalent in Scandinavia (South of 62°N). Traavik (1978) pointed out that in part of Norway situated at latitude 70-71°N CE arbovirus was isolated from *A. hexodontus,* and that the human potential pathogenicity of the virus was demonstrated.

III. CURRENT STATUS ON ARBOVIRUSES

A. Mosquito Vectors

The experimental injection of snowshoe hare subtype of CEV to *Culiseta inornata* mosquitoes provokes infection. Viral replication is detected after 27 days of incubation at 0°C or 6 days at 13°C. Demonstration of viral antigens and virions in the salivary glands of the infected mosquitoes was possible by direct immunoflurescence. In field conditions the infection rates for infected species of *A. canadensis* and *A. communis* are 1:503 and 1:2564, respectively. Neutralizing antibodies to the snowshoe hare subtype of CEV are detectable in 14% of mammals in the northern Yukon Territory of Canada. One of the means of overwintering of CEV in the Canadian Arctic is by transovarial transfer of the infection as suggested earlier. Experimental transmission of arboviruses has also been demonstrated by intrathoracic injection and feeding or arctic *A. communis* mosquitoes. Intracytoplasmic virions are demonstrated by immunoelectron microscopy. These data demonstrate their potential as natural vectors in polar regions (McLean *et al.*, 1970; McLean, 1978).

B. Tick Vectors

1. *Togaviridae-Flavivirus*

Tyuleniy virus, one of the important Togaviruses isolated from *Ixodes putus* ticks principally on Tyuleniy and Commodore Islands in USSR, is pathogenic for suckling mice after intracerebral injection and can also replicate in chick embryos and BHK-21 cells. These viruses can experimentally infect *Aedes aegypti* and *Culex molestus* mosquitoes and several species of birds. In man, Tyuleniy virus can cause a febrile illness after visit to colonies of birds. The infection rate of adults of

I. putus varies from 1:35 to 1:600. Serological data demonstrated that the natural reservoirs of Tyuleniy virus are birds, mainly Brünnich's guillemots, common murres, pelagic and red-faced cormorants, and fulmars. Detection of antibodies against this virus in fur seals, cows and man suggest their involvement in the dissemination of Tyuleniy virus (Lvov, 1978).

2. *Bunyaviridae-Uukuniemi and Sakhalin Groups*

Two viruses of these groups were isolated at high latitudes in the USSR: Zaliv Terperniya virus (Uukuniemi group) and Sakhalin virus (Sakhalin group), both *I. putus* ticks. Another Bunyavirus (ungrouped) isolated is Paramushir virus, which infects *I. putus* and *I. signatus* ticks. The infection rates of ticks range from 1:50 to 1:10,000 (Lvov, 1978).

Zaliv Terpeniya virus can infect experimentally suckling mice by intracerebral inoculation and chick and duck embryo cell cultures. Under natural conditions rodents can be infected by the ingestion of tick and dead bodies of infected birds.

Sakhalin virus as Zaliv Terpeniya virus infects suckling mice and can replicate in primary duck and human embryo cell cultures, as well as in BHK-21, HeLa, and Vero cell lines. Viruses related to Sakhalin virus were isolated on Gull Island (Alaska) and on the Newfoundland coast (Canada). The presence of antibodies against viruses related to Sakhalin virus in several species of birds in the Far East part of the Soviet Union suggests their important role in the dissemination of the virus.

Paramushir virus, isolated on Tyuleniy and Commodore Islands, present several properties similar to Bunyaviruses. Among the laboratory animals only suckline mice are sensitive to the virus by intracerebral inoculation. Replication of the virus with cytopathogenic effect (CPE) occurs in L-cells, HeLa, and pig embryo kidney cell cultures. Without CPE, replication can be detected in chick, duck, human embryo fibroblast cultures, and in BHK-21 cells (Lvov, 1978; Lvov *et al.*, 1973; Ritter and Felz, 1974).

IV. CONCLUSIONS

The role and importance of arctic and subarctic arboviruses, as well as influenza and rabies viruses, in human and veterinary medicine is not sufficiently known. The ICPV recommends generating new research on all the aspects of these viruses and the infections they can cause in circumpolar regions, which hitherto have been insufficiently investigated. This new knowledge will contribute directly to the upgrading of public health measures in polar and subpolar regions, thereby improving the work output and general well-being of human beings both settled and transient in these developing areas, through the prevention of virus infections.

Acknowledgments

Work on this chapter was sponsored by the Ministry of Indian and Northern Affairs of Canada and by the Université de Montréal.

References

Artsob, H., and Spence, L. (1978). *In* "Arctic and Tropical Arboviruses" (E. Kurstak, ed.). Academic Press, New York.

Calisher, C. H., Lindsey, H. S., Ritter, D. G., and Sommerman, K. M. (1974). *Can. J. Microbiol. 20,* 219-223.

Kurstak, E. (1977). "Considerations on Arctic Arboviruses." *Int. Symp. Microbial Ecol.* Dunedin, New Zealand.

McLean, D. M. (1978). *In* "Arctic and Tropical Arboviruses" (E. Kurstak, ed.). Academic Press, New York.

McLean, D. M., and Bryce-Larke, R. P. (1963). *Can. Med. Assoc. J. 88,* 182-185.

McLean, D. M., Crawford, M. A., Ladyman, S. R., Peers, R. R., and Purvin-Good, K. W. (1970). *Am. J. Epidemiol. 92,* 266-272.

McLean, D. M., Grass, P. N., Judd, B. D., Cmiralova, D., and Stuart, K. M. (1977). *Can. J. Publ. Health 68,* 61-73.

Lvov, D. K. (1978). *In* "Arctic and Tropical Arboviruses" (E. Kurstak, ed.). Academic Press, New York

Lvov, D. K., Timopheera, A. A., Gromashevsky, V. L., Gostinshchikova, G. V., Veselovskaya, O. V., Chervonsky, V. I., Fomina, K. B., Gromov, A. I., Progrebenko, A. G., and Zhermer, V. Y. (1973). *Arch. Ges. Virusforsch. 41,* 165-169.

Ritter, D. G., and Feltz, E. (1974). *Can. J. Microbiol. 20,* 1359-1366.

Rossier, E., Harrisson, R. J., and Lemieux, B. (1974). *Can. Med. Assoc. J. 110,* 1173-1175.

Traavik, T. (1978). *In* "Arctic and Tropical Arboviruses" (E. Kurstak, ed.). Academic Press, New York

Chapter 2

ARBOVIRUS VECTORS IN THE CANADIAN ARCTIC

D. M. McLean

1. OVERVIEW

Mosquito-borne arboviruses in Canada are distributed in nature principally according to vegetation zones. Thus the two agents with high encephalitis-inducing incidence, St. Louis (SLE) (Flavivirus) and western equine encephalomyelitis (WEE) (Alphavirus) are found south of latitude 53°N in cultivated grassland and parkland in western Canada or cultivated and irrigated portions of the southeastern mixed forest zone of southern Ontario (McLean, 1975). In the boreal forest regions north of 53°N on the prairies, or north of 45°N in Ontario, the main mosquito-borne agent is the snowshoe hare subtype of California encephalitis (CE) virus (McLean, 1975; McLintock and Iversen, 1975) but Northway (NOR) virus was also identified recently (McLean *et al.*, 1977a). Both of these are Bunyaviruses, whose prevalence extends northward from the boreal forest through open woodlands to the tundra region.

California encephalitis virus (snowshoe hare subtype) was first isolated in Canada from blood of sentinel rabbits near Ottawa, Ontario (45°N, 75°W) during summer 1963 (McKiel *et al.*, 1966) by which time its prevalence was also recognized through

ISBN 0-12-429765-X

serological surveys of snowshoe hares (*Lepus americanus*) collected near North Bay, Ontario 46°N, 79°W) (McLean *et al.*, 1964), Manitoulin Island, Ontario (46°N, 82°W), Kamloops, B. C. (51°N, 120°W) (Newhouse *et al.*, 1963). Extensive ecological investigations in the boreal forest at Rochester, Alberta (54°N, 113°W) showed clearly the presence of a natural cycle of infection during successive summers from 1964 to 1968 involving *Aedes* mosquitoes, principally *A. communis* as vectors and small mammals, especially *Lepus americanus* as reservoirs (Iversen *et al.*, 1969). Virus was first isolated from mosquitoes in British Columbia near Penticton (49°N, 120°W) in 1969 and in Saskatchewan near Lac La Ronge (55°N, 105°W) during 1972 (Iversen *et al.*, 1973). In subarctic western Canada north of 60 N, CE virus was first isolated from *A. canadensis* mosquitoes collected at Marsh Lake (61°N, 134°W) near Whitehorse, Yukon Territory in June 1971 (McLean *et al.*, 1972), and it has been isolated repeatedly from *A. communis* and three other *Aedes* species in boreal forest areas as far north as mile 123 Dempster Highway (66°N, 138°W) (McLean *et al.*, 1975). In the Mackenzie District of the Northwest Territories, where the terrain is largely open woodland, CE virus has been isolated from *Aedes* mosquitoes collected at Fort Smith (60°N, 112°W) and Fort Simpson (62°N, 122°W) in the south, northwards to Arctic tidewater at Inuvik (69°N, 135°W) (McLean *et al.*, 1977a).

II. CURRENT FIELD STATUS

A. Yukon Territory

In the Yukon Territory during successive summers 1971 through 1975, isolation of 26 strains of the snowshoe hare subtype of CE virus were achieved from 41, 167 mosquitoes of five species that were processed in 910 pools (McLean *et al.*, 1977). Minimum field infection rates for the two commonly infected mosquito species were 1:503 *A. canadensis* and 1:2564 *A. communis*. At two important collection sites, virus isolations were achieved during three successive summers, 1971 to 1973 at Marsh Lake and 1972 to 1974 along the Dempster Highway (Table I).

Neutralizing antibody to the snowshoe hare subtype was detected in sera from 704 to 4913 (14%) mammals collected throughout the Yukon between 1971 and 1974, including 430 of 1076 (40%) *Lepus americanus* (snowshoe hares), 266 of 3610 (7%) *Citellus undulatus* (arctic ground squirrels), and 9 of 227 red squirrels (*Tamiasciurus hudsonicus*) (McLean *et al.*, 1975). Persistently high antibody rates in snowshoe hares collected at each site during successive summers suggests their importance as natural

TABLE I
California Encephalitis Virus Isolations from Yukon Mosquitoes, 1971-1974[a]

Location	*Year*	*Week no.*	*Species*	*Ratio*[b]	*MFIR*[c]
Marsh Lake	1971	26-29	A.canadensis	6/34	1:282
(61°N, 134°W)	1972	27-31	A.communis	4/40	1:668
	1973	28	A.canadensis	1/6	1:295
	1973	27	A.communis	1/20	1:117
Fish Lake	1971	29	A.canadensis	5/16	1:160
(61°N, 135°W)	1973	23	Cs.inornata	1/9	1:179
Mayo Road (61.5°N, 135°W)	1971	27	A.canadensis	1/6	1:650
Hunker Creek (64°N, 138°W)	1972	30	A.communis	1/21	1:1654
Dempster Hwy	1972	30	A.communis	1/19	1:2854
(66°N, 138°W)	1973	23	A.cinereus	1/5	1:370
	1974	29	A.communis	3/24	1:316
Carmacks (62°N, 135°W)	1974	23	A.canadensis	1/8	1:175

[a]*A total of 30,686 unengorged female mosquitoes were tested in 648 pools, of which 26 yielded virus.*
[b]*Ratio: number of mosquito pools that yielded virus/number of pools tested.*
[c]*MFIR: minimum field infection rate.*

reservoirs of infection, but the repeated detection of antibody in ground squirrels, even at sites where no snowshoe hares were collected, suggests that they also serve as natural reservoirs (Table II).

Attempts were also made to isolate CE virus from wild-caught larvae, which were processed in pools of about 100. During May and early June of 1974, CE virus was isolated from 1 of 84 pools of *Aedes sp.* larvae collected at Kusawa Lake, and from late April to early June 1975, CE virus was recovered from 2 of 218 pools of larvae, both of which were collected at Marsh Lake (McLean *et al.*, 1977). These findings strongly suggest transovarial transfer as an additional means of overwintering of CE virus in the Canadian Arctic, comparable to findings with other CE subtypes in Maryland (Le Duc *et al.*, 1975), Wisconsin *et al.*, 1974), and Minnesota (Balfour *et al.*, 1975) (Table III).

TABLE II
California Encephalitis Neutralizing Antibodies in Small Mammals, Yukon, 1971-1974

	Species			
Location	*La*[a]	*Cu*[b]	*Th*[c]	*Total*
Fish Lake 61°N, 135°W	76/268[d]	42/305	0/13	118/586
Mayo Road 61.5°N, 135°W	16/92	28/572	0/3	44/667
Carcross etc 60.5°N, 135°W	77/162	102/1422	6/120	185/1704
Marsh Lake 61°N, 134°W	155/337	39/429	3/68	197/834
Haines Jn etc 61°N, 138°W	58/129	38/723	0/17	96/869
Carmacks 62°N, 135°W	-	1/19	-	1/19
Hunker Ck 64°N, 138°W	46/86	2/51	0/4	48/141
Dempster Hwy 66°N, 138°W	2/2	14/89	0/2	16/93
Total				
No.	430/1076	266/3610	9/227	704/4913
%	40	7	4	14
All				
No.		705/4913		
%		14		

[a]*La: Lepus americanus.*
[b]*Cu: Citellus undulatus.*
[c]*Th: Tamiasciurus hudsonicus.*
[d]*Numerator, number of animals whose sera neutralized California encephalitis virus; denominator, number of animal sera tested.*

B. Mackenzie District of Northwest Territories

In the Mackenzie District of the Northwest Territories during summer 1976, eight isolations of CE virus (snowshoe hare subtype) and one isolation of Northway virus were achieved from 23,747 mosquitoes of five species, with all positive results

TABLE III
CE Virus Isolations from Yukon Larvae 1974-1976

Location		*1974*	*1975*	*1976*	*Total*
Kusawa Lake	(61°N, 136°W)	1/13	0/57	0/39	1/109
Marsh Lake	(61°N, 134°W)	0/13	2/67	0/8	2/88
Carcross	(60.5°N, 135°W)			0/23	0/23
Fish Lake	(61°N, 135°W)	0/7	0/16		0/23
Mayo Road	(61.5°N, 135°W)	0/4	0/52		0/56
Carmacks	(62°N, 135°W)	0/29	0/10	0/5	0/44
Hunker Creek	(64°N, 138°W)		0/16		0/16
Dempster Hwy	(66°N, 138°W)	0/18			0/18
Total		1/84	2/218	0/75	3/377

being obtained from *A. Hexodontus* or *A. punctor* collected between 10 and 24 July (McLean *et al.*, 1977a). At various localities minimum field infection rates for CE virus in *A. hexodontus* from 1:256 at Fort Simpson to 1:3662 at Inuvik, and also at Inuvik 1:3662 *A. punctor* yielded Northway virus. To date, insufficient mammalian sera have been collected in the Mackenzie District to suggest possible natural reservoirs (Table IV).

III. ARCTIC MOSQUITOES AS BUNYAVIRUS VECTORS

During successive summers, live arctic mosquitoes have been infected either after feeding on suspensions of the 74-Y-234 strain of CE virus in defibrinated blood or by intrathoracic injection of this and other Yukon CE isolates from larval and adult mosquitoes. For the 74-Y-234 strain, virtually identical results have been achieved with comparable doses of virus either without laboratory passage, or after one or occasionally two intracerebral mouse passages.

A. *Culiseta inornata*

After injection of *Culiseta inornata* mosquitoes with 0.1 to 100 mouse LD_{50} of first mouse passage 74-Y-234 virus, viral replication was detected by infectivity assays of mosquito salivary glands following 27 or more days of incubation at 0°C and after 6 or more days of incubation of 13°C (McLean *et al.*,

TABLE IV
Bunyavirus Isolations from Northwest Territories Mosquities 1976[a]

Location	*Week No.*	*Species*	*Sera type*	*Ratio*[b]	*MFIR*[c]
Fort Smith	29	*A.hexondontus*	CE	1/18	1:922
60^oN 112^oW	29	*A.punctor*	CE	1/17	1:911
Fort Simpson	29	*A.hexondontus*	CE	2/11	1:256
62^oN 122^oW	28	*A.punctor*	CE	1/33	1:1611
	28	*A.communis*	CE	2/106	1:2734
Inuvik	27	*A.hexodontus*	NOR	1/65	1:3662
69^oN 135^oW	27	*A.hexondontus*	CE	1/65	1:3662

[a]*A total of 23,747 unengorged female mosquitoes were tested in 475 pools, of which eight yielded CE virus (snowshoe hare subtype) and one yielded Northway virus.*

[b]*Ratio: number of mosquito pools that yielded virus/number of pools tested.*

[c]*MFIR: minimum field infection rate.*

1977a). Virus transmission was attained 27 days after injection with 1 LD_{50} and 96 days after injection with 10 or 100 LD_{50} when mosquitoes were incubated at 0^oC. Transmission was also demonstrated 20 days after injection with 10 LD_{50} and 27 days after injection with 0.1 LD_{50} when mosquities were incubated at 13^oC. Consistent demonstration of viral antigen in salivary glands by direct immunofluorescence was attained only with mosquitoes injected with 100 mouse LD_{50} (Table V).

B. *Aedes communis*

Virus transmission by *Aedes communis* has been demonstrated after injection of 0.01 to 100 mouse LD_{50} of first mouse passage 74-Y-234 virus when mosquitoes were incubated at 0 or 13^oC for 13 or 20 days and after injection of 1 or 10 mouse LD_{50} when mosquitoes were incubated at 23^oC (McLean *et al.*, 1977a). Immunofluorescence was less consistent for detection of virus antigen in salivary glands than infectivity assays in mice. Virus transmission has also been demonstrated by *A. communis* that were infected by feeding on 0.01, 1, 100, and 1000 mouse LD_{50} after incubation at 13^oC for 13 days, after incubation at 0^oC for 20 days after imbibing 100 mouse LD_{50}, and after holding at 23°C for 13 days after imbibing 100 mouse LD_{50}, and for 20 days after feeding on 0.1 and 0.01 mouse LD_{50} (Table VI).

When *A. communis* were fed 1 mouse LD_{50} unpassaged 74-Y-234 virus, transmission was observed after 13 and 20 days of incubation at 13°C and after 13 days incubation at 23°C (McLean *et al.*, 1977b). When 1 or 0.1 mouse LD_{50} virus was injected into *A. communis*, transmission was noted after 20 days incubation at 13 and 23°C (Table VII).

These results have confirmed and extended previous findings with the 74-Y-234 strain before and after one mouse passage, where viral replication was demonstrated regularly in salivary glands and thoraces of *A. communis* mosquitoes following intrathoracic injection of 0.1 mouse LD_{50} or higher virus doses and incubation at 0, 13, and 23°C (McLean *et al.*, 1976). However, virus antigen was demonstrated more reliably by direct immunoperoxidase reactions in mosquito salivary glands than by direct immunofluorescence. Virons with particle diameters 45-53 nm surrounded by electron-dense peroxidase granules were demonstrated intracytoplasmically in thin sections of salivary glands of *A. communis* after 21 days of incubation at 13°C when glands were treated with CE antibody conjugated to horseradish peroxidase before fixation and osmium staining.

A. communis mosquitoes have also supported the replication of the prototype Northway virus in its third mouse passage following intrathoracic injection of 100 mouse LD_{50} and incubation at 13 and 23°C for 7 and 14 days, and after injection of 10 mouse LD_{50} and incubation at 13°C for 14 and 21 days (McLean *et al.*, 1976). Virons 89 nm diameter, which are morphologically typical of Bunyaviruses, were visualized in sections of cell pellets of BHK-21 cells infected with NOR virus (McLean *et al.*, 1978.

Murray Valley encephalitis virus in its fifteenth mouse passage multiplied in *A. communis* mosquito salivary glands after injection of 0.03 to 300 mouse LD_{50} and incubation for 7, 14, or 21 days at 0, 13, and 23°C. Virus antigen was detected regularly in salivary glands of infected mosquitoes by the direct immunoperoxidase test after mosquitoes received 3 mouse LD_{50} or higher virus doses (McLean *et al.*, 1976).

IV. CONCLUSIONS

Isolation of CE virus from six *Aedes* mosquito species collected throughout boreal forest and open woodland portions of the western Canadian Arctic during successive summers points strongly toward their importance as summertime vectors. Transmission of arctic mosquito CE isolates, including an unpassaged strain, by arctic *A. communis* mosquitoes incubated at tempera-

TABLE V

Transmission of Mosquito CE Isolate 74-Y-234 by C. inornata Mosquitoes[a]

Dose mouse LD_{50}		0°C					13°C				
		6	13	20	27	96	6	13	20	27	96
100	F	$\frac{2}{2}$	$\frac{1}{2}$	$\frac{2}{4}$		$\frac{1}{2}$ tr $\frac{2}{2}$	$\frac{4}{4}$	$\frac{3}{4}$	$\frac{6}{6}$ tr $\frac{1}{1}$		$\frac{1}{1}$
INJ	SG	$\frac{1}{2}$(2.0)	$\frac{0}{2}$	$\frac{3}{4}$(2.0)		$\frac{2}{2}$(2.0)	$\frac{4}{4}$(1.8)	$\frac{4}{4}$(2.5)	$\frac{6}{6}$(3.0)		$\frac{1}{1}$(2.5)
	TH	$\frac{2}{2}$(2.0)	$\frac{0}{2}$	$\frac{4}{4}$(3.0)		$\frac{2}{2}$(2.5)	$\frac{4}{4}$(2.0)	$\frac{4}{4}$(2.5)	$\frac{6}{6}$(3.0)		$\frac{1}{1}$(3.5)
10	F	$\frac{1}{2}$	$\frac{2}{4}$ tr $\frac{0}{1}$	$\frac{2}{6}$	$\frac{0}{2}$	$\frac{0}{2}$ tr $\frac{1}{1}$	$\frac{0}{2}$	$\frac{2}{4}$	$\frac{0}{6}$ tr $\frac{3}{3}$	$\frac{0}{2}$	
INJ	SG	$\frac{2}{2}$(1.8)	$\frac{1}{4}$(1.8)	$\frac{4}{6}$(2.5)	$\frac{2}{2}$(1.8)	$\frac{2}{2}$(3.0)	$\frac{2}{2}$(2.0)	$\frac{4}{4}$(2.0)	$\frac{6}{6}$(3.5)	$\frac{2}{2}$(3.5)	
	TH	$\frac{2}{2}$(1.8)	$\frac{4}{4}$(1.5)	$\frac{5}{6}$(3.2)	$\frac{2}{2}$(1.8)	$\frac{2}{2}$(3.2)	$\frac{2}{2}$(2.0)	$\frac{4}{4}$(2.0)	$\frac{6}{6}$(3.2)	$\frac{2}{2}$(2.5)	
1	F	$\frac{0}{2}$	$\frac{2}{2}$	$\frac{0}{2}$	$\frac{0}{1}$ tr $\frac{1}{1}$		$\frac{0}{2}$	$\frac{2}{2}$	$\frac{0}{2}$	$\frac{0}{1}$	

INJ SG	$\frac{2}{2}$(1.8)	$\frac{0}{2}$	$\frac{0}{2}$	$\frac{1}{1}$(2.0)		$\frac{2}{2}$(2.0)	$\frac{2}{2}$(1.8)	$\frac{1}{2}$(2.8)	$\frac{1}{1}$(2.0)
TH	$\frac{2}{2}$(1,8)	$\frac{0}{2}$	$\frac{0}{2}$	$\frac{1}{1}$(2.8)		$\frac{2}{2}$(2.0)	$\frac{2}{2}$(2.0)	$\frac{2}{2}$(2.0)	$\frac{1}{1}$(2.8)
0.1 F	$\frac{0}{2}$	$\frac{2}{2}$	$\frac{0}{2}$	$\frac{2}{2}$	$\frac{1}{1}$	$\frac{0}{2}$	$\frac{1}{2}$	$\frac{2}{2}$	$\frac{0}{2}$ tr $\frac{1}{2}$
INJ SG	$\frac{2}{2}$(1.8)	$\frac{0}{2}$	$\frac{0}{2}$	$\frac{2}{2}$(2.5)	$\frac{1}{1}$(4.0)	$\frac{2}{2}$(2.0)	$\frac{1}{2}$(1.8)	$\frac{1}{2}$(2.0)	$\frac{2}{2}$(4.0)
TH	$\frac{2}{2}$(1.8)	$\frac{0}{2}$	$\frac{0}{2}$	$\frac{2}{2}$(2.8)	$\frac{1}{1}$(4.0)	$\frac{2}{2}$(2.0)	$\frac{2}{2}$(3.0)	$\frac{2}{2}$(3.5)	$\frac{2}{2}$(3.5)
100 F	$\frac{0}{1}$	$\frac{1}{2}$	$\frac{0}{2}$			$\frac{1}{1}$	$\frac{1}{2}$	$\frac{1}{2}$ tr $\frac{0}{2}$	
FED SG	$\frac{1}{1}$(2.0)	$\frac{2}{2}$(1.8)	$\frac{0}{2}$			$\frac{1}{1}$(2.0)	$\frac{2}{2}$(1.8)	$\frac{0}{2}$	
TH	$\frac{1}{1}$(1.8)	$\frac{2}{2}$(1.8)	$\frac{0}{2}$			$\frac{1}{1}$(1.5)	$\frac{2}{2}$(2.0)	$\frac{0}{2}$	

[a]*F: proportion of salivary glands with cytoplasmic immunofluorescent foci in acinar cells; SG: proportion of salivary glands containing virus (log mouse* LD_{50} *titers per gland in parentheses); TH: proportion of thoraces containing virus; tr: proportion of mosquitoes that transmitted virus while feeding on defibrinated blood in pledgets; INJ: intrathoracic injection; FED: feeding on virus in defibrinated blood.*

TABLE VI
Transmission of First Mouse Passage Mosquito CE

Mouse LD_{50} fed		*0°C* 3	6	13	20
	F				
1000	SG				
	TH				
	F	$\frac{0}{2}$	$\frac{2}{2}$	$\frac{0}{.2}$	$\frac{0}{2}$ tr $\frac{1}{1}$
100	SG	$\frac{0}{2}$	$\frac{2}{2}$(1.5)	$\frac{2}{2}$(2.0)	$\frac{2}{2}$(2.0)
	TH	$\frac{0}{2}$	$\frac{2}{2}$(1.8)	$\frac{2}{2}$(2.0)	$\frac{2}{2}$(2.5)
	F				
1	SG				
	TH				
	F				
0.1	SG				
	TH				
	F				
	SG				
	TH				

[a]*See Table V for meaning of abbreviations.*

Isolate 74-Y-234 by A. communis Mosquitoes after Feeding[a]

13°C					23°C	
3	6	13	20	27	13	20
		$\frac{1}{4}$ tr $\frac{1}{1}$	$\frac{0}{4}$ tr $\frac{2}{2}$		$\frac{0}{2}$ tr $\frac{2}{2}$	
		$\frac{4}{4}$(2.0)	$\frac{4}{4}$(2.0)		$\frac{2}{2}$(2.0)	
		$\frac{4}{4}$(2.5)	$\frac{4}{4}$(2.0)		$\frac{2}{2}$(1.8)	
$\frac{0}{2}$	$\frac{1}{2}$	$\frac{2}{5}$ tr $\frac{1}{1}$	$\frac{1}{4}$ tr $\frac{2}{3}$	$\frac{0}{3}$		
$\frac{0}{2}$	$\frac{2}{4}$(1.8)	$\frac{4}{5}$(3.0)	$\frac{4}{4}$(3.3)	$\frac{2}{3}$(2.0)		
$\frac{0}{2}$	$\frac{2}{4}$(1.8)	$\frac{4}{5}$(3.5)	$\frac{4}{4}$(3.3)	$\frac{2}{3}$(2.0)		
		$\frac{0}{2}$ tr $\frac{1}{1}$	$\frac{0}{1}$			$\frac{0}{1}$
		$\frac{2}{2}$(2.0)	$\frac{0}{1}$			$\frac{1}{1}$(2.0)
		$\frac{2}{2}$(2.0)	$\frac{1}{1}$(1.8)			$\frac{1}{1}$(1.5)
		$\frac{2}{2}$	$\frac{0}{2}$	$\frac{0}{2}$		$\frac{0}{1}$ tr $\frac{1}{1}$
		$\frac{2}{2}$(1.8)	$\frac{2}{2}$(1.8)	$\frac{2}{2}$(1.8)		$\frac{1}{1}$(1.8)
		$\frac{2}{2}$(1.8)	$\frac{2}{2}$(1.8)	$\frac{2}{2}$(1.8)		$\frac{1}{1}$(1.5)
		$\frac{0}{2}$ tr $\frac{1}{2}$	$\frac{0}{2}$	$\frac{0}{2}$		$\frac{0}{2}$ tr $\frac{2}{2}$
		$\frac{2}{2}$(2.0)	$\frac{2}{2}$(1.8)	$\frac{2}{2}$(2.0)		$\frac{2}{2}$(1.8)
		$\frac{2}{2}$(2.0)	$\frac{2}{2}$(2.0)	$\frac{2}{2}$(2.0)		$\frac{2}{2}$(1.8)

TABLE VII
Transmission of Zero Passage California Encephalitis Virus (74-Y-234 Strain) and Localization of Antigen in Salivary Glands by Direct Immunofluorescence, in A. communis Mosquitoes Infected by Feeding or Intrathoracic Injection[a]

Dose mouse LD_{50}		13 C, days: 6	13 C, days: 13	13 C, days: 20	23 C, days: 6	23 C, days: 13	23 C, days: 20
1 FED	F		0/2 (TR 2/2)	0/2 (TR 2/2)		0/2 (TR 2/2)	
	SG		2/2 (1.8)	2/2 (1.8)		2/2 (1.8)	
1 INJ	F	0/2	1/2	2/2 (TR 1/1)	0/2	1/2	2/2 (TR 1/2)
	SG	2/2 (1.8)	2/2 (1.8)	2/2 (1.8)	2/2 (3.3)	2/2 (3.0)	2/2 (2.0)
0.1 INJ	F	2/2	1/2	0/2 (TR 2/2)	0/2	2/2	0/2 (TR 2/2)
	SG	2/2 (1.5)	2/2 (1.5)	2/2 (1.8)	2/2 (1.8)	2/2 (1.8)	2/2 (1.8)
Total	F	2/4	2/6	2/6	0/4	3/6	2/4
	SG	4/4	6/6	6/6	4/4	6/6	4/4

[a]*See Table V for meaning of abbreviations.*

tures of 0, 13, and 23°C following infection both by feeding and by intrathoracic injection of 1 mouse LD_{50} or smaller virus doses, demonstrates clearly their high potential as natural vectors. Furthermore, the isolation of CE virus from *Aedes sp.* larvae, collected at two Yukon locations during spring of two successive years, before emergence of adult mosquitoes, points strongly toward a transovarial mechanism for maintenance of virus through winter to the following summer. Arctic mosquitoes have shown equal susceptibility to intrathoracic injection with a nonendemic Flavivirus, Murray Valley encephalitis, as to the endemic Bunyavirus, CE virus.

References

Balfour, H. H., Edelman, C. K., Cook, F. E., Barton, W. I., Buzicky, A. W., Seim, R. A., and Bauer, H. (1975). *J. Infect. Dis. 131,* 712-716.

Iversen, J. O., Hanson, R. P., Papadopoulos, O., Morris, C. C., and De Foliart, G. R. (1969). *Am. J. Trop. Med. Hyg. 18,* 735-742.

Iverson, J. O., Wagner, R. J., De Jong, C., and McLintock, J. R. (1973). *Can. J. Publ. Health 64,* 590-594.

Le Duc, J. W., Suyemoto, W., Eldridge, B. F., Russell, P. K., and Barr, A. R. (1975). *Am. J. Trop. Med. Hyg. 24,* 124-126.

McKiel, J. A., Hall, R. R., and Newhouse, V. F. (1966). *Am. J. Trop. Med. Hyg. 15,* 98-102.

McLean, D. M. (1975). "Arboviruses and Human Health in Canada." Publ. No. 14106, National Research Council of Canada.

McLean, D. M., De Vos, A., and Quantz, E. J. (1964). *Am. J. Trop. Med. Hyg. 13,* 747-753.

McLean, D. M., Crawford, M. A., Ladyman, S. R., Peers, R. R., and Purvin-Good, K. W. (1970). *Am. J. Epidemiol. 92,* 266-272.

McLean, D. M., Grass, P. N., Judd, B. D., Stolz, K. J. and Wong, K. K. (1978). Transmission of Northway and St. Louis Encephalitis Viruses by Arctic Mosquitoes. *Arch. Virol. 57,* 315-322.

McLean, D. M., Goddard, E. J., Graham, E. A., Hardy, G. J., and Purvin-Good, K. W. (1972). *Am. J. Epidemiol. 95,* 347-355.

McLean, D. M., Bergman, S. K. A., Gould, A. P., Grass, P. N., Miller, M. A., and Spratt, E. E. (1975). *Am. J. Trop. Med. Hyg. 24,* 676-684.

McLean, D. M., Grass, P. N., Judd, B. D., and Wong, K. S. K. (1976), *Can. J. Microbiol. 22,* 1128-1136.

McLean, D. M., Grass, P. N., Judd, B. D., Cmiralova, D., and Stuart, K. M. (1977). *Can. J. Publ. Health 68,* 69-73.

McLean, D. M., Grass, P. N., Judd, B. D., Ligate, L. V., and Peter, K. K. (1977a). *J. Hyg. (Camb.), 79,* 61-71.

McLean, D. M., Grass, P. N., and Judd, B. D. (1977b).

McLintock, J., and Iversen, J. (1975). *Can. Ent. 107,* 695-704.

Newhouse, V. F., Burgdorfer, W., McKiel, J. A., and Gregson, J. D. (1963). *Am. J. Hyg. 78,* 123-129.

Watts, D. M., Thompson, N. H., Yuill, T. M., De Foliart. G. R., and Hanson, R. P. (1974). *Am. J. Trop. Med. Hyg. 23,* 694-700.

Chapter 3

ARBOVIRUSES OF HIGH LATITUDES IN THE USSR

D. K. Lvov, V. L. Gromashevski, T. M. Skvortsova, L. K. Berezina, Y. P. Gofman, V. M. Zhdanov, A. S. Novokhatski, S. M. Klimenko, A. A. Sazonov, N. V. Khutoretskaya, V. A. Aristova, N. G. Kondrashina, K. B. Fomina, and B. G. Sarkisyan

I. INTRODUCTION

Ixodid ticks are absent from the continental territory of the USSR to the north of the range of *Ixodes persulcatus* ticks, where a warm season with temperatures effective for virus replication is very short. Therefore, mosquitoes, though numerous in this region, cannot as a rule provide circulation of viruses; only under certain conditions do they probably take part in the functioning of seasonal foci. First of all, this possibility appears when some viruses are capable of transovarial transmission in mosquitoes. Apparently it is just the case with some viruses of the California group (Le Duc *et al.*, 1975; Watts *et al.*, 1973) and perhaps also Bunyamwera viruses. The virus of snowshoe hares was repeatedly isolated in Alaska, the

ISBN 0-12-429765-X

U.S.A. and the Yukon (McLean, 1975; McLean *et al.*, 1972) and northwestern areas of Canada (Newhouse *et al.*, 1963; Wagner *et al.*, 1975) from *Aedes canadensis* and *A. communis* mosquitoes, hares, and lemmings. Isolation of the virus from larvae of mosquitoes indicates the transovarial transmission of viruses or the alimentary route of infection of larvae in sites of their breeding. Northway virus of the Bunyamwera group was isolated in the southeastern and central parts of Alaska from "sentinel" rabbits, voles, and mosquitoes of the *Aedes* and *Culiseta* genera (Calisher *et al.*, 1974; Ritter and Felt, 1974). The existence of natural foci of these or ecologically similar arboviruses in the north of the USSR has not been established for the present but is very likely.

Over the littoral and insular territories of the north of the temperate zone in the Subarctic and the Artic, there are huge populations of colonial seabirds on their nesting grounds. Two species of Ixodid ticks , *Ixodes (Ceratixodes) putus (I. uriae)* and *I. signatus*, are obligatory parasites of these birds. *Ixodes putus* ticks infest 17 species of birds in the northern hemisphere and 16 species of birds in the southern hemisphere (Roberts, 1970; Zumpt, 1952; Belopolskaya, 1952). However, alcids in the northern hemisphere and penguins in the southern hemisphere are the major hosts of *I. putus*. In the southern hemisphere ticks are distributed over Cape Horn and Tierra del Fuego in South America (Argentina), in the south of Africa (Republic of South Africa), on southern islands of New Zealand, Tasmania, subantarctic islands (Kerguelen, Antipodes, Macquarie, Tristan da Cunha, Marion, Campbell, Herd, and others), on the southern coast of Australia, and Victoria Land in the Antarctic (Roberts, 1970). In the northern hemisphere, the range covers northern Europe (Great Britain, France, Ireland, Norway, Finland, Iceland), the coast of Greenland, and areas to the north of latitude 45^{0}north in the U.S.A. and Canada (Arthur, 1963; Bequaert, 1946; Clifford *et al.*, 1970). In the USSR, ticks are distributed on islands of the Barents Sea--Novaya Zemlya, Kharlov, Kuvshin (Belopolskaya, 1952), the northern coast of the Kola Peninsula (Flint and Kostyrko, 1967; Karpovich and Pilipas, 1972), the Kuril Islands, Sakhalin, Tyuleniy, Moneron Islands in the Sea of Okhotsk, and Commodore Islands. Our data also indicate the presence of ticks on Iona Island in the Sea of Okhotsk and on the southeastern coast of Chukotka (Lvov *et al.*, 1975; Timofeeva *et al.*, 1974). Thus, within the borders of the country, the range covers littoral and insular territories in the north of the temperate zone (islands of the Sea of Okhotsk, the Kuril Islands), the Subarctic (Commodore Islands, the southeastern coast of Chukotka, islands of the Barents Sea), and the Arctic (Novaya Zemlya Island).

The range of *I. signatus* ticks is considerably narrower and includes Hokkaido Island and the northern part of Honshu Island (Japan), Aleutian Islands and the coast of California (Zumpt, 1952). In the USSR, ticks were discovered in islands of the Rimski-Korsakov Archipelago in the Sea of Japan, Tyuleniy Island in the Sea of Okhotsk, and Kuril Islands. We established an enormous infestation rate of birds with ticks on Commodore Islands. *I. signatus* ticks are obligatory parasites of cormorants: pelagic cormorants *Phalocrocorax pelagicus*, red faced cormorants *P. urile*, European cormorants *P. carbo*, Temminck's cormorants *P. filamentosus*. Individual ticks are caught on fulmars *Fulmaris glacialis* (USSR), common murres *Uria aalige* and gulls *Larus crassirostris* in Japan. Thus, the range of the species covers the northern part of the temperate zone reaching the border with the Subarctic in the north (Commodore Islands). In 1969 the existence of active natural foci of arboviruses associated with *I. putus* and *I. signatus* ticks was hypothesized (Lvov, 1969).

II. Places of Material Collection

During the course of investigations in the period 1969-1976, about 70,000 ticks were used for virological studies. Field material was collected in 11 regions of the north of the Far East and European part of the USSR (Table I). Most essential collections of ticks with a subsequent isolation of the virus were made on Tyuleniy and Iona Islands in the Sea of Okhotsk, on Commodore Islands in the Bering Sea in the Far East, and on Kuvshin Island in the Barents Sea in the European part of this country.

A. The Far East of the USSR

Tyuleniy Island is in Zaliv Terpeniya, the Sea of Okhotsk, and is adjacent to the Terpeniya Peninsula at the southeastern coast of Sakhalin. One of the world's three largest harem lairs of fur seals *Callorhinus ursinus* is on the island. There is a considerable colony of common murre *Uria aalge* on the plateau of the island. *I. putus* ticks are concentrated in clefts of friable rocks at a depth of up to 20 cm. Up to 7000 ticks can be collected from a plot of 1 m^2. Ticks have also been collected from common murres and fur seal calves. Cases have been recorded of ticks sticking to man. Iona Island presents above-water rocks on the fringe of the mainland in the Sea of Okhotsk 200 km north of Sakhalin. On the island there

TABLE I

Collection of Material in Different Regions in the North of the Far East and European Part of the USSR

Site of collection of material	*Detection of ticks*		*Annual sum of T>10°C*	*Climate zone*
	I. putus	*I. signatus*		
European:				
Basin of the Barents Sea				
Murmansk region, Coast of the Kola Peninsula, Kuvshin Island	+	-	200	Subarctic
Arkhangelsk Region, southeastern part of Novaya Zemlya Island	-	-	0	Arctic
Far Eastern:				
Basin of the Sea of Japan				
Primorye territory, Rimski-Korsakov Archipelago, Furugelm Island	-	+	2300	Temperate
Basin of Sea of Okhotsk				
Sakhalinsk region, South Kuril Islands	-	+	800-1200	Temperate
North Kuril Islands	-	+	200-800	North of temperate zone
Tyuleniy Island	+	-	600	
Iona Island	+	-	400	
Basin of the Bering Sea				
Kamchatsk region, Commodore Islands	+	+	200	Border of temperate zone and subarctic
Magadansk region, south eastern Coast of Chukotka	+	-	0	Subarctic
Ratmanov Island	-	-	0	Arctic
Basin of the Chukotsk Sea				
Magadansk region, Vrangel Island	-	-	0	Arctic

is a large colony of seabirds, the main population of which consists of Brünnich's guillemots *Uria lomvia* fulmars *Fulmaris glacialis*.

On Commodore Islands, collection of ticks was carried out on Ariy Kamen Island, a rock at 20 km from the Pacific coast of Bering Island. A very large mixed colony of seabirds (alcids, gulls, cormorants) lives on the island, among which *U. lomvia* and *U. aalge, Lunda cirrhata,* and *Phalacrocorax pelagicus* prevail. *I. signatus* ticks were collected in cracks of rocks, in holes of puffins, and on common murres and puffins. Ticks actively attacked man. *I. signatus* ticks were collected from cormorants and their nests. Sometimes large epizootics are observed among birds.

B. European Part of the USSR

Kuvshin Island is situated at the southeastern coast of the Kola Peninsula, where there is a large colony of seabirds among which *U. lomvia* is the main species.

III. ISOLATION OF VIRUSES

In all, 241 strains of five viruses were isolated: Tyuleniy (Flaviviruses, Togaviridae), Okhotskiy (Orbibiruses, Reoviridae), Zaliv Terpeniya (the Uukuniemi group, Bunyaviridae), Sakhalin (the Sakhalin group, Bunyaviridae), Paramushir (Bunyaviridae). A total of four strains of viruses was isolated from 20, 048 *I. signatus* ticks: Okhotskiy and Zaliv Terpeniya viruses on Commodore Islands, Paramushir and Sakhalin viruses on South Kuril Islands. The infection rate of *I. signatus* ticks with these viruses was only about 1:10,000-1:20,000. The isolates came from females (two strains), males (one strain), and nymphs (two strains), and never from larvae. In these studies 3200, 3200, 6800, and 6600 specimens were examined, respectively. The infection rate of males, females and nymphs was approximately the same but about 20 times lower than that of *I. putus* ticks. Perhaps this phenomenon can be explained by a small range of the tick species and rare species of host birds. It cannot be excluded that the susceptibility of *I. signatus* ticks to virus infection is considerably lower than that of *I. putus*, the overall infection rate of which with viruses in different areas varies only within the narrow limits of 1:150-1:350. In the north of the European part, the infection rate of *I. putus* ticks with Zaliv Terpeniya and Okhotskiy viruses is the highest, whereas that with Tyuleniy virus is 10-20 times less. On the other hand, in the Far East, the highest infection rate of ticks is with Sakhalin (on Tyuleniy Island) and Tyuleniy viruses (on Iona and Commodore Islands), the lowest is with Paramushir virus (on Iona and Commodore Islands), the lowest is with Paramushir virus (Table II).

IV. TYULENIY VIRUS (FLAVIVIRUSES, TOGAVIRIDAE)

Tyuleniy virus was first isolated in the USSR from *I. putus* ticks on Tyuleniy Island in 1969 (lvov *et al.*, 1971), and then on the Commodore Islands (Lvov *et al.*, 1972a), Murmansk coast (Gaydamovich *et al.*, 1971), and the Oregon coast in the U.S.A. (Clifford *et al.*, 1971; Thomas *et al.*, 1973). The virus is pathogenic for suckling mice and 2-3-week-old mice upon intracerebral inoculation. The virus replicates in chick embryos and BHK-21 cell cultures without cytopathogenic effect (CPE).

Electron-microscopic studies of the virus in both suckling mouse brain cells and tissue cell cultures have revealed crystal-like formations and a "loopy net" along with typical virons, about 40 nm in size (Fig. 1). Apparently, these structures are associated with synthesis and formation of virions.

On experimental infection of *Aedes aegypti* mosquitoes, the virus was demonstrated from the fourth to the thirty-first day of observation. Virus titers reached 1.5-2.0 log LD_{50}/0.01 ml from day 4 up to day 17, increased to 3.0-3.5 at days 23-27, and finally decreased to 1.5 at day 31. Transmission of the virus through bite was established from the seventh to the nineteenth postinfection day. In *Culex molestus* mosquitoes, the virus was detected from the fifth up to the twenty-first postinfection day (the observation period) in titers of 1.0-2.0 log LD_{50}/0.01 ml. On experimental infection of birds (kittiwakes *Rissa tridactyla*, herring-gulls *Larus argentatus* Brünich's guillemots *Uria lomvia)*, an apparent clinical infection with lesions in the central nervous system develops, sometimes with fatal issue (Berezina *et al.*, 1974). The virus can cause a general febrile condition in man; under natural conditions the illness develops after visit to colónies of common murres without the presence of any ticks on these persons (Votyakov *et al.*, 1974).

The infection rate of imagoes (males or females) if *I. putus* is actually the same in all parts of the range, 1:350-1:600, whereas the infection rate of nymphs and larvae is approximately 2 and 20 times less, respectively. It is likely that ticks are mainly infected during the stages of larvae and nymphs. The virus is then transmitted in the course of metamorphosis, which is indicated by the same infection rate in females

TABLE II

Frequency of Isolation of Strains of Different Viruses from I. Putus Ticks

	European:	*Far East*					
Sites of tick collection	*Kuvshin Island*	*Tyuleniy Island*	*Iona Island*	*Commodore Islands*	*Southeastern coast of Chukotka*	*Total*	*Total*
Viruses:							
Tyuleniy	2(3%)	16(20%	5	30(42%)	0	51(30%)	53(21%)
Okhotskiy	23(34%)	11(14%)	2	8(11%)	3	24(14%)	47(21%)
Zaliv Terpeniya	42(63%)	12(16%)	6	25(35%)	0	43(24%)	85(37%)
Sakhalin	0	32(42%)	2	6(9%)	3	43(29%)	43(19%)
Paramushir	0	6(8%)	0	2(3%)	0	8(3%)	8(2%)
Total	67(100%)	77(100%)	15	71(100%)	6	169(100%)	236(100%)

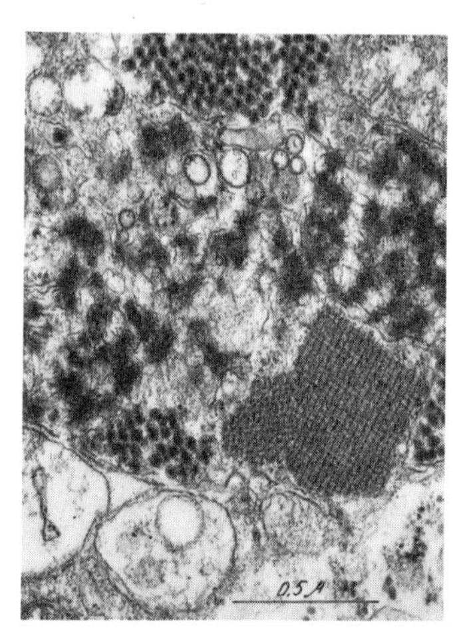

FIGURE 1. Tyuleniy virus. Ultrathin section of the brain of an infected suckling mouse. Crystallike formations and inclusion in the form of "loopy net" are revealed in the cytoplasm along with virions. (×90,000)

and males not feeding on blood. However, transovarial transmission of the virus occurs in less than 5% of infected females. The virus has not been isolated from *I. signatus* ticks.

According to data of serological examination of 2500 birds in the Far East, the main range covers the Commodore, Sakhalin, Iona, and Tyuleniy Islands. Of major importance for the circulation of the virus are Brünnich's guillemots *U. lomvia* and common murres *U. aalge*, puffins *L. cirrhata* and, some parts of the range, pelagic cormorants *P. pelagicus* , red-faced cormorants *P. urile* (Commodore Islands), gulls *L. glaucescens*, kittiwakes *R. tridactyla*, and fulmars *F. glacialis*. Detection of antibodies in snipes during autumn migration suggests the importance of some species of snipes in the possible spreading of the virus in the southern hemisphere. Antiviral immunity among fur seals on Commodore and Tyuleniy Islands reaches 10% for adults and 20% for the young of the individuals, indicating the active involvement of these animals in the circulation of the virus. Isolated findings of antibodies against the virus on Commodore Islands in sera of cows and man suggest the participation of mosquitoes in the dissemination of the virus over the territory.

The presence of complement-fixing antibodies in the indigenous population (Saami, Komi, Nenets) in the tundra zone on the Barents coast of the Cola Peninsula, as well as in reindeer, snipes *Phalaropus lobatus*, snow bunting *Plectrophenax nivalis*, snipes *Philomachus pugnax*, and root voles *Myerotus aeconomus* would suggest the possibility of an outburst of the virus during summer on the continent, where mosquitoes are probably involved in its circulation. Here birds are attacked by *Aedes communis*,

A. punctor, and *A. excrucians.* The infection rate of mosquitoes at the end of July and beginning of August reaches 0.3% in the area of nesting grounds of sea birds and 0.1% on the coast of the Cola Peninsula.

V. OKHOTSKIY VIRUS (ORBIVIRUSES, REOVIRIDAE)

Okhotskiy virus was first isolated from *I. putus* ticks collected in 1969 on Tyuleniy Island (Lvov *et al.*, 1973a) and afterwards on the Murmansk coast of the Barents sea, Commodore Islands, and the southeastern coast of Chukotka. Related viruses were isolated in the north of the American continent, Yaquina Head virus on the Oregon coast (Yunker *et al.*, 1973), Great Island and Bauline viruses on the coast of Newfoundland (Main *et al.*, 1973), Poovot virus on St. Lawrence Island and Kenai virus on Gull Island at Alaska (Ritter and Feltz, 1974), and Cape Wrath virus in the north of Scotland (Main *et al.*, 1976b). and on Macquarie Island in the south Atlantic Ocean (Doherty *et al.*, 1975). Among laboratory animals, only suckling mice are found to be sensitive to the virus on intracerebral inoculation. The replication with CPE occurs in primary embryo cell cultures of chicks, ducks, and man and in continuous L, Rh, and pig embryo kidney cell cultures. The virus does not agglutinate goose erythrocytes.

The regularities of Okhotskiy virus replication were studied on chick embryo fibroblast cultures. The virus is characterized by a short replication cycle (6 hours), a high percentage of virus release into the culture medium (80%), and a lack of sensitivity to actinomycin D. In these cultures the virus induces production of moderate amounts of interferon and possesses a low sensitivity to the inhibiting action of interferon and its inducer, a complex of synthetic polyribonucleotides (Fig. 2). The property of double-strandedness of the viral RNA was exploited for obtaining information on the state of the genetic material of Okhotskiy virus. The RNA of the virus, obtained by the method of phenol extraction from concentrated and purified virus preparations and treated with RNase at a concentration of 20 μg/ml for 30 minutes 37 C, still possessed a pronounced ability to induce antiviral resistance in a chick embryo fibroblast culture, with respect to vesicular stomatitis virus (VSV) Fig. 3).

On centrifugation in a cesium sulfate denisty gradient, RNA of Okhotsk virus gives a peak in the zone corresponding to a density of 1.60 g/cm^3 which is characteristic of double-stranded

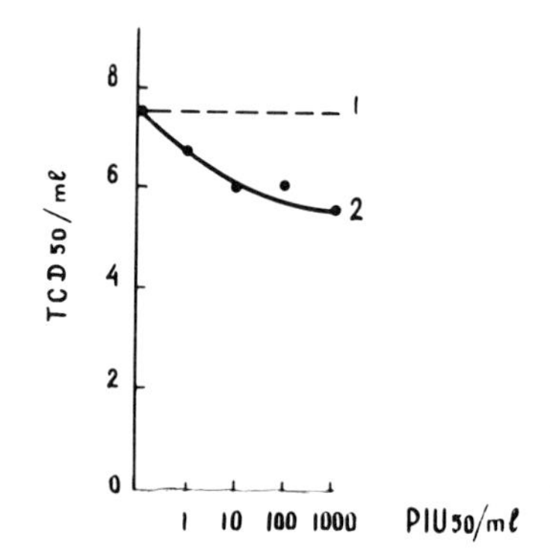

FIGURE 2. Inhibition of production of infectious Okhotsk virus in chick embryo fibroblasts by increasing doses of interferon. (1) Control level of Okhotsk virus production. (2) Production of Okhotskiy virus in interferon-treated chick embryo fibroblast culture. The abscissa represents the activity of interferon (PIU_{50}/ml), the ordinate represents the activity of viruses in log TCD_{50}/ml.

RNA (Fig. 4). Material from this fraction also induces inhibition of VSV replication in cell culture (per 1.4 PFU/ml as compared with control) and inhibits production of CPE.

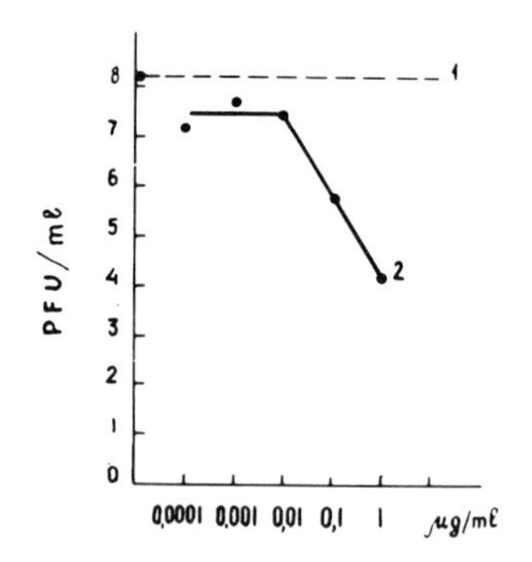

FIGURE 3. Inhibition of the replication of VSV in chick embryo fibroblast culture treated with RNA isolated from Okhotsk virus, in the presence of 100 μg/ml DEAE-dextran. The abscissa represents the dose of RNA in μg/ml, the ordinate represents the virus titer in log PFU/ml.

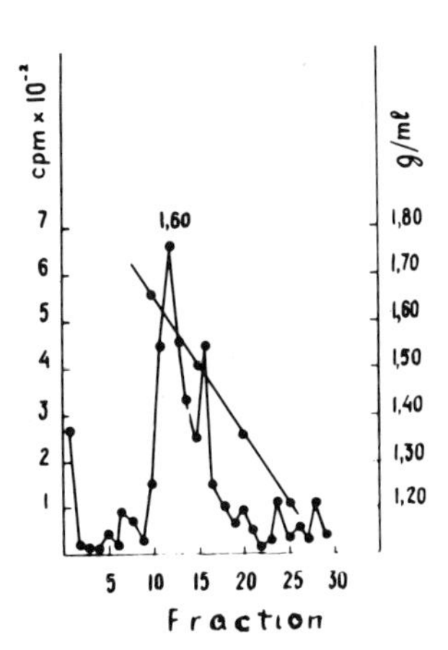

FIGURE 4. Distribution RNA of Okhotsk virus labeled with ^{3}H-uridine in a cesium sulfate density gradient.

On examination of ultrathin sections of the brain tissue of suckling mice, morphologic changes characteristic of arbovirus-affected cells were revealed (Fig. 5). Virions are formed in matrices that are localized in the paranuclear zone of the cytoplasm and have a peculiar appearance. The most characteristic components in the structure of the matrix are (1) virus particles with a diameter of 50 to 65 nm and a nucleoid, about 40 nm in size, (2) tubular structures, 40-50 nm in diameter, (3) strands of different lengths and diameters.

In different regions the infection rate of imagoes and nymphs of *I. putus* varies within the limits of 1:500-1:2500 and that of larvae is 5-10 times less. One virus strain was isolated from *I. signatus* on Commodore Islands. Two strains were recovered from kittiwakes *R. tridactyla* on the coast of the Cola Peninsula (Votyakov *et al.*, 1974). Complement-fixing antibodies (according to the data of a serologic survey of about 1500 birds) were found in 15% in fulmars *F. glacialis* (Iona Island), in 4-6% in common murres *U. aalge* (Tyuleniy and Commodore Islands) and in 6% in pelagic cormorants *P. pelagicus* (Commodore Islands). Only in one case were antibodies demonstrated by the indirect complement fixation test, in all other cases the direct complement fixation test was used. Antibodies were also detected in man (up to 12%), permanent residents of Commodore Islands. However, cases of diseases associated with virus infection were not revealed. The results of the complement fixation test were confirmed by the examination of the positive sera in the neutralization test in chick and duck embryo cell cultures and in suckling mice.

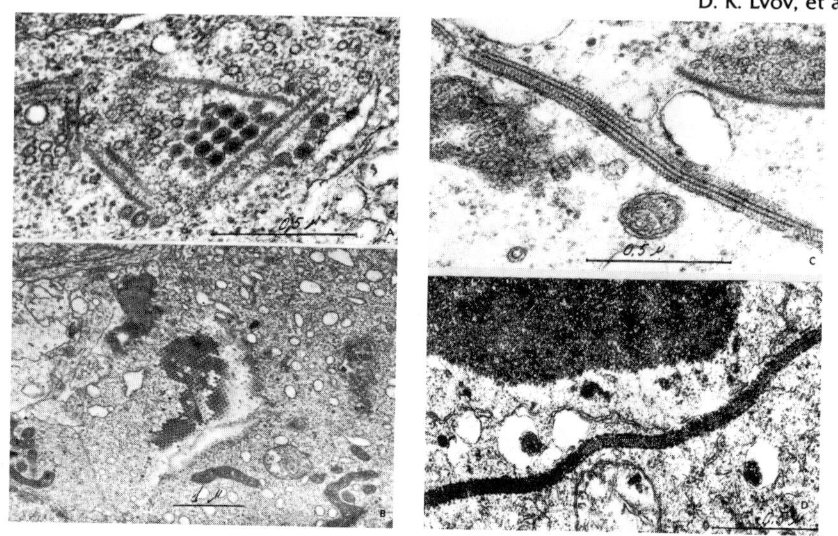

FIGURE 5. Okhotskiy virus. Ultrathin sections of the brain of an infected suckling mouse. (A) An area of the cytoplasm where virions, tubular structures in cross and longitudinal sections, and nonstructural strands are localized (×120,000) (B) Crystallike accumulation of viruses (×30,000) (C, D) Strands possessing a complex organization (×90,000).

VI. ZALIV TERPENIYA VIRUS (THE UUKUNIEMI GROUP, BUNYAVIRIDAE)

Zaliv Terpeniya virus was first isolated from *I. putus* ticks collected in 1969 on Tyuleniy Island (Lvov *et al.*, 1973b) and later on on Iona Island, Commodore Island, the southeastern coast of Chukotka, and the Murmansk coast of the Barents Sea. Related viruses were isolated on the Oregon coast (Thomas *et al.*, 1973) and California (Yunker, 1975). Among laboratory animals only suckling mice are found to be sensitive to the virus on intracerebral inoculation. Replication without CPE occurs in chick and duck embryo cell cultures. The virus does not agglutinate goose erythrocytes.

In the north of the European part of this country the virus is most frequently isolated as compared with other viruses. The infection rate of male and female ticks in this region reaches 1:50, that of nymphs and larvae is 5 and 13 times lower, respectively. A relatively high infection rate of nonengorged larvae indicates a high frequency (8-10%) of transovarial transmission of the virus by infected female ticks. In the Far East, the infection rate of *I. putus* with the virus is 15-20 times lower. The virus was isolated on two occasions from *I. signatus* on Commodore Islands. However, the mean infection rate of these ticks with the virus does not exceed 1:10,000.

Complement-fixing antibodies against the virus on Commodore Islands were found only in common murres *U. aalge* (1%) and in the north of the European part of this country (Murmansk region) in Brünnich's guillemots *U. lomvia* (6%), sea gulls *L. marinus* (7%), kittiwakes *R. tridactyla* (4%), herring gulls *L. argentatus P.* (0%), root voles *M. oeconomus* (1%). Rodents can be infected by the alimentary route by the ingestion of ticks and bodies of birds dead of infection. Also, their infection by mosquito bite cannot be excluded. Thus, the data of the serological survey of birds corroborate the results of the virological examination of ticks on a considerable general activity of natural foci of the virus in the north of the European part, as compared with the Far Eastern part, of the range.

VII. SAKHALIN VIRUS (THE SAKHALIN GROUP, BUNYAVIRIDAE)

Sakhalin virus was first isolated from *I. putus* ticks collected in 1969 on Tyuleniy Island, and then on Iona Island, Commodore Islands, and the southeastern coast of Chukotka (Lvov *et al.*, 1972b). Related viruses were isolated in the North of the American continent--Sakhalin virus on the Oregon coast and northern California (Yunker, 1975), on Gull Island at Alaska (Ritter and Feltz, 1974), Avalon virus on the Newfoundland coast (Main *et al.*, 1976a), and Taggert virus on Macquarie in the south Atlantic Ocean (Doherty *et al.*, 1975). Among laboratory animals, only suckling mice are sensitive to the virus on intracerebral inoculation. Replication without CPE occurs in primary duck and human embryo cell cultures and in continuous Vero, HeLa, and BHK-21 cell cultures.

A continuous line of BHK-21 cells was used for a study of the properties of Sakhalin virus. The virions of the Sakhalin virus have a density of 1.23 g/cm^3 in sucrose density gradient (Fig. 6A). Treatment of the virions with NP-40 results in the release of structures with a density of 1.30 g/cm^3 in sucrose and cesium chloride gradients, which appear to be nucleoids (Fig. 6B, C), and the structures with a density of 1.36 g/cm^3 (Fig. 6C), which are the products of a subsequent deproteinization. Viral RNA, with a sedimentation coefficient of 39 S, was isolated both from virions and nucleoids (Fig. 7).

An electron-microscopic study of ultrathin sections of the brain of infected suckling mice has demonstrated that this virus was similar in its morphology and morphogenesis to virions of the Bunyaviridae family (Fig. 8). The virions, round and oval in shape, have a diameter of about 90-95 nm and are localized in vacuoles. The envelop possesses distinct outlines, and spikes covering the particle can be distinguished in some areas.

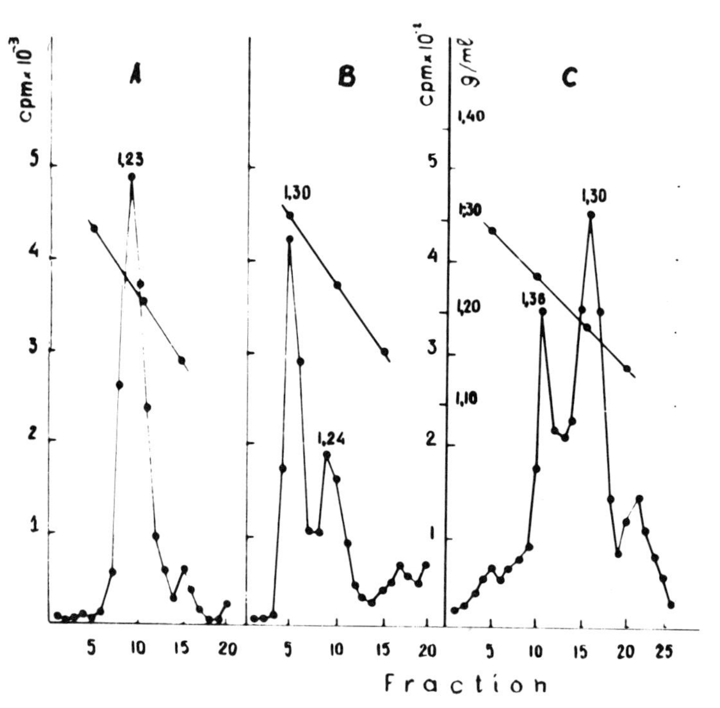

FIGURE 6. Distribution of preparations of Sakhalin virus labeled with ^{3}H-uridine in sucrose (A,B) and cesium sulfate (C) density gradients. Preparations in some experiments (B,C) were treated with a 1% solution of the nonionic detergent NP-40.

In the inner cavity of the particle, small granules (of about 10 nm) can be observed, which apparently present a cross section of ribonucleoprotein strands.

Parallel studies with the direct and indirect complement fixation tests of sera of about 1800 birds caught in the northern regions of the European part of this country and in the Far East revealed antibody only by the indirect test. Antibodies were found in snipes (17%) on Sakhalin Island (*P. lobatus*, sandpipers *Calidris alba, C. subminuta)*, fulmars *F. glacialis* (5%) on Iona Island, tufted puffins *L. cirrhata* (5%), Atlantic murres *U. aalige* (2-4%) on Commodore, Tyuleniy, and Iona Islands, Temminck's cormorants *P. filamentosus* (1%) on South Kuril Islands, and kittiwakes *R. tridactyla* (0.6%) on Iona Islands. According to the findings of the serological survey, the range of the virus is limited by the southern part of the basin of the Sea of Okhotskiy and in the north it does not exceed the bounds of the southeastern coast of the Chukotsk Peninsula.

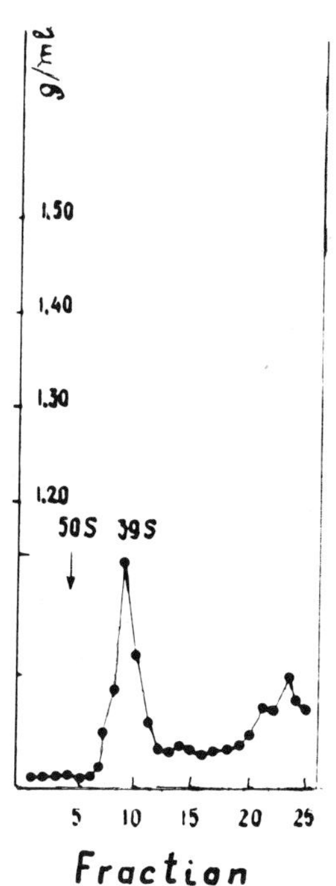

FIGURE 7. Sedimentation pattern of RNA of ^{3}H-uridine-labeled Sakhalin virus from structures with a density of 1.30 g/cm^3.

VIII. PARAMUSHIR VIRUS (BUNYAVIRIDAE, UNGROUPED)

Five strains of Paramushir virus were isolated in 1969-1974 from *I. signatus* (1 strain) and *I. putus* ticks on Tyuleniy Island and Commodore Islands (Lvov *et al.*, 1976). Among laboratory animals, only suckling mice are sensitive to the virus on intracerebral inoculation. Replication of the virus without CPE was detected in chick, duck, and human embryo fibroblast cultures and of continuous cell cultures in BHK-21. Replication of the virus with CPE was established in continuous L, A, HeLa, and pig embryo kidney cell cultures. The virus does not agglutinate goose erythrocytes. Preliminary results of an electron-microscopic study of ultrathin sections of the brain of infected suckling mice revealed the presence of viruslike particles, about 100 nm in size, morphologically similar to virions of the Bunyaviridae family.

The infection rate of *I. putus* ticks with Paramushir virus is the lowest as compared with the other four viruses: 1:3,300 among imago, 1:11,000 among nymphs. The virus was not isolated from larvae.

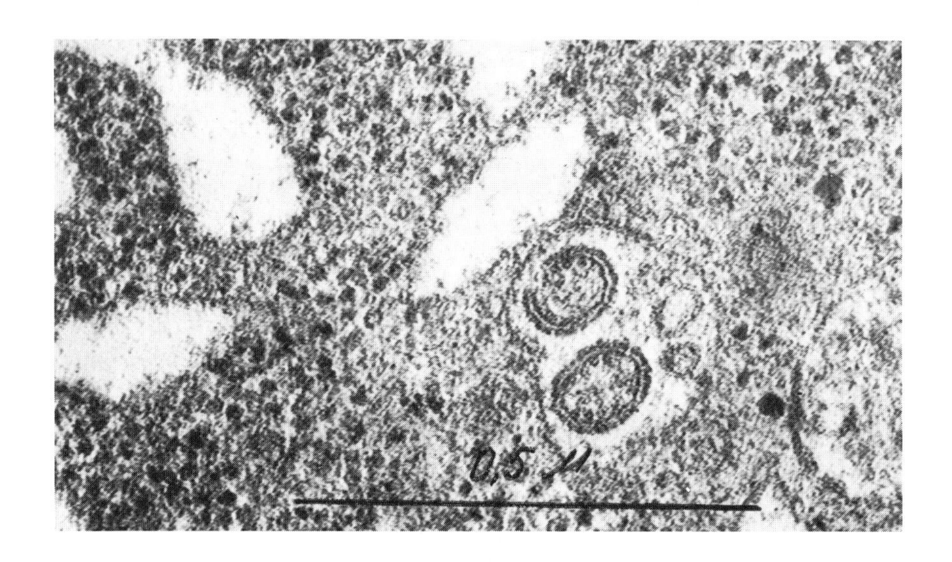

FIGURE 8. Sakhalin virus (C 2). An ultrathin section of the brain of an infected suckling mouse. Two virions localized in the cytoplasmic vacuole. In the inner cavity of the particles, a longitudinal section of the ribonucleoprotein strand can be seen. (×150,000)

References

Arthur, D. R. (1963). "British Ticks." Butterworths, London.

Belopolskaya, M. M. (1952). "Parasitophauna of Seabirds." Uchenie Zapiski, Leningrad. *Gosudarstv. Univ., Ser. Biol. 141*(28), 127-180.

Bequaert, J. C. (1946). *Ent. Am. 24*, 73-120.

Berezina, L. K., Smirnov, V. A., and Zelenskiy, V. A. (1974). *In* "Ecology of Viruses," pp. 13-17. Moscow.

Calisher, C. H., Lindsey, H. S., Ritter, D. G., and Sommerman, R. M. (1974). *Can. J. Microbiol. 20*, 219-223.

Clifford, C. M., Yunker, C. E., Easton, E. R., and Keirans, J. E. (1970). *J. Med. Entomol.* 7, 438-445.

Clifford, C. M., Yunker, C. E., Thomas, L. A., Easton, E. R., and Corwin, D. (1971). *Am. J. Trop. Med. Hyg. 20*, 461-468.

Doherty, R. L., Carley, J. G., Murray, M. D., Main, A. J., Kay, B. H., and Domrow, R. (1975). *Am. J. Trop. Med. Hyg. 24.*, 521-526.

Flint, V. E., and Kostyrko, I. N. (1967). *Zool. J. 66*(8), 1253-1256.

Gaydamovich, S. Ya., Melnikova, E. E., Bekleshova, A. Yu., Terskich, I. I., and Obuchova, V. R. (1971). *Vopr. Med. Virusol., Part 2,* 68-69.

Karpovich, V. N., and Pilipas, N. I. (1972). *Proc. 6th Symp. Study of Viruses Ecologically Connected with Birds,* pp. 91-94. Moscow.

LeDuc, J. M., Suyemoto, W., Eldridge, B. F., Russel, P. K., and Barr, A. R. (1975). *Am. J. Trop. Med. Hyg. 24,* 124-126.

Lvov, D. K. (1969). *In* "Virological Investigations in the Far East, pp. 15-49. Vladivostok.

Lvov, D. K., Timopheeva, A. A., Chervonsky, V. I., Gromashevsky, V. L., and Klisenko, G. A. (1971). *Vopr. Med. Virusol. 1,* 128.

Lvov, D. K., Chervonsky, V. I., Gostinshcikova, I. N., *et al.* (1972a). *Arch. Ges. Virusforsch. 38,* 139-142.

Lvov, D. K., Timopheeva, A. A., Gromashevsky, V. L., *et al.* (1972b). *Arch. Ges. Virusforsch. 38,* 133-138.

Lvov, D. K., Timopheeva, A. A., Gromashevsky, V. L., *et al.* (1973a). *Arch. Ges. Virusforsch. 41,* 160-164.

Lvov (D. K., Timopheeva, A. A., Gromashevsky, V. L., *et al.* (1973b). *Arch. Ges. Virusforsch. 41,* 165-169.

Lvov, D. K., Timopheeva, A. A., Smirnov, V. A., *et al.* (1975). *Med. Biol. 53,* 325-330.

Lvov, D. K., Sazonov, A. A., Gromashevsky, V. L., *et al.* (1976). *Arch. Virol. 51,* 157-161.

McLean, D. M. (1975). *Med. Biol. 53,* 264-270.

McLean, D. M., Clarke, A. M., Goddard, E. J., Manes, A. S., Montalbetti, C. A., and Pearson, R. E. (1972). *J. Hyg. Camb. 71,* 391-402.

Main, A. J., Downs, W. G., Shope, R. E., and Wallis, R. C. (1973). *J. Med. Entomol. 10,* 229-235.

Main, A. J., Downs, W. G., Shope, R. E., and Wallis, R. C. (1976a). *J. Med. Entomol. 13,* 309-315.

Main, A. J., Shope, R. E., and Wallis, R. C. (1976b). *J. Med. Entomol. 13,* 304-308.

Newhouse, V. F., Burgdorfer, W., McKiel, J. A., and Gregson, J. D. (1963). *Am. J. Hyg. 78,* 123-129.

Ritter, D. G., and Feltz, E. (1974). *Can. J. Microbiol. 20,* 1359-1366.

Roberts, F. H. S. (1970). "Australian Ticks." Commonwealth Sci. and Industrial Res. Organ., Melbourne, Australia.

Thomas, L. A., Flifford, C. M., Yunker, C. E., Keirans, J. E., Patzer, E. R., Monk, G. E., and Easton, E. R. (1973). *J. Med. Entomol. 10,* 165-168.

Timopheeva, A. A., Lvov, D. K., Pogrebenko, A. G., and Gromashevsky, V. L. (1974). *Zool. J. 53,* 906-911.

Votyakov, V. I., Voinov, I. N., Samoylova, T. I., Leshko, S. T., Gembitskiy, A. S., and Smirnov, V. A. (1974). *Proc. Symp. Ecology of Viruses Connected with Birds*, pp. 42-44. Minsk.
Wagner, R. J., De Long, C., Leung, M. K., McLintock, J., and Iverson, J. O. (1975). *Can. J. Microbiol. 21*, 574-576.
Watts, D. M., Pantuwatana, S., DeFoliart, G. R., Yuill, T. M., and Thompson, W. H. (1973). *Science 182*, 1140-1141.
Yunker, C. E. (1975). *Med. Biol. 53*, 302-311.
Yunker, C. E., Clifford, C. M., Keirans, J. E., Thomas, L. A., and Cory, J. (1973). *J. Med. Entomol. 10*, 264-269.
Zumpt, F. (1952). *In Austral. Nat. Antarctic Res. Expedition Rep. Bl*, 12-20.

Chapter 4

ARBOVIRUSES IN CANADA

Harvey Artsob and Leslie Spence

ISBN 0-12-429765-X

I. INTRODUCTION

Arbovirus investigations in Canada have resulted in the isolation of 16 arboviruses, including nine mosquito transmitted, six tick transmitted, and one culicoides transmitted (Table I). Evidence also exists for past or current activity of vesicular stomatitis and Bluetongue viruses. [The classification of vesicular stomatitis as an arbovirus is questionable. However, a Subcommittee on Evaluation of Arthropod-borne Status has categorized different members of the VSV group either as arboviruses or as probable or possible arboviruses ("International Catalogue of Arboviruses," 2nd Edition, edited by T. O. Berge, U. S. Dept. Health, Education and Welfare, 1975). For purposes of this review, the VSV group are being considered as arboviruses.] It exists less strongly for the possible occurrence of Buttonwillow and other flaviviruses in Canada.

This chapter will review studies concerning the natural occurrence of these arboviruses in Canada as well as their documented role in human and veterinary disease.

II. NATURAL OCCURRENCE OF ARBOVIRUSES IN CANADA

A. Western Equine Encephalitis

Since the first Canadian isolation of Western equine encephalitis (WEE) virus from horse brain in 1935 (Fulton, 1938), the virus has been found to be endemic in the prairie provinces. The virus has also been isolated from western Ontario (Mitchell and Walker, 1941) and British Columbia (Kettyls *et al.*, 1972) and one human infection has been diagnosed in Quebec (Pavilanis *et al.*, 1957).

Numerous studies have been undertaken to define the natural cycle(s) of WEE virus in Canada. Potential reservoirs have been examined including birds, mammals, reptiles, and amphibians.

Studies of Saskatchewan birds in 1962 and 1963 yielded 25 isolates of WEE virus including 23 isolates from English sparrows and one isolate each from a Swainson's hawk and a mourning dove (Burton *et al.*, 1966b; McLintock *et al.*, 1966). The large number of isolates from nestling English sparrows suggests that they are an important source of WEE virus. The isolate from a mourning dove occurred at a time of the year when mosquito activity was lacking, providing suggestive evidence for latent infection of the bird (Burton *et al.*, 1966b).

TABLE I
Arboviruses Isolated in Canada

Virus	*Taxonomic status*[a]	*Vector*
Eastern equine encephalitis	Alphavirus	Mosquito
Western equine encephalitis	Alphavirus	Mosquito
St. Louis encephalitis	Flavivirus	Mosquito
California encephalitis		
snowshoe hare	Bunyavirus	Mosquito
Jamestown Canyon	Bunavirus	Mosquito
Cache Valley	Bunyavirus	Mosquito
Northway	Bunyavirus	Mosquito
Turlock	Bunyavirus-like	Mosquito
Flanders-Hart Park	Rhabdovirus	Mosquito
Powassan	Flavivirus	Tick
Silverwater	Bunyavirus-like	Tick
Colorado Tick Fever	Orbivirus	Tick
Great Island	Orbivirus	Tick
Bauline	Orbivirus	Tick
Avalon	Unclassified	Tick
Epizootic hemorrhagic disease	Orbivirus	Culicoides

[a]*According to the "International Catalogue of Arboviruses," 2nd edition, edited by T. O. Berge, U. S. Dept. Health, Education and Welfare, 1975.*

Neutralizing antibodies have been demonstrated in several wild duck species (Burton *et al.*, 1961), chickens (Hoff *et al.*, 1970), turkeys (Burton and McLintock, 1970), and in other bird species in the prairie provinces including Brewer's blackbird, common red-wing, rock dove, English sparrow, barn swallow, starling, ruffed grouse, crow, Swainson's hawk, Franklin's gull, ring-billed gull, magpie, robin, and sharptailed grouse (Burton *et al.*, 1966b).

The Richardson ground squirrel *Spermophilus richardsonii* has been implicated as a possible reservoir of WEE on the prairies. Several virus isolations have been obtained (Burton *et al.*, 1966b; Gwatkin and Moynihan, 1942; Leung *et al.*, 1975) and analysis of antibody rates of Richardson ground squirrels in Saskatchewan from 1964 to 1973 revealed a high rate of seropositive squirrels (11.6%) in the epidemic year 1965 but low rates (less than 2%) in subsequent years (Leung *et al.*, 1975).

Viremias of 3 to 5 days duration and of sufficient magnitude to infect numerous mosquito species could be experimentally induced, emphasizing the potential of this rodent to serve as an amplifying agent of WEE virus (Leung *et al.*, 1976). The possibility that Richardson ground squirrels can be latently infected with WEE virus has also been raised following an early spring isolation of virus from one squirrel (Burton *et al.*, 1966b).

Studies of antibodies to WEE virus in snowshoe hare *Lepus americanus* in Alberta revealed evidence of late spring epizootics in 1963 (Yuill and Hanson, 1964) and in 1965 (Yuill *et al.*, 1969). Both epizootics preceded by about two months outbreaks of WEE infecting man and horses in Alberta, and it was suggested that hares may be effective amplifying hosts during epidemic times (Yuill and Hanson, 1964).

Neutralizing antibodies have been found in other mammals including Franklin ground squirrels (Yuill and Hanson, 1964; Yuill *et al.*, 1969; Hoff *et al.*, 1970) red foxes, pigs, skunks, reindeer and bison (Burton and McLintock, 1970), moose (Trainer and Hoff, 1971), pronghorns (Barrett and Chalmers, 1975), red-backed voles, weasles, cattle, and domestic rabbits (Hoff *et al.*, 1970) and in muskrats (Burton *et al.*, 1966b).

Reptiles and amphibians have been implicated as possible reservoirs of WEE virus. Virus isolations have been reported from nine naturally infected garter snakes and six leopard frogs (Burton *et al.*, 1966a) and neutralizing antibodies have been demonstrated in these species (Spalatin *et al.*, 1964; Burton *et al.*, 1966a; Prior and Agnew, 1971).

Western equine encephalitis virus has been isolated from *Culex, Culiseta,* and *Aedes* genera mosquitoes in the prairie provinces (McLintock, 1947; McLintock *et al.*, 1966, 1970; Norris, 1946; Shemanchuk and Morgante, 1968; Spalatin *et al.*, 1963). Whereas *Culex tarsalis* is undoubtedly the principal epidemic transmitter, *Culiseta inornata* may be a significant transmitter of virus in horse epidemics (McLintock *et al.*, 1970).

McLintock (1976) has suggested that early spring *Aedes* species such as *A. campestris, A. spencerii,* and *A. flavescens* may be involved in early summer amplification of WEE virus before *C. tarsalis* emerges from hibernation. In northern areas where *C. tarsalis* is not abundant, virus transmission is likely by *C. inornata* or *Aedes* mosquitoes (Burton and McLintock, 1970).

B. Eastern Equine Encephalitis

Eastern equine encephalitis (EEE) virus was first isolated in Canada in 1938 from the blood of a horse at St. George, Ontario (Schofield and Labzoffsky, 1938). Other isolations of EEE virus have been made in Canada from the blood of a migrating junco captured at Long Point, Ontario in 1961 (Karstad, 1965), from two snowshoe hares captured in central Alberta in 1962 (Yuill *et al.*, 1969), and from horse brain in the Eastern Townships of Quebec in 1972 (Bellavance *et al.*, 1973).

Neutralizing antibodies have been detected in snowshoe hares taken in north-central Alberta in 1961 and 1969 (Hoff *et al.*, 1970), in ruffed grouse captured in Alberta (Hoff *et al.*, 1970), and in pronghorns from southwestern Saskatchewan and southeastern Alberta in 1971 and 1972 (Barrett and Chalmers, 1975).

Antibodies to EEE virus have also been detected in migratory birds caught in Ontario at Long Point (Karstad, 1965) and at Prince Edward Point (Stewart, 1977). Apparent EEE virus activity was detected in Quebec province in 1976 as evidenced by hemagglutination inhibition antibody conversions in sentinel chickens (S. Belloncik, personal communication).

Eastern equine encephalitis virus has never been isolated from mosquitoes in Canada. However, potentially important vectors, *Culiseta melanura* and *Coquillettidia perturbans* are present in Quebec (Harrison *et al.*, 1975). The distribution of *C. perturbans* extends from Nova Scotia to British Columbia (Steward and McWade, 1960).

C. St. Louis Encephalitis

St. Louis encephalitis (SLE) virus was first isolated in Canada in 1971 from a pool of *C. tarsalis* collected in the Weyburn area of southern Saskatchewan (Burton *et al.*, 1973). The only other SLE virus isolate reported to date in Canada has been from human brain during a 1975 outbreak in Ontario (Spence *et al.*, 1977).

A low incidence of hemagglutination inhibition (HI) antibodies to SLE antigen has been observed in migratory birds captured at Long Point, Ontario in 1961 and 1962 (Karstad, 1965) and at Prince Edward Point in 1976 (Stewart, 1977). High HI antibody rates were observed in nonmigratory English sparrows captured in the Windsor-Essex region following the SLE outbreak in southern Ontario in 1975 (R. Dorland, H. Artsob, and L. Spence unpublished data). Thus, it would appear that English sparrows were an important reservoir during the outbreak.

No isolations of virus were obtained from a limited number of mosquito pools screened following the 1975 outbreak. However, the most important vector of SLE virus was probably *Culex pipiens*, which occurs fairly commonly in southern Ontario, only being unrecorded from central and northern areas (Steward and McWade, 1960) where no cases were diagnosed. *Culex pipiens* is also found in the province of Quebec (Harrison and Cousineau, 1973).

Antibodies to SLE virus have been found in western Canada including Alberta, where positive serology was obtained in one moose (Trainer and Hoff, 1971), two bull snakes (Spalatin *et al.*, 1964), and in snowshoe hares (Hoff *et al.*, 1970). In British Columbia neutralizing antibodies were observed in sera from wild rodents, marmots, three nonmigratory birds, and snowshoe hares (McLean *et al.*, 1968, 1969, 1970, 1971).

D. California Encephalitis

As shown in Table II, California encephalitis (CAL) virus has a widespread distribution in Canada, having been isolated in Quebec, Ontario, Saskatchewan, Alberta, British Columbia, the Yukon, and the Northwest Territories. In addition, neutralizing antibodies to CAL virus have been detected in Nova Scotia (Embree *et al.*, 1976) and Manitoba (McKiel *et al.*, 1970). Most CAL isolations have been of the snowshoe hare serotype but the Jamestown Canyon serotype has been isolated in Saskatchewan and Alberta (Table II).

Most mammalian surveys in Canada for the natural distribution of CAL virus have centered around the snowshoe hare, *L. americanus*, which is a primary reservoir. Virus has been isolated from a juvenile hare captured in Alberta (Hoff *et al.*, 1969) and antibodies have been detected in hares throughout Canada including Nova Scotia (Embree *et al.*, 1976), Quebec (Belloncik *et al.*, 1975), Ontario (Newhouse *et al.*, 1963, 1964; McLean *et al.*, 1964a, 1966, 1968), Alberta (Yuill and Hanson, 1964; Yuill *et al.*, 1969; Hoff *et al.*, 1969, 1970), British Columbia (Newhouse *et al.*, 1963; McLean *et al.*, 1968, 1971), and the Yukon (McLean *et al.*, 1972, 1973, 1974, 1975).

Other small mammals appear to play a role in the natural occurence of CAL virus in Canada. These include chipmunks, *Eutamias amoenus* (McLean *et al.*, 1970), ground squirrels, *Citellus franklinii* (Hoff *et al.*, 1970) and *C. undulatus* (McLean *et al.*, 1972, 1973, 1974, 1975), and red squirrels, *Tamiasciurus hudsconicus* (McLean *et al.*, 1968, 1971, 1973, 1974, 1975; Hoff *et al.*, 1970). *Citellus undulatus* was shown to be the probable main reservoir for snowshoe hare virus in the Yukon in 1974 at a time when snowshoe hares were virtually absent (McLean *et al.*, 1975).

Neutralizing antibodies have been detected in many other mammals including marmots (Newhouse *et al.*, 1964; McLean *et al.*, 1968, 1970, 1971), porcupines (Newhouse *et al.*, 1964; Hoff *et al.*, 1970), racoons (Hoff *et al.*, 1970), coyotes (Hoff *et al.*, 1970), lynx (Hoff *et al.*, 1970), moose (Hoff *et al.*, 1970), cattle (Hoff *et al.*, 1970), and horses (Hoff *et al.*, 1970; H. Artsob and L. Spence unpublished data).

California encephalitis virus has been isolated from *C. inornata* and several *Aedes* species of mosquitoes (Table II). In addition to isolations from adult mosquitoes, the snowshoe hare serotype has been isolated from *Aedes* sp. larvae and pupae collected in the Yukon (McLean *et al.*, 1975, 1977) and from *A. implicatus* reared from larvae collected in Saskatchewan (Mclintock *et al.*, 1976). These isolations strongly suggest the possibility of snowshoe hare virus overwintering by transovarial transfer.

E. Powassan

Isolations of Powassan (POW) virus have been made in Ontario from human brain (McLean and Donohue, 1959), tick pools (McLean and Bryce Larke, 1963; McLean *et al.*, 1964b, 1966, 1967), squirrel blood (McLean and Bryce Larke, 1963) and groundhog blood (McLean *et al.*, 1964b). Human infections due to POW virus have been diagnosed in Quebec (Section III, E) and serological evidence has been found to suggest the natural occurrence of POW virus in Nova Scotia (Embree *et al.*, 1976), Alberta (Hoff *et al.*, 1970), and British Columbia (McLean *et al.*, 1968, 1969, 1970, 1971).

Studies for neutralizing antibodies to POW virus have revealed evidence of the exposure of numerous small mammals to POW including squirrels (McLean *et al.*, 1960, 1961, 1962, 1964a, b, 1967, 1968, 1969; McLean and Bryce Larke, 1963), chipmunks (McLean *et al.*, 1960, 1961 1962, 1968, 1969), snowshoe hares (McLean *et al.*, 1961, 1968, 1969, 1970, 1971; Hoff *et al.*, 1970), groundhogs (McLean *et al.*, 1964b 1966, 1967), porcupines (McLean *et al.*, 1964a), and woodchucks (McLean *et al.*, 1964a)

McLean and co-workers undertook studies of the natural cycle of POW virus in northern Ontario and presented evidence to implicate the red squirrel *T. hudsonicus*, and the groundhog *Marmota monax* as important reservoirs of POW virus. Virus isolations were achieved from *Ixodes marxi* ticks collected from one red squirrel and from the blood of a second squirrel (McLean and Bryce Larke, 1963). These isolations as well as high antibody rates in squirrels in the north bay-Powassan region of Ontario led to the suggestion that squirrels are important reservoirs in the maintenance cycle of POW virus (Mc-

TABLE II
Isolations of California Encephalitis Virus in Canada

Province or territory	*Year*	*Source*	*Serotype*	*Reference*
Quebec	1974	*A. communis*	Type identity pending	J. Woodall and M. Grayson, personal communication
Ontario	1963	Indicator rabbits	Snowshoe hare	McKiel *et al.*, 1966
	1975	*A. fitchii, A. triseriatus*	Snowshoe hare	Wright *et al.*, 1977
Saskatchewan	1972	*A. fitchii gr., A. cataphylla, A. punctor gr., A. excrucians*	Snowshoe hare and Jamestown Canyon	Iversen *et al.*, 1973
	1975	*A. implicatus*	Snowshoe hare	McLintock *et al.*, 1976
Alberta	1965	*C. inornata*	Snowshoe hare, La-Crosse BFS 283 subgroup	Morgante and Shemanchuk 1967
	1964-1968	*A. communis gr.* *A. stimulans gr.*	Snowshoe hare	Hoff *et al.*, 1969; Iversen *et al.*, 1969
	1964-1965	*Aedes sp.* *A. communis gr.*	Jamestown Canyon	Hoff *et al.*, 1969; Iversen *et al.*, 1969
	1968	Snowshoe hare	Snowshoe hare	Hoff *et al.*, 1969
British Columbia	1969	Mixed *A. vexans-A. canadensis* pool	Snowshoe hare	McLean, 1970; McLean *et al.*, 1970
	1973	*A. fitchii*	Snowshoe hare	McLean *et al.*, 1974

Yukon	1971	*A. canadensis*	Snowshoe hare	McLean *et al.*, 1972, 1975
	1972	*A. communis*	Snowshoe hare	McLean *et al.*, 1973, 1975
	1973	*A. canadensis*, *A. communis*, *A. cinerus*, *C. inornata*	Snowshoe hare	McLean *et al.*, 1974, 1975
	1974	*A. canadensis*, *A. communis*	Snowshoe hare	McLean *et al.*, 1975
	1975	*Aedes* sp. larvae	Snowshoe hare	McLean *et al.*, 1977
Northwest Territories	1973	*A. hexodontus-punctor* gr.	Snowshoe hare	Wagner *et al.*, 1975

Lean and Bryce Larke 1963; McLean *et al.*, 1967). Similar virus isolations and antibody data were obtained from groundhogs (McLean *et al.*, 1964b, 1966, 1967).

Powassan virus has been isolated from *Ixodes* ticks collected in northern Ontario including *I. marxi* (McLean and Bryce Larke, 1963) and *I. cookei* (McLean *et al.*, 1964b, 1966, 1967). The distribution of *I. cookei* ticks is known to extend to southern Ontario (Ko, 1972) and to the Eastern Townships of Quebec (Rossier *et al.*, 1974; Harrison *et al.*, 1975b).

F. Colorado Tick Fever

Isolations of Colorado Tick Fever (CTF) virus have been made from *Dermacentor andersoni* ticks collected in Alberta (Brown 1955; Edkund *et al.*, 1955) and in southeastern British Columbia (Hall *et al.*, 1968b). In addition, neutralizing antibodies have been detected in five of 49 (9.8%) of snowshoe hares captured in Richmond, Ontario, providing suggestive evidence for the natural occurrence of CTF virus in eastern Canada (Newhouse *et al.*, 1964).

G. Flanders-Hart Park

Two isolates of the Flanders-Hart Park group were obtained from pools of *C. tarsalis* collected in Saskatchewan in 1967 (Hall *et al.*, 1969). Subsequent isolations were made in Saskatchewan from *C. tarsalis, Aedes vexans,* and *A. spencerii* mosquitoes collected in 1970 and 1971 and from birds including one house sparrow and two barn swallows (McLintock and Iverson, 1975).

Finally, an isolation of Flanders virus was made from a pool of *A. triseriatus* collected near London, Ontario in 1976 (Wright *et al.*, 1977).

H. Turlock

Turlock virus was initially isolated in Canada from a pool of *C. tarsalis* collected in Saskatchewan in 1962 (Hall *et al.*, 1968a). Additional isolations were achieved in 1965 from a pool of *C. inornata* collected in Alberta (Hall *et al.*, 1968a) and in 1970 in Saskatchewan from three *C. tarsalis* pools, from the blood of a house sparrow, and from the brain of a barn swallow (McLintock and Iversen, 1975).

I. Cache Valley

A bunyamwera group virus, probably Cache Valley (CV) was isolated from a pool of 63 unidentified mosquitoes collected in Alberta in 1965 (Hall *et al.*, 1968a). Another isolation of CV virus was made from a pool of 75 *C. inornata* collected in the Weyburn area of southern Saskatchewan in 1971 (Burton *et al.*, 1973). Subsequent CV isolations were made in Saskatchewan primarily from *C. inornata* but also from *C. tarsalis* (McLintock and Iversen, 1975).

J. Northway

Northway virus was originally isolated from sentinel rabbits and mosquitoes in Alaska and the authors speculated that Northway virus activity might extend to contiguous portions of Canada (Calisher *et al.*, 1974). This speculation was confirmed by a recent isolation of Northway virus in Canada (McLean, 1978).

K. Silverwater

Silverwater (SIL) virus was first isolated from a pool of *Haemophysalis leporis-palustris* ticks collected from snowshoe hares at Manitoulin Island in 1960 (McLean 1961; McLean *et al.*, 1961). Further studies in northern Ontario resulted in other SIL virus isolations from HLP ticks (McLean *et al.*, 1962, 1964a, 1966; McLean and Bryce Larke, 1963) and high rates of neutralizing antibodies were detected in snowshoe hares (McLean *et al.*, 1962, 1964a, 1966, 1967; McLean and Bryce Larke, 1963).

The importance of snowshoe hares as reservoirs of SIL virus was confirmed by studies conducted in Alberta. Three SIL virus isolates were obtained from viremic hares and 15 isolations were made from HLP tick pools collected from hares during the spring and summer from 1962 to 1965 (Yuill *et al.*, 1969; Hoff *et al.*, 1969, 1971a). In addition, neutralizing antibodies to SIL virus were detected in the hare population from 1961 to 1969 (Yuill *et al.*, 1969; Hoff *et al.*, 1970, 1971b).

Antibodies to SIL virus were demonstrated in five other vertebrate species in Alberta including red squirrels, chipmunks, coyotes, domestic rabbits, and domestic cattle (Hoff *et al.*, 1970, 1971b).

L. Great Island

Great Island (GI) virus, a Kemerovo group arbovirus, was isolated from *Ixodes uriae* ticks collected from a puffin (*Fratercula arctica)* colony on Great Island, Newfoundland (Main *et al.*, 1973). Six isolates were obtained in 1971 and one isolate in 1972. Neutralizing antibodies to GI virus were found in 19.8% of avian sera tested including Atlantic Puffins (37.0%) and Leach's Petrels (5.8%) (Main *et al.*, 1976b).

M. Bauline

Bauline (BAU) virus is another Kemerovo group arbovirus that was isolated from *I. uriae* ticks collected from a puffin (*F. arctica*) colony on Great Island, Newfoundland (Main *et al.*, 1973). Six isolations of BAU virus were achieved including one isolate in 1971 and five isolates in 1972. Neutralizing antibodies were detected in 18.3% of avian sera tested including Atlantic Puffins (37.3%) and Leach's Petrels (3.9%) (Main *et al.*, 1976b).

N. Avalon

A Sakhalin group arbovirus, Avalon (AVA) virus, was isolated from three lots of *I. uriae* ticks and from the blood of a Herring Gull *Larus argentatus* chick collected in a common puffin *F. arctica* colony on Great Island, Newfoundland in 1972 (Main *et al.*, 1972a). Neutralizing antibodies to AVA virus were also demonstrated in 27.6% of avian sera tested including Atlantic Puffins, Leach's Petrels, and Herring Gulls (Main *et al.*, 1976b).

O. Epizootic Hemorrhagic Disease

An epizootic of epizootic hemorrhagic disease (EHD) of wild ungulates was recognized in Alberta during the summer of 1962. Mortalities included approximately 450 white-tailed deer *Odocoileus virginianus*, 20 mule deer *O. hemionus*, and 15 pronghorn antelopes *Antilocapra americana* (Chalmers *et al.*, 1964). The causative agent was recognized as being serologically identical with the New Jersey strain of EHD virus (Ditchfield *et al.*, 1964). Evidence of EHD virus acitivity in British Colombia has also been reported (Trainer, 1964).

Subsequent serological studies of pronghorns captured in southwestern Saskatchewan in 1970 and in southeastern Alberta in 1971 and 1972 (Barrett and Chalmers, 1975) and of moose from southeastern Alberts (Trainer and Hoff, 1971) revealed no evidence of antibodies to EHD virus.

P. Buttonwillow

Buttonwillow (BUT) virus has never been isolated in Canada. However, HI antibodies were found to BUT virus in one marmot from Ontario and in 21 of 117 snowshoe hares from Alberta, leading to speculation that BUT, or a closely related Simbu group virus, may be circulating in these areas (Reeves *et al.*, 1970). Neutralizing antibodies to BUT virus have also been detected in snowshoe hares captured in Alberta in 1968 and 1969 (Hoff *et al.*, 1970).

Q. Other Flaviviruses

Serological surveys employing flavivirus antigens in Newfoundland (Main *et al.*, 1976b), British Columbia (McLean *et al.*, 1969, 1970), and the Yukon (McLean *et al.*, 1972) yielded evidence of HI positive sera that did not neutralize the corresponding flaviviruses used. Assuming these HI reactions to be specific, circumstantial evidence has thus been provided to support speculation that other flaviviruses may be circulating in Canada.

III. EXPOSURE OF HUMANS TO ARBOVIRUSES IN CANADA

A. Western Equine Encephalitis

Historically, WEE has been the most important arbovirus in Canada with regard to human disease. Its recognized role in causing human infections in Canada dates back to 1938, when 27 encephalitis cases were reported in Manitoba (Donovan and Bowman, 1942 a, b) and 29 in Saskatchewan (Gareau, 1941). In 1941 a large epidemic occurred in the prairie provinces, with 509 cases in Manitoba (Jackson, 1942), 543 cases in Saskatchewan (Davidson, 1942), and 42 cases in Alberta (McGugan, 1942).

The 1941 epidemic was studied extensively with reports concerning clinical findings of WEE infection (Adamson and Dubo, 1942 a, b) as well as reports on WEE infection in infants (Medovy, 1942) and on the pathology of 18 fatal cases (Quong,

1942). During the epidemic, WEE was isolated from the CSF of one six-week-old boy in Manitoba (Chown and Norris, 1942) and an adult in Alberta (Gwatkin and Moynihan, 1943). Both patients recovered. A survey of 1013 Manitoba residents following the 1941 outbreak revealed neutralizing antibodies in 18.95% of the population (Mitchell and Pullin, 1943).

Subsequent outbreaks have occurred in Canada as shown in Table III. In addition to the prairie provinces, one case was diagnosed in Quebec in 1955 and human infection due to WEE virus was recognized in British Columbia in 1971 and 1972. Sporadic cases of WEE infection have occurred in the prairie provinces without much documentation, including 15 cases in Manitoba between 1960 and 1975 (Medovy, 1976). Circumstances relating to the most recent human outbreak of WEE in Canada, which occurred in Manitoba in 1975, have been well documented in a recent monograph (Waters, 1976).

B. Eastern Equine Encephalitis

No clinical cases of EEE have ever been diagnosed in Canada. A study of 208 human sera from north-central Alberta tested between 1961 and 1968 revealed no evidence of neutralizing antibodies (Hoff *et al.*, 1970). However, a survey of 4706 human sera collected from all parts of the province of Quebec between 1971 and 1974 yielded three HI positive reactors (Artsob *et al.*, 1978).

C. St. Louis Encephalitis

Diagnostic tests conducted on patients with central nervous system disease revealed neutralizing antibodies to SLE virus in one 71-year-old female in Manitoba in 1941 (Donovan and Bowman, 1942a, b) and complement-fixing antibodies to SLE in another patient in Saskatchewan in 1954 (Dillenberg *et al.*, 1956). However, no human cases of SLE infection were diagnosed in Canada until 1975, when 66 cases were reported from Ontario (Davidson *et al.*, 1976; Spence *et al.*, 1977) and one case each from Quebec (Davidson *et al.*, 1976) and Manitoba (Davidson *et al.*, 1976; Sekla and Stackiw, 1976).

The 1975 outbreak in Ontario occurred in the southern part of the province encompassing the Windsor-Sarnia-Chatham area, the Niagara region, and metropolitan Toronto. Fatalities occurred in four laboratory-confirmed cases and virus was recovered from the brain of one fatal case (Spence *et al.*, 1977). St. Louis encephalitis reappeared in Canada in 1976 with four

TABLE III
Human Infections in Canada Due to Western Equine Encephalitis Virus

Year	*Province*	*Comments*	*Reference*
1935+ 1937	Saskatchewan	Large equine outbreaks; increasing number of acute encephalitis in humans	Fulton, 1941
1938	Manitoba	27 encephalitis cases with 6 deaths	Donovan and Bowman, 1942a,b
	Saskatchewan	29 cases with 4 deaths	Gareau, 1941
1939	Saskatchewan	Few cases; WEE recovered from brain of one fatal case.	Fulton, 1941
1941	Manitoba	Epidemic with 509 cases and 78 deaths; outbreak predominantly rural; 70% of cases males; concurrent polio outbreak.	Jackson, 1942; Donovan and Bowman 1942a,b
	Saskatchewan	543 cases with 44 deaths; 66% of cases males; attempted treatment with immune horse sera.	Davidson, 1942
	Alberta	42 reported cases with 8 deaths; concurrent polio outbreak	McGugan, 1942
1947	Manitoba	81 human cases	Snell, 1966
	Saskatchewan	Sizable outbreak with 68 cases	Dillenberg, 1965
1953	Saskatchewan	Localized outbreak in Radville; 53 cases including some polio infections	Dillenberg *et al.*, 1956
1955	Quebec	Serologically diagnosed case in 20-month-old infant on Montreal Island	Pavilanis *et al.*, 1957
1963	Saskatchewan	Estimate 90-100 cases; 33 serologically confirmed with 3 deaths; simultaneous enterovirus infections	Medical News, 1963; Epid. Bull., 1964; Dillenberg, 1965
	Alberta	6 cases of WEE occurred simultaneously with echovirus 9 outbreak	Morgante *et al.*, 1968a
1965	Saskatchewan	72 cases with 8 deaths; WEE from CSF, throat washing, and brain specimens	Rozdilsky *et al.*, 1968
1971	British Colombia	5 human cases with 2 deaths	Kettyls *et al.*, 1972
1972	British Colombia	14 encephalitis cases with 7 laboratory confirmed as WEE	Kettyls and Bowmer, 1975
1975	Manitoba	14 laboratory diagnosed cases with no deaths; permanent sequelae in four cases	Sekla and Stackiw, 1976

human cases diagnosed from southern Ontario (Weekly Bulletin, Community Health Protection Branch, Epidemiology Service, Ontario Ministry of Health, Oct. 8, 1976).

Serological evidence is also available to suggest human infection with SLE, or a closely related virus, in western Canada. Neutralizing antibodies to SLE have been demonstrated in residents of south Saskatchewan (McLintock, 1976), north-central Alberta (Hoff *et al.*, 1970), and British Columbia (McLean *et al.*, 1969; Kettyls *et al.*, 1972).

D. California Encephalitis

No clinical cases of CAL infection have been diagnosed in Canada. However, evidence does exist to document the exposure of humans in different parts of Canada to CAL virus. Hemagglutination inhibition and neutralizing antibodies to snowshoe hare virus were detected in 25 of 4706 sera collected in the province of Quebec between 1971 and 1974 (Artsob *et al.*, 1978). Antibodies to CAL virus have also been detected in residents of Ontario (Newhouse *et al.*, 1963; McKiel *et al.*, 1970; Spence *et al.*, 1977), Manitoba (McKiel *et al.*, 1970), north central Saskatchewan (McLintock and Iversen, 1975), Alberta (Hoff *et al.*, 1970; Iversen *et al.*, 1971), British Columbia (Kettyls *et al.*, 1968, 1972), and the Northwest Territories (McLintock and Iversen, 1975).

In view of the widespread distribution of CAL virus in Canada (see Section II, D) and the documented exposure of residents in several provinces to CAL virus, diagnostic tests should be emphasized for suspect CAL cases, particularly in children. It should be stressed that severe encephalitis occurs in only a low porportion of infected individuals and that clinically apparent infections with CAL group viruses are generally manifested as mild, undifferentiated febrile illnesses, influenzalike syndromes, or acute central nervous system diseases (Lindsey *et al.*, 1976). Thus far, the predominant CAL serotype circulating in Canada, snowshoe hare, has not been definitely implicated in human disease, but continued vigilance is warranted.

E. Powassan

The first diagnosed human case of POW infection occurred in 1958 in a 5-year-old boy from Powassan, Ontario, who developed acute encephalitis and died (McLean and Donohue, 1959). Perivascular cuffing typical of viral encephalitis was observed in brain sections and the agent, isolated from brain material, was identified as a new flavivirus.

Three subsequent cases of POW infection have been diagnosed in Canada. The second case involved an 8-year-old boy in the Eastern Townships of Quebec in 1972 (Rossier *et al.*, 1974; Harrison *et al.*, 1975b). The patient displayed a severe case of encephalitis but survived suffering moderate sequelae. In 1975 another severe case of POW encephalitis was diagnosed in a 3½-year-old boy from western Quebec (Conway *et al.*, 1976). The virus also affected the upper cervical cord of the patient, resulting in the involvement of shoulder girdle muscles. Some residual sequelae was observed. Finally, a fourth case of POW infection was diagnosed in a 15-year-old female in Ontario in 1976 (Rossier, 1976). Symptoms were mild and no sequelae was observed.

Studies by McLean and co-workers from 1959 to 1961 revealed evidence of neutralizing antibodies to POW virus in northern Ontario residents, particularly in areas around Powassan, North Bay, and Manitoulin Island (McLean *et al.*, 1960, 1961, 1962). Hemagglutination inhibition and complement fixing antibodies to POW virus have also been detected in human sera in British Columbia (Kettyls *et al.*, 1968, 1972).

F. Colorado Tick Fever

No human cases of CTF virus infection have been diagnosed in Canada. However, serological surveys in British Columbia revealed eight complement fixation positive reactors as well as one reactor who displayed neutralizing antibodies to CTF virus (Kettyls *et al.*, 1968, 1972).

Vigilance of physicians and diagnostic laboratories to possible CTF virus infections, particularly in Alberta and British Columbia, should be stressed.

IV. ARBOVIRUSES OF VETERINARY IMPORTANCE IN CANADA

A. Western Equine Encephalitis

The first definite recognition of WEE virus in Canada occurred in 1935 when virus was isolated from horse brain (Fulton, 1938). Prior to this time diseases of horses likely due to WEE virus occurred in western Canada and were described by such terms as forage poisoning, corn-stalk disease, sleeping sickness, blind staggers and cerebrospinal meniingitis.

Horse outbreaks due to WEE virus have been documented in Canada since 1935 with cases observed mainly in the prairie provinces, especially Manitoba and Saskatchewan, but also in western Ontario and British Columbia. The most extensive horse epizootics occurred between 1935 and 1938, involving over 60,000 horses. A subsequent decline was observed both in the morbidity and mortality of the disease following the introduction of a horse vaccine in 1938. However, due to the endemicity of WEE virus, small outbreaks still occur in western Canada.

A summary of WEE horse epizootics that have occurred in Canada is presented in Table IV.

B. Eastern Equine Encephalitis

A small outbreak of equine encephalitis occurred at St. George, Ontario in 1938 with 12 reported cases and two deaths (Schofield and Labzoffsky, 1938). Virus isolated from the blood of a horse was suggested by the authors as being EEE virus and this was confirmed in a related publication (Schoening *et al.*, 1939). Six or seven apparent cases of encephalitis also occurred in the St. Catharines area of Ontario in 1938 with some fatalities. No virus was demonstrated but histopathological changes characteristic of encephalitis were observed (Schofield and Labzoffsky, 1938).

Another equine outbreak occurred in the Eastern Townships of Quebec in 1972. Virus was isolated from the brains of five horses and it is believed that about 25 other horses died from the disease (Bellavance *et al.*, 1973; Harrison *et al.*, 1975a).

C. Vesicular Stomatitis

An outbreak due to vesicular stomatitis virus (VSV) was reported in cattle and horses in Manitoba in 1937. The disease affected livestock in wooded areas east of the Red River and along Lake Manitoba and Lake Winnipeg but failed to spread to livestock on farms situated on the open plain (Hanson, 1952).

A subsequent outbreak occurred in Manitoba in 1949 covering almost exactly the same territory that was invaded in 1937. The outbreak started near St. Paul, Minnesota in June and spread slowly east and west reaching into Manitoba in September. It is estimated that 500 horses and cattle were affected in Manitoba before the disease died out in October (Hanson, 1952) The New Jersey serotype of VSV was isolated in the United States and is the likely serotype to have invaded into Manitoba.

There are no published reports of VSV activity in eastern Canada or British Columbia. Twenty-three moose captured in southeastern Alberta were tested for antibodies to VSV but all were negative (Trainer and Hoff, 1971).

D. Bluetongue

Complement fixation tests conducted on 13,210 sheep sera and 13,486 bovine sera collected across Canada between 1970 and 1974 revealed no definite evidence for antibodies to Bluetongue (BLU) virus although three questionable reactors were obtained (Carrier and Boulanger, 1975). Similarly, no antibodies were detected in pronghorns captured in southwestern Saskatchewan in 1970 and in southeastern Alberta in 1971 and 1972 (Barrett and Chalmers, 1975) nor in moose captured in southeastern Alberta (Trainer and Hoff, 1971).

However, serological evidence for BLU virus activity in Canada was obtained in 1976. Complement-fixing antibodies were detected in sera from over 1000 cattle, sheep, goats, and deer from the Similkameen and Okanagon Valley areas of British Columbia. Confirmatory neutralizing antibodies were demonstrated in many of these positives. No apparent deaths or clinical illnesses were observed and attempted virus isolations from cattle blood, fetuses of seropositive dams, spleens from nursing calves of seropositive dams, and pools of culicoides flies have thus far proven negative (F. C. Thomas, personal communication).

V. DISCUSSION

Most arbovirus studies have been undertaken in central and western Canada. Few studies have been reported from the Atlantic provinces where little is known about the natural distribution of arboviruses including those of human health significance such as EEE, SLE, and CAL viruses.

Undoubtedly, the best studied arbovirus in Canada to date has been WEE virus. In addition, extensive investigations have been made concerning the natural occurrence of CAL virus in various parts of Canada including subarctic regions (McLean, 1975) and of POW virus in northern Ontario.

Eastern equine encephalitis has been observed only sporadically in Canada. However, the most recent evidence for EEE activity in Quebec in 1976 (Section II, B) with no reports of human or horse disease does underline the importance of constant

TABLE IV
Horse Epizootics in Canada Due to Western Equine Encephalitis Virus

Year	*Location*	*Comments*	*Reference*
1935	Manitoba and Saskatchewan	Large WEE outbreak; first isolation of WEE in Canada from horse brain	Fulton, 1938
1937	Ontario	Virus isolated from brain of horse infected in Rainy River District of western Ontario	Mitchell and Walker, 1941
	Manitoba	Large outbreak involving approximately 12,000 horses with 27% mortality rate	Jackson, 1942; Savage, 1942
	Saskatchewan	Extensive outbreak involving all sections of Saskatchewan except the extreme northwest	Fulton, 1938
1938	Manitoba	WEE outbreak extending into northwestern Manitoba	Savage, 1942
	Saskatchewan	52,500 horse cases with over 15,000 deaths	Davidson, 1942
1938-1940	Manitoba and Saskatchewan	Chick embryo vaccine introduced in 1938 with decline in disease during the following two years.	Fulton, 1941; Savage, 1942
1941	Prairie Provinces	Estimate between 250 and 350, mostly mild, cases in each province with low mortality rate	Cameron, 1942
1953-1954	Saskatchewan	76 reported cases in southeastern Saskatchewan	Dillenberg *et al.*, 1956

1962	Saskatchewan	Extensive horsebreak; no human cases	Spalatin *et al.*, 1963
1963	Manitoba	173 clinical cases	Lillie *et al.*, 1976
	Saskatchewan	279 cases with 47 deaths	Dillenberg, 1965
1964	Manitoba	73 clinical cases	Lillie *et al.*, 1976
1965	Manitoba	75 horse cases with nine deaths	Snell, 1966
	Alberta	63 serologically diagnosed and 73 presumptive positives; also five WEE isolations from brain	Morgante *et al.*, 1968b
1966	Manitoba	51 equine cases	McKay *et al.*, 1968
1967-1969	Manitoba	8 cases in 1967, 14 cases in 1968, and 41 cases in 1969	Lillie *et al.*, 1976
1971	British Columbia	60 cases in Okanagan Valley with 15 deaths	Kettyls *et al.*, 1972
1972-1973	British Columbia	17 equine cases in 1972 and 7 cases in 1973	Kettyls and Bowmer, 1975
1975	Ontario	Serological evidence for up to 4 cases in northwestern Ontario	Lillie *et al.*, 1976
	Manitoba	261 clinically suspect cases; 65 confirmed and 80 presumptive positives	Lillie *et al.*, 1976

vigilance for this potentially dangerous virus as well as the need for more intesnive studies concerning the natural distribution of this agent, particularly in eastern Canada.

The recent outbreak of SLE in southern Ontario in 1975 followed by four cases in 1976 as well as evidence for SLE activity in Quebec and western Canada point out the importance for further studies on this agent concerning distribution in Canada, mosquito vectors, animal reservoirs, and potential for endemicity.

An awareness should also be maintained for the possible introduction of new arboviruses into Canada. In this respect the potential for Japanese B encephalitis virus to become established in Canadian mosquitoes and wildlife has been stressed (West, 1958; McLintock and Iversen, 1975).

The observation of BLU virus antibodies in British Columbia in 1976 (Section IV, D) may reflect a new arbovirus introduction into Canada. However, no virus isolations were obtained and it is possible that this agent has been in Canada previously but was only serologically recognized in 1976 (F. C. Thomas, personal communication).

Addendum

Since preparing this review neutralization tests have been performed on the hemagglutination inhibition (HI) conversions demonstrated in sentinel chickens in the province of Quebec (Sections II, B and V). No neutralizing antibodies could be demonstrated and it would appear that the HI conversions were nonspecific.

References

Adamson, J. D., and Dubo, S. (1942a). *Can. Med. Assoc. J. 46,* 530-537.

Adamson, J. D., and Dubo, S. (1942b). *Can. Publ. Health J. 33,* 288-300.

Artsob, H., Spence, L., Th'ng, C., and West, R. (1978). Manuscript in preparation.

Barrett, M. W., and Chalmers, G. A. (1975). *J. Wildl. Dis. 11,* 157-163.

Bellavance, R., Rossier, E., LeMaitre, M., and Willis, N. G. (1973). *Can. J. Publ. Health 64,* 189-190.

Belloncik, S., Artsob, H., Trudel, C., and Spence, L. (1975). ACFAS, Moncton, New Brunswick (Abstract).

Brown, J. H. (1955). *Can. J. Zool. 33,* 389-390.

Burton, A. N., and McLintock, J. (1970). *Can. Vet. J. 11,* 232-235.

Burton, A. N., Connell, R., Rampel, J. G., and Gollop, J. B. (1961). *Can. J. Microbiol. 7*, 295-302.
Burton, A. N., McLintock, J., and Rempel, J. G. (1966a). *Science, 154*, 1029-1031.
Burton, A. N., McLintock, J., Spalatin, J., and Rempel, J. G. (1966b). *Can. J. Microbiol. 12*, 133-141.
Burton, A. N., McLintock, J., and Francy, D. B. (1973). *Can. J. Pub. Hlth. 64*, 368-373.
Calisher, C. H., Lindsey, H. S., Ritter, D. G., and Sommerman, K. M. (1974). *Can. J. Microbiol. 20*, 219-223.
Cameron, C. D. W. (1942). *Can. Publ. Hlth. J. 33*, 383-387.
Carrier, S. P., and Boulanger, P. (1975). *Can. J. Comp. Med. 39*, 231-233.
Chalmers, G. A., Vance, H. N., and Mitchell, G. J. (1964) *Wildl. Dis. 42*, 1-6.
Chown, B., and Norris, M. (1942). *JAMA 120*, 116-117.
Conway, D., Rossier, E., Spence, L., and Artsob, H. (1976) *Can. Dis. Weekly Rep. 2-22*, 85-87 Health and Welfare Canada.
Davidson, R. O. (1942). *Can. Pub. Hlth. J. 33*, 388-398.
Davidson, W. G., Snell, E., Joshua, J. M., and West, R. (1976). *Can. Dis. Weekly Rep. 2-21*, 81-83 Health and Welfare Canada.
Dillenberg, H. (1965). *Can. J. Publ. Health 56*, 17-20.
Dillenberg, H. O., Acker, M. S., Belcourt, J. P., and Nagler, F. P. (1956). *Can. J. Publ. Health 47*, 6-14.
Ditchfield, J., Debbie, J. G., and Karstad, L. (1964). *Trans. N. A. Wildl. Nat. Res. Conf. 29*, 196-201.
Donovan, C. R., and Bowman, M. (1942a). *Can. Med. Assoc. J. 46*, 525-530.
Donovan, C. R., and Bowman, M. (1942b). *Can. Publ. Health J. 33*, 246-257.
Eklund, C. M., Kohls, G. M., and Brennan, J. M. (1955). *JAMA 157*, 335-337.
Embree, J. E., Rozee, K. R., and Embil, J. A. (1976). *Abstr. 44th Annu. Meeting, Can. Publ. Health Assoc., Montreal, Quebec.*
Epidemiological Bulletin (1964). *Med. Serv. J. Can. 20*, 103-105.
Fulton, J. S. (1938). *Can. J. Comp. Med. 2*, 39-46.
Fulton, J. S. (1941). *Can. Publ. Health J. 32*, 6-12.
Gareau, U. (1941). *Can. Publ. Health J. 32*, 1-5.
Gwatkin, R., and Moynihan, I. W. (1942). *Can. J. Res. 20*, 321-337.
Gwatkin, R., and Moynihan, I. W. (1943). *Can. J. Publ. Health 34*, 42-43.
Hall, R. R., McKiel, J. A., and Brown, J. H. (1968a). *Can. J. Publ. Health 59*, 159-160.
Hall, R. R., McKiel, J. A., and Gregson, J. D. (1968b). *Can. J. Publ. Health 59*, 273-275.

Hall, R. R., McKiel, J. A., McLintock, J., and Burton, A. N. (1969). *Can. J. Publ. Health 60,* 486-488.
Hanson, R. P. (1952). *Bacterial Rev. 16,* 179-204.
Harrison, R. J., and Cousineau, G. (1973). *Ann. Soc. Ent. Quebec 18,* 138-146.
Harrison, R. J., Rossier, E., and Lemieux, B. (1975a). *Ann. Soc. Ent. Quebec 20,* 27-32.
Harrison, R. J., Rossier, E., and Lemieux, B. (1975b). *Ann. Soc. Ent. Quebec 20,* 48-52.
Hoff, G. L., Yuill, T. M., Iversen, J. O., and Hanson, R. P. (1969). *Bull. Wildlife Dis. Assoc. 5,* 254-259.
Hoff, G. L., Yuill, T. M., Iversen, J. O., and Hanson, R. P. (1970). *J. Wildlife Dis. 6,* 472-478.
Hoff, G. L.,Iversen, J. O., Yuill, T. M., Anslow, R. O., Jackson, J. O., and Hanson, R. P. (1971a). *Am. J. Trop. Med. Hyg. 20,* 320-325.
Hoff, G. L., Yuill, T. M., Iversen, J. O., and Hanson, R. P. (1971b). *Am. J. Trop. Med. Hyg. 20,* 326-330.
Iversen, J., Hanson, R. P., Papadopoulos, O., Morris, C. V., and DeFoliart, G. R. (1969). *Am. J. Trop. Med. Hyg. 18,* 735-742.
Iversen, J. O., Seawright, G., and Hanson, R. P. (1971). *Can. J. Publ. Health 62,* 125-132.
Iversen, J. O., Wagner, R. J., DeJong, C., and McLintock, J. (1973). *Can. J. Publ. Health 64,* 590-594.
Jackson, F. W. (1942). *Can. Med. Assoc. J. 47,* 364-365.
Karstad, L. (1965). *Ontario Bird Banding 1,* 1-9.
Kettyls, G. D., and Bowmer, E. J. (1975). *Epidemiol. Bull.* Health and Welfare Canada, *19,* 24.
Kettyls, G. D., Verrall, V. M., Hopper, J. M. H., Kokan, P., and Schmitt, N. (1968). *Can. Med. Assoc. J. 99,* 600-603.
Kettyls, G. D., Verrall, V. M., Wilton, L. D., Clapp, J. B., Clarke, D. A., and Rublee, J. D. (1972). *Can. Med. Assoc. J. 106,* 1175-1179.
Ko, R. C. (1972). *Can. J. Zool. 50,* 433-436.
Leung, M. K., Burton, A., Iversen, J., and McLintock, J. (1975). *Can. J. Microbiol. 21,* 954-958.
Leung, M. K., Iversen, J., McLintock, J., and Saunders, J. R. (1976). *J. Wildlife Dis. 12,* 237-246.
Lillie, L. E., Wong, F. C., and Drysdale, R. A. (1976). *Can. J. Publ. Health 67 (Suppl. 1),* 21-27.
Lindsey, H. S., Calisher, C. H., and Mathews, J. H. (1976). *J. Clin. Microbiol. 4,* 503-510.
McGugan, A. C. (1942). *Can. Publ. Health J. 33,* 148-151.
McKay, J. F. W., Stackiw, W., and Brust, R. A. (1968). *Manitoba Med. Rev. 48,* 56-57.
McKiel, J. A., Hall, R. R., and Newhouse, V. F. (1966). *Am. J. Trop. Med. Hyg. 15,* 98-102.

McKiel, J. A., Hall, R. R., Valentine, G. H., and Tusz, L. J. (1970). *Abstr. 20th Annu. Meeting, Can. Soc. Microbiol., Halifax, Nova Scotia.*

McLean, D. M. (1961). *Fed. Proc. 20,* 443.

McLean, D. M. (1970). *Mosquito News, 30,* 144-145.

McLean, D. M. (1975). *Med. Biol. 53,* 264-270.

McLean, D. M. (1978). *In* "Arctic and Tropical Arboviruses" (E. Kurstak, ed.). Academic Press, New York.

McLean, D. M., and Bryce Larke, R. P. (1963). *Can. Med. Assoc. J. 88,* 182-185.

McLean, D. M., and Donohue, W. L. (1959). *Can. Med. Assoc. J. 80,* 708-711.

McLean, D. M., Macpherson, L. W., Walker, S. J., and Funk, G. (1960). *Am. J. Publ. Health 50,* 1539-1544.

McLean, D. M., Walker, S. J., MacPherson, L. W., Scholten, T. H., Ronald, K., Wyllie, J. C., and McQueen, E. J. (1961). *J. Infect. Dis. 109,* 19-23.

McLean, D. M., McQueen, E. J., Petite, H. E., McPherson, L. W., Scholten, T. H., and Ronald, K. (1962). *Can. Med. Assoc. J. 86,* 971-974.

McLean, D. M., deVos, A., and Quantz, E. J. (1964a). *Am. J. Trop. Med. Hyg. 13,* 747-753.

McLean, D. M., Best, J. M., Mahalingam, S., Chernesky, M. A., and Wilson, W. E., (1964b). *Can. Med. Assoc. J. 91,* 1360-1362.

McLean, D. M., Smith, P. A., Livingstone, S. E., Wilson, W. E., and Wilson, A. G. (1966). *Can. Med. Assoc. J. 94,* 532-536.

McLean, D. M., Cobb, C., Gooderham, S. E., Smart, C. A., Wilson, A. G., and Wilson, W. E. (1967). *Can. Med. Assoc. J. 96,* 660-664.

McLean, D. M., Ladyman, S. R., and Purvin-Good, K. W. (1968). *Can. Med. Assoc. J. 98,* 946-949.

McLean, D. M., Chernesky, M. A., Chernesky, S. J., Goddard, E. J., Ladyman, S. R., Peers, R. R., and Purvin-Good, K. W. (1969). *Can. Med. Assoc. J. 100,* 320-326.

McLean, D. M., Crawford, M. A., Ladyman, S. R., Peers, R. R., and Purvin-Good, K. W. (1970). *Am. J. Epidemiol. 92,* 266-272.

McLean, D. M., Bergman, S. K. A., Goddard, E. J., Graham, E. A., and Purvin-Good, K. W. (1971). *Can. J. Publ. Health 62,* 120-124.

McLean, D. M., Goddard, E. J., Graham, E. A., Hardy, G. J., and Purvin-Good, K. W. (1972). *Am. J. Epidemiol. 95,* 347-355.

McLean, D. M., Clarke, A. M., Goodard, E. J., Manes, A. S., Montalbetti, C. A., and Pearson, R. E. (1973). *J. Hyg. 71,* 391-402.

McLean, D. M., Bergman, S. K. A., Graham, E. A., Greenfield, G. P., Olden, J. A., and Patterson, R. D. (1974). *Can. J. Publ. Health 65:* 23-28.

McLean, D. M., Bergman, S. K. A., Gould, A. P., Grass, P. N., Miller, M. A., and Spratt, E. E. (1975). *Am. J. Trop. Med. Hyg. 24,* 676-684.

McLean, D. M., Grass, P. N., Judd, B. D., Cmiralova, D., and Stuart, K. M. (1977). *Can. J. Publ. Health 68,* 69-73.

McLintock, J. (1947). *Manitoba Med. Rev. 27,* 635-637.

McLintock, J. (1976). *Can. J. Publ. Health 67 (Suppl. 1),* 8-12.

McLintock, J., and Iversen, J. (1975). *Can. Ent. 107,* 695-704.

McLintock, J., Burton, A. N., Dillenberg, H., and Rempel, J. G. (1966). *Can. J. Publ. Health 57,* 561-575.

McLintock, J., Burton, A. N., McKiel, J. A., Hall, R. R., and Rempel, J. G. (1970). *J. Med. Ent. 7,* 446-454.

McLintock, J., Curry, P. S., Wagner, R. J., Leung, M. K., and Iversen, J. O. (1976). *Mosquito News, 36,* 233-237.

Main, A. J., Downs, W. G., Shope, R. E., and Wallis, R. C. (1973). *J. Med. Ent. 10,* 229-235.

Main, A. J., Downs, W. G., Shope, R. E., and Wallis, R. C. (1976a). *J. Med. Ent. 13,* 309-315.

Main, A. J., Downs, W. G., Shope, R. E., and Wallis, R. C. (1976b). *J. Wildlife Dis. 12,* 182-194.

Medical News (1963). *Mod. Med. Can. 18,* 59.

Medovy, H. (1942). *Can. Publ. Health J. 33,* 307-312.

Medovy, H. (1976). *Can. J. Publ. Health 67 (Suppl. 1),* 13-14.

Mitchell, C. A., and Pullin, J. W. (1943). *Can. J. Publ. Health 34,* 419-420.

Mitchell, C. A., and Walker, R. V. L. (1941). *Can. J. Comp. Med. 5,* 314-319.

Morgante, O., and Shemanchuk, J. A. (1967). *Science 157,* 692-693.

Morgante, O., Barager, E. M., and Herbert, F. A. (1968a). *Can. Med. Assoc. J. 98,* 1170-1175.

Morgante, O., Vance, H. N., Shemanchuk, J. A., and Windsor, R. (1968b). *Can. J. Comp. Med. 32,* 403-408.

Newhouse, V. F., Burgdorfer, W., McKiel, J. A., and Gregson, J. D. (1963). *Am. J. Hyg. 78,* 123-129.

Newhouse, V. F., McKiel, J. A., and Burgdorfer, W. (1964). *Can. J. Publ. Health 55,* 257-261.

Norris, M. (1946). *Can. J. Res. 24,* 63-70.

Pavilanis, V., Wright, I. L., and Silverberg, M. (1957). *Can. Med. Assoc. J. 77,* 128-130.

Prior, M. G., and Agnew, R. M. (1971). *Can. J. Comp. Med. 35,* 40-43.

Quong, T. L. (1942). *Can. Publ. Health J. 33,* 300-306.

Reeves, W. C., Scrivani, R. P., Hardy J. L., Roberts, D. R., and Nelson, R. L. (1970). *Am. J. Trop. Med. Hyg. 19,* 554-551.

Rossier, E. (1976). *Can. Dis. Weekly Rep. 2-51,* 202-203 Health and Welfare Canada.
Rossier, E., Harrison, R. J., and Lemieux, B. (1974). *Can. Med. Assoc. J. 110,* 1173-1175.
Rozdilsky, B. Robertson, H. E., and Chorney, J. (1968). *Can. Med. Assoc. J. 98,* 79-86.
Savage, A. (1942). *Can. Publ. Health J. 33,* 258.
Schoening, H. W., Giltner, L. T., and Shohan, M. S. (1939). *Proc. U. S. Livestock Sanitary A,* 145-150.
Schofield, F. W., and Labzoffsky, N. (1938). *Rep. Ont. Dept. Agric., Ont. Vet. Coll. 29,* 25-29.
Sekla, L. H., and Stackiw, W. (1976). *Can. J. Publ. Health 67 (Suppl. 1),* 33-39.
Schemanchuk, J. A., and Morgante, O. (1968). *Can. J. Microbiol. 14,* 1-5.
Snell, E. (1966). *Manitoba Med. Rev. 46,* 23-29.
Spalatin, J., Burton, A. N., McLintock, J., and Connell, R. (1963). *Can. J. Comp. Med. Vet. Sci. 27,* 283-289.
Spalatin, J., Connell, R., Burton, A. N., and Gollop, B. J. (1964). *Can. J. Comp. Med. 28,* 131-142.
Spence, L., Artsob, H., Grant, L., and Th'ng, C. (1977). *Can. Med. Assoc. J. 116,* 35-36.
Steward, C. C., and McWade, J. W. (1960). *Proc. Ent. Soc. Ont. 91,* 121-188
Stewart, R. B. (1977). *Blue Bill Suppl. 24,* 8-10.
Trainer, D. O. (1964). *J. Wildlife Manag. 28,* 377-381.
Trainer, D. O., and Hoff, G. L. (1971). *J. Wildlife Dis. 7,* 118-119.
Wagner, R. J., DeJong, C., Leung, M. K., McLintock, J., and Iversen, J. O. (1975). *Can. J. Microbiol. 21,* 574-576.
Waters, J. R. (1976). *Can. J. Publ. Health 67 (Suppl. 1),* 28-32.
West, A. S. (1958). *Proc. 10th Int. Congr. Ent. (1956) 3,* 581-586.
Wright, R. E., Artsob, H., and Shipp, L. (1977). Manuscript in preparation.
Yuill, T. M., and Hanson, R. P. (1964). *Zoonoses Res. 3,* 153-164.
Yuill, T. M., Iversen, J. O., and Hanson, R. P. (1969). *Bull. Wildlife Dis. Assoc. 5,* 248-253.

Chapter 5

ARBOVIRUSES IN NORWAY

T. Traavik

I. INTRODUCTION

This chapter reviews the research on arboviruses that has taken place in Norway since 1973/1974. Prior to this date, no allusions to this group of potential pathogens are found in the literature, with a few noticeable exceptions connected to *Ixodes ricinus*. In 1959, Tambs-Lyche (1959) tried to provoke an interest in tick-borne viruses, evidently with little success. In 1962, Brennaas and Raeder (1962) described an outbreak of meningoencephalomyeloradiculitis on the Western coast, and

ISBN 0-12-429765-X

suggested that it was an "endemic virus disease transmitted by the castor bean tick." Efforts to demonstrate hemagglutinating antibodies to tick-borne encephalitis virus (TBE) in these patients failed, which may in part explain why the authors' interesting observations were not followed up by virological and epidemiological research.

Our interest in the arbovirus field (Traavik, 1970) was aroused by the discrepancy between the lack of information in Norway and the knowledge accumulated in Finland, Denmark (Freundt, 1963), and Sweden (Svedmyr *et al.*, 1958). An analysis of the ecological circumstances in Norway, compared to our neighboring countries, and the results of the extensive Finnish approaches to tick-borne (Brummer-Korvenkontio *et al.*, 1973a; Oker-Blom, 1956; Saikku, 1974) and mosquito-borne viruses (Brummer-Korvenkontio, 1969; Brummer-Korvenkontio *et al.*, 1973b) formed the basis of the Norwegian research program.

II. VECTORS

A. Ticks

A considerable pool of information concerning the distribution and ecology of Ixodidae is available, mainly due to the investigations of Tambs-Lyche (1943a,b) and Mehl (1968, 1970, 1972). Presumably the most important vector within this group, *I. ricinus*, is abundant in the coastal areas from the southern parts of the country up to Helgeland, near the Arctic circle. It is well-known, under various local names, to the public, mainly because of the connection to bovine piroplasmosis (Tambs-Lyche, 1943b). The tick of small rodents, *I. trianguliceps*, is found within the distribution area of *I. ricinus*, but it is also numerous in inland localities where the latter is absent (see Section III,2). *Ixodes hexagonus* has been found in Southern Norway, both within and outside the distribution area of *I. ricinus*, and is probably active throughout the whole year (Tambs-Lyche, 1943a). *Ixodes lividus* was demonstrated by Mehl (1970) in Sand Martin colonies. *Ixodes caledonicus* was found on a starling in 1965 (Mehl, 1970). Another bird tick, *I. arboricola*, was reported by Tambs-Lyche in 1943 (1943a). The seabird tick *I. uriae*, which has been firmly established as an arbovirus vector during recent years (Lvov *et al.*, 1975; Traavik and Mehl, 1975; Yunker, 1975), is abundant in the vast Norwegian seabird colonies. Its ability to attack man has been repeatedly demonstrated (Mehl, 1968;

Tambs-Lyche, 1959). Finally, on various occasions *Hyalomma marginatum* has been picked off passerine migratory birds arriving in the spring.

B. Mosquitoes

In spite of the extensive pioneering work of Natvig (1948), the knowledge of Norwegian mosquitoes is still scanty. This lack of ecological information is unsatisfactory, particularly when considering mosquitoes as potential arbovirus vectors, as nothing is known about the seasonal density variations of the various species, their host preference, or biting habits. Mehl (1976) recently summed up the mosquito species demonstrated in Norway and produced a list comprising two *Anophelini*, three *Culicini,* six *Culisetini,* and 21 *Aedini,* as well as another 11 mosquito species that have been found in the neighboring countries, some of which "may probably also be found in Norway." In our collections, as in those of Natvig and Brummer-Korvenkotio *et al.*, (1973) in Finland, the most numerous widely distributed mosquito species belong to the *Aedes communis* group.

C. Other Potential Vectors

During our field expeditions, some fleas from small rodents and several thousands of *Simulidae* and *Culicoides* have been collected and processed for virus isolation: no virus isolates have been found so far.

III. TICK-BORNE VIRUSES

A. Collection of Materials

Ticks and trapped small mammals were collected from the Autumn of 1973 through the year 1976. In some locations passerine birds were also captured by mist-netting. Selected biotopes were to be visited at regular intervals during the season, and the collection was to be performed in a way that would have made it possible to record virus activity compared to ecological parameters.

B. The Tick-Borne Encephalitis (TBE) Complex

1. *Virus Isolations*

From the thousands of ticks collected in 1973-1975 no TBE virus was isolated. However, these field expeditions mainly took part during late summer and fall and it has been suggested that the seasonal fluctuation of TBE virus may make isolations in early summer more likely (Chumakov *et al.*, 1973). Accordingly, in 1976 ticks were collected in some selected biotopes in early June. The localities were chosen in accordance with the results of a serological screening of cattle (Traavik, 1973). The spring and summer of 1976 were extremely dry and it proved to be hard to catch *I. ricinus* by the blanket drag method. The expedition was terminated after three days, during which only 358 *I. ricinus* were caught. This was less than half the number usually collected in one day during the preceding years. The ticks were divided into 16 pools according to developmental stage and processed for virus isolation attempts by intracerebral inoculations into newborn mouse litters aged 1-3 days. Five pools yielded mouse-pathogenic agents passing 220 nm filters and two of these virus strains have been shown to be antigenically related to the TBE complex (T. Traavik, R. Mehl, and R. Wiger, 1978a). Pool E 672 was composed of nine female *I. ricinus* collected at Lavik in the Sogn og Fjordane county on June 9, while pool 674 was made of 12 male *I. ricinus* collected at Dragsvik in the same county on the following day. For serological characterization, micro-hemaglutination (Clarke and Casals, 1958), complement fixation, closed-hexagon immunodiffusion (Traavik *et al.*, 1972; Traavik and Mehl, 1977), and immunoelectroosmophoresis (IEOP) (Traavik and Mehl, 1977; Traavik *et al.*, 1977; Traavik, 1977a) were used. The Czech Hypr strain and a hyperimmune mouse antiserum to this virus were used as references. No antigenic differences between the two Norwegian strains, or between the newly isolated viruses and the Hypr strain, have been revealed so far. However, homologous antisera to the new viruses have not yet been employed in serological tests. No attempts to characterize the remaining three *I. ricinus* isolates have been made so far (see Section III,F).

2. *Serological Screenings*

Sera from 103 cattle, 341 human patients from *I. ricinus* invaded areas, and more than 1000 wildlife animals have been screened from various parts of Norway by the traditional hemagglutination method. Positive reactions were also verified by immunodiffusion and, to some extent, also after concentration by Lyphogel (Aschavai and Peters, 1971). The modified immunoelectroosmophoresis method that was successfully used for Uukuniemi, Runde, Tribec, and Tahyna viruses could not be applied to TBE serology, probably because of the lack of a suitable, highly concentrated antigen preparation (Traavik, 1977a). The overall prevalence of seropositive individuals was very similar within the distribution areas of *I. ricinus* for all the species tested and was around 20%: in cattle 18% (Traavik, 1973 and unpublished results), in humans 20%, and in small rodents 24% were seropositive. Analysis of the seropositive human subjects revealed some interesting features related to sex and age. In men, the prevalence of antibody reactors was essentially unchanged between the ages 10 and 60 years (25-29%), but dropped to 19% for individuals over 60 years of age. In women, a noticeable drop in antibody reactors was demonstrated from the age group 20-30 years (32%) to 30-40 years (17%). Among women over 60 years, only 6-8% had antibodies to TBE. The overall prevalence for men was 24%, for women 16%. Some implications of the serological screening data are discussed in Section III, B,4.

3. Human Disease

Very few patients with CNS symptoms have been examined for TBE in Norway in the past, mainly because physicians and research workers in hospitals and laboratories have not been aware of this potential etiological factor. Consequently, in the few cases where TBE has finally been considered, samples have not been adequately taken. Only four patients, with an acute disease, have been demonstrated to have antibodies to TBE and in all cases sera were taken too late to demonstrate significant rises in antibody titers.

The 341 patients used for screening (Section III,B,2) were chosen from the files of the Department of Microbiology, Haukeland University Hospital, Bergen. Sera had been remitted to the virus laboratory requesting diagnostic or serological controls. The only criterion for inclusion in our panel was that the patient should live within the known distribution area of *I. ricinus*. The samples were matched with regard to age and sex and no clinical information was obtained about the patients chosen. After the screening was concluded, information was requested concerning 19 patients with "high" antibody titers (320) and two patients who demonstrated significant increases in titers in paired sera. As the present manuscript was in preparation, only some of the required information was received for 18 of the patients, i.e., only the clinical information received by the laboratory with the serum specimens. Fever, fever and myalgia, or only myalgia were the only indications given for ten of these patients, pneumonia was mentioned for three, encephalitis/multiple sclerosis for two, and polyradiculitis with pareses for one.

4. *Comments*

Much more work is needed in order to gain a comprehensive view of the extension, ecology, and clinical significance of TBE viruses in Norway. The explanation of 25% seropositive *C. glareolus* in a locality where *I. ricinus* is absent strongly suggests the existence of alternative vector(s). The animals collected at Kviteseid in Telemark county are often strongly infested by *I. trianguliceps*, which has been proposed as an alternative vector. The serological screening of 383 bank voles captured at Kviteseid poses the question of alternative or tangential transmission cycles for TBE virus. Another problem is presented by the age and sex differences in the human subjects revealed by the anti-TBE antibody screenings mentioned in Section III,B,2. This phenomenon has not been observed in Finland (Salminen *et al.*, 1961) or in Czechoslovakia (Gresikova *et al.*, 1967). In addition, a lack of correspondence between hemagglutination titers and the ability of the antisera to react in immunodiffusion has been revealed in the present study. One speculative hypothesis may be that there are local, antigenic variants of TBE and that the avidity of antibodies for the Hypr strain decreases with time after infection. According to this interpretation, the age and sex differences could be due to reinfections.

C. The Uukuniemi (UUK) Virus Group

1. *Virus Isolations*

From *I. ricinus* ticks collected in 1973, two UUK virus strains were isolated: By E 50 was isolated from a pool composed of 40 *I. ricinus* nymphs collected in the vegetation at Bygstad in the Sogn og F jordane county in late August, while SF El originated from a pool composed of eight engorged *I. ricinus* nymphs picked off passerine migratory birds at Store Faerder in the Oslofjord (a small island with an ornithological station) in early May (Traavik *et al.*, 1974). These two virus strains were shown to have the same biological characteristics. They seemed antigenically identical by the methods employed and also identical to the Finnish prototype strain S 23 (Traavik and Mehl, 1977). A third virus strain, Ru E82, belonging to the UUK group, was isolated from *I. uriae* collected in the seabird colonies at Runde, an island in the Møre og Romsdal county. This was the same collection from which the two Runde virus strains were isolated (see Section III,E,1). The actual pool was composed of five female *I. uriae*. The Ru E82 strain demonstrated some divergent characteristics and also differed antigenically from S 23 and the two Norwegian strains (Traavik and Mehl, 1977). All three strains were only weakly active in hemagglutination, but satisfactory antigens were obtained by employing Aerosil-treatment and sonication of sucrose-acetone-extracted mouse brains or cell culture fluids concentrated with polyethylene glycol 6000/NaCl (Traavik, 1977b). During 1974 and 1975 no UUK viruses were isolated from several thousands of *I. ricinus* ticks.

2. *Serological Screenings*

The same serum panel which was used for the screening for TBE (bovine, human, small mammal, and passerine birds) was also tested for antibodies to the By E50 UUK strain by a modified IEOP method (Traavik and Mehl, 1977) and by hemagglutination inhibition (Clarke and Casals, 1958; Traavik, 1977b). It was evident that the UUK activity within the actual biotopes was weaker than the TBE activity, since antibodies were demonstrated only in 13.5% of the sera of cattle, 5% of humans, and approximately 10% of small mammals (Traavik and Mehl, 1977; R. Wiger, T. Traavik, and R. Mehl, unpublished). There was good agreement between the results of the two tests.

3. *Comment*

It has been established by Saikku (1974) and it has also been demonstrated for the Norwegian strains that resistance to the acute UUK virus disease is initiated early in life for the laboratory mouse (Traavik and Mehl, 1977). However, persistent infection to the UUK group virus Ru E82 with a resulting chronic disease can be established by intracerebral infection of 2-week-old mice. The capability of other UUK strains in this respect should be further investigated.

D. The Kemorovo Virus Group

No virus belonging to this group of tick-borne viruses has been isolated so far. Preliminary screenings of human and cattle sera with Tribec virus have demonstrated 3-4% seropositives by the IEOP (Traavik, 1977a) and complement fixation tests. These investigations will be continued. However, since these methods probably demonstrate strain-specific antibodies, the results may not give a fair evaluation of the extent of Kemorovo group infections and activity (T. Traavik, R. Wiger, and R. Mehl, unpublished).

E. "Runde" Virus

1. Virus Isolations

During September 25-27, 1973, *I. Uriae* was collected in the vast seabird colonies at the island Runde in the Møre og Romsdal county. Parts of the colonies inhabited by the Common Puffin (*Fratercula arctica*) were chosen due to relatively safe access. The birds had migrated south about three weeks earlier and the ticks were found unengorged, resting in diapause between rocks.

From the collected material 206 *I. uriae*, divided into 24 pools, were processed for virus isolation in newborn mice. Three pools yielded virus isolates, one of which turned out to be the UUK group virus Ru E82 (Section III,C,1, 3). The additional two strains seemed antigenically identical, but could not be related to any major arbovirus group (Traavik *et al.*, 1977). In the electron microscope the viruses displayed a coronavirus-like morphology, although the virion size was larger (around 170 nm diameter on average) than that of the other members of the coronavirus group. The two strains, Ru E81 and Ru E85, were found serologically unrelated to avian infectious bronchitis virus. Their ability to multiply in the presence of BUdR points to a single-stranded RNA genome and an inner helical material was seen in partly disrupted particles. These characteristics were reported in detail elsewhere (Traavik and Brunvold, 1978). Although the viruses were isolated only after one blind passage in mice, control experiments have proved that they originated in the tick pools. From a scientific expedition to the seabird colonies in Hernyken, Røst in Lofoten, four out of 19 seabird sera had antibodies to Runde virus.

2. Persistent Infections

Runde virus displayed a low degree of pathogenicity in mice, as well as in BHK 21/cl3 cell cultures. In the cell cultures persistent, productive infections were easily established, while in mice an age-dependent infection was observed. Newborn mice contracted an actue CNS disease that was regularly fatal, whereas at the age of 2-3 weeks moderate virus doses produce persistent infections with or without chronic disease. Virus was reisolated from the brains of sick or symptomless mice, up to 150 days after the infection. In newborn mice the chronic disease and virus persistence could be provoked by intracerebral injection of either virus from the undiluted cell culture supernatants of serial passages, or by mixtures of virus and hyperimmune mouse antiserum. Addition of unheated guinea pig serum to the mixtures strengthened the tendency toward virus persistence, and even unheated guinea pig serum alone, modified the clinical expression of Runde virus infection in newborn mice.

It was also demonstrated that Runde virus multiplied in the amnion of embryonated fowl eggs, that the virus persisted until hatching, and that the chickens were still infected. The chickens had episodes of epilepsy-like disease, but seemed to behave normally in the intervals between these episodes (Traavik, 1978a,b,c).

3. *Comment*

In order to explain our interest in investigating seabirds-colonies for arbovirus infections, the following considerations can be listed:

(a) The seabird colonies represent confined ecosystems with very few components in the supposed virus transmission cycles. Knowledge of the ecological circumstances for the virus may provide valuable information of general importance.

(b) Seabirds are at the end of ecological food chains and readily reflect alterations in the environment. The high mortality of chicks, which was one of the reasons for choosing Runde among the many Norwegian seabird colonies, is still unexplained. Pesticide residues were below the "danger limit" in dead chicks. Since some of the chemicals released into the environment in industrial wastes are affecting the metabolism and enzyme systems at the cellular level, it might not be too surprising that they also influence the balance between macro- and micro-organisms.

(c) Since Norwegian seabird colonies are often located on inhabited islands that are used as pastures, or for recreational and scientific purposes, as well as for egg-picking and bird-hunting, the implications of arboviruses in public health should not be totally ignored. *Ixodes uriae* readily attacks man and domestic animals (Tambs-Lyche, 1959).

F. Unidentified Virus Isolates from *I. ricinus*

From *I. ricinus* collected in 1976, three agents that are passing through 220 nm filters were isolated. For these isolates, the short incubation time in suckling mice (2-5 days) makes it unlikely that they belong to the UUK virus group.

G. Discussion

Although much work still has to be done in order to have a detailed knowledge of the extension and activity of tick-borne arboviruses in Norway, the field work and collection of material during the next few years should be concentrated on mosquito-borne viruses and attention should be focused only on some specefic problems concerning tick-borne viruses. Some of these problems have already been mentioned in the preceding sections. Since it is evident that at least three different arboviruses circulate within the same biotopes by *I. ricinus* and that antibodies to more than one virus have been detected in some individuals during the screenings, the possibilities of mixed infections in nature might be considered as far from hypothetical (Traavik, 1973; Traavik and Mehl, 1975). In recent experiments with mixed infections mice were inoculated with two or three viruses (TBE, UUK, Tribec) simultaneously or at different intervals. The ability of mixed infections to modulate the expression of a single virus, both for the benefit or the disadvantage of the host animal, has been demonstrated in these experiments. The ability of Norwegian arboviruses to produce persistent infections should be further investigated and related to the symptoms of selected groups of patients.

The finding of antibodies to TBE virus outside the distribution area of *I. ricinus* has been mentioned earlier. Interestingly, antibodies to UUK virus were not demonstrated in the same sera. The possibility of *I. trianguliceps* as an alternative vector for TBE should be given fair attention, although it should not be overlooked that another flavivirus transmitted by mosquitoes might account for the peculiar findings.

Finally, a more thorough biological biochemical classification of our virus isolates will be necessary.

VI. MOSQUITO-BORNE VIRUSES

A. Collection of Mosquitoes

During 1975 and 1976, approximately 12,000 mosquitoes were collected. As already stated, relatively little information on the relevance of mosquitoes to arbovirus research is presently available. Consequently, in 1975 a typical forest biotope was chosen in Trandum in southeastern Norway, situated close to Oslo, which was used extensively for military training and recreational purposes. The area (Romerike) contains some of the best agricultural and livestock districts of the country. Mosquitoes were also collected at Øyern, the largest delta area in Norway, which has now become a preserved national park and is well-known for its wildlife population, particularly birds. Finally, a subarctic tundra biotope, Masi in Finmark county, situated within an area of great importance for the native Lap population and their reindeer, was also used for the collection of mosquitoes.

Collection of mosquitoes was continued in these localities in 1976, in addition to some other biotopes, among which the two providing virus isolates (see below). Sjusjøen in Opland county has some of the most intensely expanded mountain cabin areas in the country, and is situated at approximately 1000 meters above sea level. Trysil in Hedmark county lies close to the Swedish border and is a typical forestry municipality.

All collections were performed using the operators and open CO_2-containers as "bait" and by catching mosquitoes in flight with butterfly nets.

Only about half of the mosquitoes collected in 1976 have so far been used for virus isolation attempts in newborn mice.

B. Viruses from *Aedes* Species

1. *California Encephalitis (CE) Group*

From the mosquitoes collected in 1975, three agents passing through 220 nm filters have been isolated that were pathogenic for mice. The first strain, S 548, was isolated from a pool composed of 45 female *A. sticticus* collected in the Øyern delta at June 18, 1975. The second strain, S 586, was from 16 male *A. diantaeus* collected at Trandum on June 17, 1975. The last strain, S 618, was from 100 female *A. hexodontus* collected at Masi, nearly 70°N, on August 20, 1975 (Traavik *et al.*, 1978b.) In the first passage in newborn mice the incubation time was 3-5 days for all strains, and this was shortened to 2-3 days during consecutive passages. Originally, only strain S 586 agglutinated chicken erythrocytes but employing Aerosil treatment and sonication (Traavik, 1977b), high hemagglutinating activity was recorded for all three strains with trypsinized chicken or human O cells. With a hyperimmune mouse serum to Tahyna virus, all three strains gave identical reactions in the hemagglutination test, while strain S 586 differed from Tahyna virus in immunodiffusion. Details of this study are in press (Traavik *et al.*, 1978b.)

2. *An Accidental Human Infection*

At the time of the inoculation experiments (from 1975) one of the staff members suddenly became ill with general fatigue, headache, and fever of around 39°C. On the second day of fever, a blood sample was taken and inoculated intracerebrally into two litters of newborn mice. After 2 days all mice were moribund.

The symptoms were rather dramatic for the first few days, but then diminished quickly, and the patient returned to work after 5 days. The agent isolated from his blood was shown to be related to, but different from, Tahyna virus and was found to be identical to strain S 586 by immunodiffusion (Traavik *et al.*, 1978b).

3. Unidentified Virus Strains

Seven additional agents passing through 220 nm filters have been isolated from *Aedes* mosquitoes collected in 1976. The isolate S 16/76 was from a pool of 170 female undetermined *Aedes*, due to a lot of *Simulidae* in the collection, taken in Trysil August 8, 1976. S 20/76 originated from a pool of 13 female *A. punctor* collected at Sjusjøen on August, 11, 1976. The last five isolates were from pools consisting of 100 female *A. communis* taken at Trandum June 2-11, 1976.

C. Virus Isolates from *Anopheles claviger*

Anopheles claviger was found in the collections from Trandum, both in 1975 and 1976, but yielded no isolates in 1975. In the 1976 material, however, two pools contained agents passing through 220 nm filters. One pool consisted of 100 female mosquitoes, the other was composed of nine male. Date of collection was June 11, 1976. No serological relationship could be established between these two isolates and Tahyna, WEE, EEE, or tick-borne viruses (T. Traavik, R. Mehl, and R. Wiger, unpublished results.

D. Discussion

As is obvious from the present study, conclusions with respect to the extent and significance of mosquito-borne viruses in Norway are still premature. It may only be stated that viruses *can* be isolated from Norwegian mosquitoes with relative ease but their implication in diseases would be highly speculative at the present time.

The three virus isolates proven to belong to the CE group were obtained from three different *Aedes* species and, consequently, it might be stated that these species support CE virus multiplication. However, this suggestion does not allow the conclusion that all, or any, of these species are capable of transmitting the virus to host animals, i.e., of being a real biological vector for the virus(es). Only further research will clarify this point.

In the course of these studies viruses have been isolated from male mosquitoes and it is tempting to associate this finding with the recent demonstrations of transovarial (Watts *et al.*, 1973) and venereal (Thompson and Beaty, 1977) transmission of La Crosse virus. Although the early date of collection of the male mosquitoes could imply that they had freshly emerged, further speculation on virus overwintering mechanisms and on the role of mosquitoes as reservoirs for arboviruses in Norway must await a sounder experimental basis.

To our knowledge, the CE isolate from *A. hexodontus* at Masi, at nearly 70^{0}N, is the northernmost arbovirus isolate so far (Casals, 1975). Evidently, mosquitoes represent a nuisance to man and animals also in the part of Norway situated at latitude 70-71^{0}N, and there are no ecological reasons why mosquitoes should not carry virus in these areas.

The potential human pathogenicity of one of our CE isolates was demonstrated by an accidental infection. Whether or not the virus is capable of producing disease under natural conditions is impossible to predict, particularly since the real vector, its biting habits, and host preferences are still unknown.

We have not yet had the opportunity to compare our strains with the Finnish CE representative Inkoo, which is, as one of our isolates seems to be, antigenically distinct from the European "prototype" Tahyna (Bardos and Danielova, 1959; Brummer-Korvenkontio *et al.*, 1973).

As pointed out by Casals (1975), the capacity of CE viruses to develop antigenic variants opens up the question of reinfection with related viruses. It should be added that the capability of CE viruses to produce persistent infections is also an open question. So far there are indications that CE viruses may indeed produce persistent infections in mice.

References

Aschavai, B. S., and Peters, R. L. (1971). Am. *J. Clin. Pathol. 55,* 262-268.

Bardos, V., and Danielova, V. (1959). *J. Hyg. Epidemiol. Microbiol. Immunol. 3,* 264-276.

Brennaas, O., and Raeder, S. (1962). *Tidsskrift norske laegeforening 11,* 739-744.

Brummer-Korvenkontio, M. (1969). *In* "Arboviruses of the California Complex and the Bunyamwera Group." Publ. House of Slov at Acad. Sci., Bratislava.

Brummer-Korvenkontio, M., Saikku, P., Korhonen, P., and Oker-Blom, N. (1973a). *Am. J. Trop. Med. Hyg. 22,* 382-389.

Brummer-Korvenkontio, M., Saikku, P., Korhonen, P., Ulmanen, I., Reunala, T., and Karvonen, J. (1973b). *Am. J. Trop. Med. Hyg. 22,* 404-413.

Casals, J. (1975). *Med. Biol. 53,* 249-258.

Chumakov, M. P., Moteynas, L. I., Bychkova, M. V., and Vargin, V. V. (1973). *Zh. Mikrobiol. Epidemiol. Immunol. 5,* 83-87.

Clarke, D. H., and Casals, J. (1958). *Am. J. Trop. Med. Hyg. 7*, 561-573.
Freundt, E. A. (1963). *Acta Path. Microbiol. Scand. 57*, 87-103.
Gresikova, M., Kozuch, O., and Molnar, E. (1967). *Bull. Wld. Hlth. Org. 36, Suppl. 1*, 81-84.
Lvov, D. K., Timopheeva, A. A., Smirnov, V. A., Gromashevsky, V. L., Sidorova, L. P., Nikiforova, L. P., Sazonov, A. A., Andreev, A. P., Skvortzova, I. M., Beresina, L. K., and Aristova, V. A. (1975). *Med. Biol. 53*, 325-330.
Mehl, R. (1968). *Fauna (Oslo) 21*, 197-198.
Mehl, R. (1970). *Norsk. Ent. Tidsskr. 17*, 109-113.
Mehl, R. (197a). *Fauna (Oslo) 25*, 186-196.
Mehl, R. (1976). Report, National Institute of Public Health, Oslo.
Natvig, L. R. (1948). *Norsk. Ent. Tidsskr., Suppl. 1.*
Oker-Blom, N. (1956). *Ann. Med. Exp. Fenn. 34*, 309-318.
Saikku, P. (1974). Uukuniemi Virus, Thesis, Univ. of Helsinki.
Salminen, A., Erikson, A., and Oker-Blom, N. (1961). *Arch. Ges. Virusforsch., 11*, 215-223.
Svedmyr, A., von Zeipel, G., Holmgren, B., and Lindal, J. (1958). *Arch. Ges. Virusforsch. 8*, 565-576.
Tambs-Lyche, H. (1943a). *Bergens Museums Arbok., Naturvitenskapelig rekke*, No. 3.
Tambs-Lyche, H. (1943b). *Norsk. Veterinaertidsskr. 55*, 337-542.
Tambs-Lyche, H. (1956). *Nordisk Medicin 62*, 1217-1222.
Thompson, W. H., and Beaty, B. J. (1977). *Science 196*, 530-531.
Traavik, T. (1970). *Norsk. Veterinaertidsskr. 82*, 705-712.
Traavik, T. (1973). *Acta Path. Microbiol. Scand., Sect. B. 81*, 138-142.
Traavik, T. (1977a). *Arch. Virol. 54*, 231-240.
Traavik, T. (1977b). *Arch. Virol. 54*, 223-229.
Traavik, T., and Mehl, R. (1975). *Med. Biol. 53*, 321-324.
Traavik, T., and Mehl, R. (1977). *Arch. Virol. 54*, 317-331.
Traavik, T., Siebke, J. C., and Kjeldsberg, E. (1972). *Acta Path. Microbiol. Scand., Sect. B. 80*, 773-774.
Traavik, T., Mehl, R., and Petterson, E. M. (1974). *Acta Path. Microbiol. Scand., Sect. B. 82*, 297-298.
Traavik, T., Mehl, R., and Wiger, R. (1977). Manuscript in preparation.
Traavik, T., Mehl, R., and Kjeldsberg, E. (1978). *Arch. Virol. 55*, 25-38.
Watts, D. M., Pantuwatana, S., De Foliart, G. R., Yuill, T. M., and Thompson, W. H. (1973). *Science 182*, 1140-1141.
Yunker, C. E. (1975). *Med. Biol. 53*, 302-311.

Additional References

Traavik, T., Mehl, R., and Wiger, R. (1978a). *Acta Path. Microbiol. Scand., Sect. B. 86*, 253-255.
Traavik, T., Mehl, R., and Wiger, R. (1978b). *Acta Path. Microbiol. Scand., Sect. B. 86*, in press.
Traavik, T., and Brunvold, E. (1978). *Acta Path. Microbiol. Scand., Sect. B.*, in press.
Traavik, T. (1978a). *Acta Path. Microbiol. Scand. Sect. B. 86*, 299-301.
Traavik, T. (1978b). *Acta Path. Microbiol. Scand. Sect. B. 86*, in press.
Traavik, T. (1978c). *Acta Path. Microbiol. Scand. Sect. B.*, in press.

Chapter 6

TICK-BORNE VIRUSES OF SEABIRDS

Carleton M. Clifford

I. INTRODUCTION

Since the isolation of Hughes virus from *Ornithodoros capensis* complex ticks from sooty tern nesting areas on Bush Key, Dry Tortugas, Florida (Hughes *et al.*, 1964), the number and diversity of viruses recovered from ticks associated with seabirds have rapidly increased. Now there are more than 32 viruses of 10 serogroups isolated from seabird ticks or from the birds themselves. The number of agents and serogroups varies somewhat, depending on which serological test an investigator emphasizes when constructing antigenic groupings. Initially, the viruses that were discovered reflected the interest and geographic location of the investigators, rather than demonstrating any natural patterns of tick-virus distribution. This is still true to some extent, but enough viruses have now been recorded from various zoogeographic regions to allow us to begin to compare the relationship among viruses, their vectors, and seabird hosts.

ISBN 0-12-429765-X

II. IXODID TICKS THAT PARASITIZE SEABIRDS

Both hard and soft ticks vector viruses associated with seabirds. The hard tick from which viruses have most frequently been isolated is *Ixodes uriae* (*I. putus* of Russian workers). *I. Uriae* has an interesting bipolar distribution and in its northern and southern distributions it is circumpolar (Fig. 1). Zumpt (1952) suggests that this bipolar distribution may result from ticks being transported by seabirds between the two regions. Birds such as Wilson's Storm Petrel (*Oceanites oceanicus*), which breeds in the Antarctic and migrates northward to the Arctic, or the Arctic tern (*Sterna paradisaea*), which breeds in the Arctic circumpolar region and then migrates to the southern regions, may act as tick hosts. *I. uriae* is associated with seabirds at nearly all major bird nesting areas and can survive under a variety of adverse climatic conditions. It feeds on almost any available seabird host, including murres, kittiwakes, penguins, gulls, petrels, cormorants, and shearwaters. In addition, there are numerous records of *I. uriae* biting persons who visit bird rookeries.

There are four stages in the life cycle, i.e., egg, larva, nymph, and adult. The males do not take blood and are found

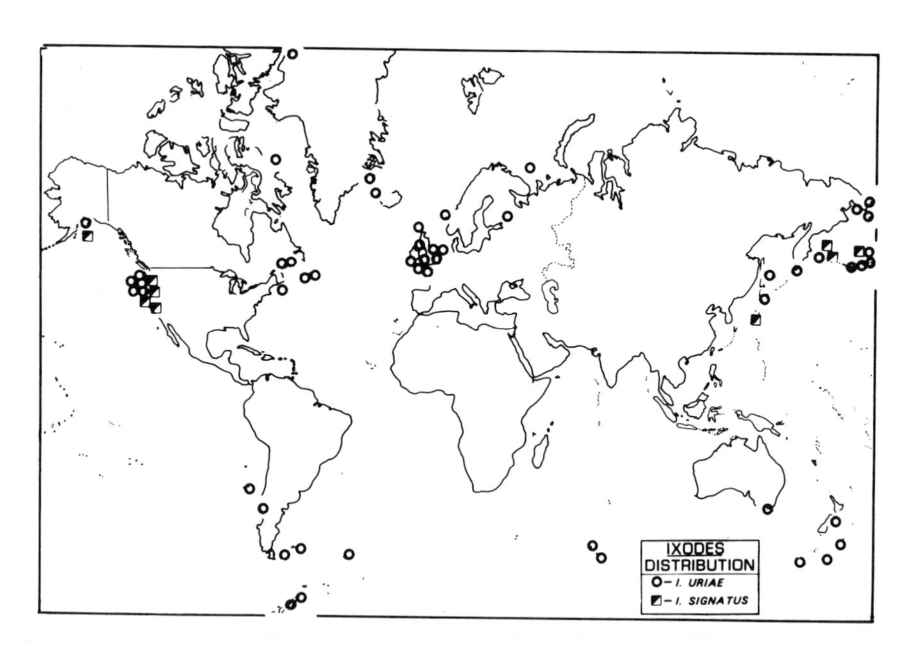

FIGURE 1. Distribution of I. uriae and I. signatus (approximate)

in the nests, whereas the larva, nymph, and female are, in addition to being collected from nests, often found on the birds. *I. uriae* is an extremely adaptive tick and its life cycle varies depending on the climate of various localities where it occurs. Flint and Kostyrko (1967) and Karpovich (1973) studied the life cycle as it occurs in the Barents Sea littoral and noted that it may take four to seven bird breeding seasons for the tick to complete its life cycle. In this area the ticks feed mostly on murres and kittiwakes. Because of the relatively short nesting period of these birds and the cold climate, only a single stage in the tick's life cycle is advanced during the bird breeding season. In contrast, Murray and Vestjens (1967) have found that on Maquarie Island, Australia, the life cycle of *I. uriae* can be completed in two bird-breeding seasons. A shorter cycle is possible here because temperatures on Marquarie Island are not particularly cold. Also, desiccation is not a factor because of the almost constant misty rains. Furthermore, *I. uriae* has a blood meal available for 6 to 7 months since its primary host there is the Royal penguin.

Of interest is the similar morphology of the ticks included under the name *I. uriae* throughout its range. Only one other closely related species, *I. jacksoni*, has been described and that was from cormorants in New Zealand. The major morphological differences are between the males of these two species. This morphological similarity of *I. uriae* from such diverse habitats indicates that either a constant transfer of genetic material among these tick populations or selection pressure on the ticks is minimal. A similar uniformity in the nature of some of the viruses isolated from *I. uriae* will be discussed later.

Another hard tick from which viruses have been isolated and that feeds on seabirds is *I. signatus*. This tick is mainly found on or in association with cormorants, but occasionally it has been recorded from gulls and nonseabirds. It has a continuous distribution from northern California in the United States along the coast of Alaska to eastern Russia and extends as far south as Japan (Fig. 1).

There are several other *Ixodes* species that are associated with sea- and shore birds that offer excellent potential for the spread and maintenance of viruses. A good example of such a species is *I. auritulus*, which ranges from southern Alaska to southern Chile (Fig. 2). Zumpt (1952) lists *I. auritulus* as a "common tick of seabirds," but Arthur (1960) records this species almost exclusively from land birds. All records in the RML collection are also from land birds. Although this species apparently feeds mainly on land birds, it is often found along coastal areas. We collected this species from sparrows on offshore islands along the northwest coast of the U. S. while conducting studies on seabird ticks. The habits

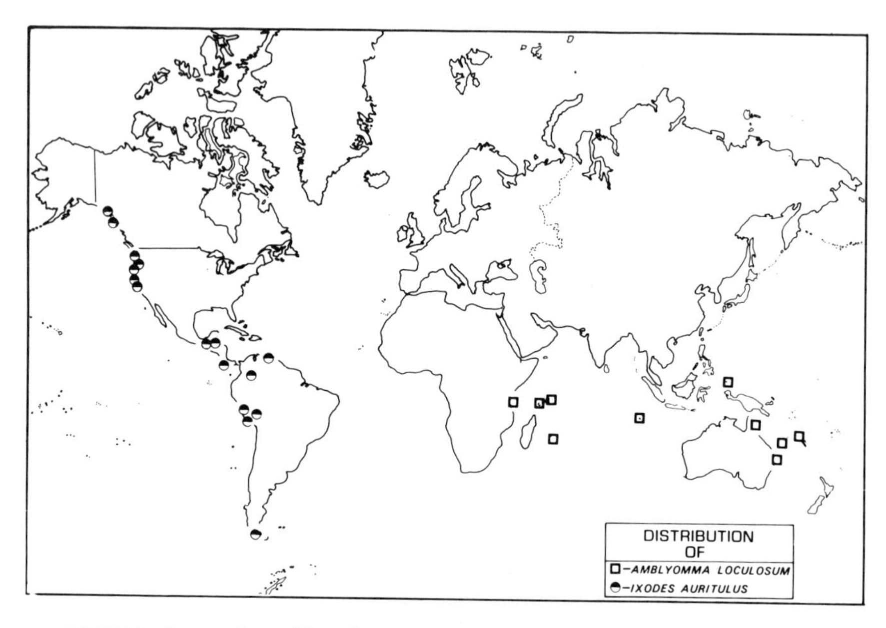

FIGURE 2. Distribution of Amblyomma loculosum and I. auritulus (approximate).

of this tick would seem to provide an excellent means for some of the seabird viruses to become introduced into inland birds.

A closely related tick, *I. auritulus zealandicus*, occurs on burrow-nesting seabirds in New Zealand (Dumbleton, 1953) and land birds in Australia (Roberts, 1970).

Another tick associated with marine birds that has yielded a virus is *Amblyomma loculosum*. Aride virus was isolated from two female ticks of this species taken off roseate terns on Bird Island of the Seychelles (Converse *et al.*, 1975). This tick parasitizes man and numerous species of marine birds, goats, and lizards in the Indian and Pacific Oceans and the Coral Sea (Hoogstraal *et al.*, 1976b). *A. loculosum* is of particular interest because it has an extensive distribution, which suggests the movement of host species between the Coral Sea and the east coast of Africa (Fig. 2). This distribution also includes areas where *O. capensis* occurs and would seem to provide an excellent mechanism for the transfer and movement of viruses between Australia and Africa. Furthermore, the broad host range of *A. loculosum* would also help with the exchange of agents between ticks, seabirds, and other vertebrate hosts.

III. ARGASID TICKS THAT PARASITIZE SEABIRDS

The only soft ticks that feed on seabirds are members of *O. capensis* group. Tick taxonomists recognize five species in this group: *O. capensis, O. denmarki, O. maritimus, O. amblus,* and *O. muesebecki*. A sixth species, *O. sawaii*, has been described from Japan (Kitaoka and Suzuki, 1973), but the distinctness of this species from *O. capensis* is open to question. Also *O. coniceps* is generally included in the *capensis* group but will not be discussed here because recent collaborative studies with Hoogstraal of NAMRU-3 (Hoogstraal *et al.*, 1976a) have demonstrated that *O. coniceps* does not feed on seabirds. Seabird ticks that were previously identified as *O. coniceps* are now recognized as a distinct species, *O. maritimus* (Hoogstraal *et al.*, 1976a). Until recently, taxonomy of species in the *capensis* complex was based on larval characters and adults generally could not be differentiated. This situation is changing as we receive more tick collections and examine them with a scanning electron microscope (SEM). These studies will eventually help to clarify the taxonomic status of the species in this group, which will in turn produce more reliable distribution records for ticks and the viruses they harbor.

The *O. capensis* complex is the counterpart of *I. uriae* in the warmer oceans and seas (Fig. 3). As extensive taxonomic studies have been undertaken, new species have been recognized such as *O. denmarki* and *O. maritimus*. The placement of *O. maritimus* in the USSR requires a comment. These ticks having a distribution as shown in Fig. 3 were sent to the RML from N. A. Filippova, who identified them as *O. capensis*. After careful study of adults and immature stages with the SEM, we have identified these ticks as *O. maritimus*. To date I have not seen any ticks from the USSR that can be identified as *O. capensis*. Other changes in the taxonomic status of this complex can be anticipated, especially in the seemingly morphologically uniform populations of *O. capensis* found in southern Africa, Australia, New Zealand, and the Pacific region.

O. capensis group ticks can usually be found in or adjacent to the nests of the host bird. The conditions under which they can exist vary considerably.

On Bush Key, Dry Tortugas, Florida, *O. denmarki* is found in old abandoned Brown Noddy tern nests 2 to 5 ft. above the ground and on living and dead bay cedar plants and in litter collected beneath growths of white mangrove and around the roots of sedge plants. Approximately 5000 ticks were recovered from one-fourth of a cubic foot of soil in this area near one dead clump of sedge (Denmark and Clifford, 1962).

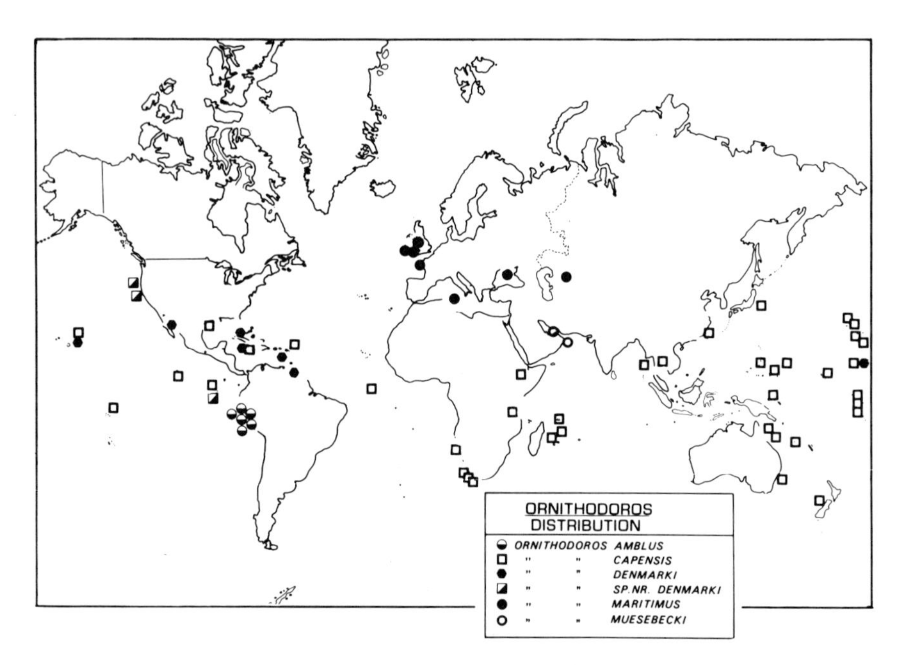

FIGURE 3. Distribution of the O. capensis complex (approximate).

On the Oregon coast where there is little or no vegetation in and around the nesting sites of murres, *Ornithodoros* ticks are found by breaking off chunks of rock and examining crevices adjacent to the bird nests.

Species of the *capensis* group feed on a great variety of seabird hosts. Some of the birds associated with the ticks from which viruses have been isolated are terns, cormorants, murres, gulls, and pelicans.

A generalized life cycle for these soft ticks includes the egg, larva, three to five nymphal stages, and adult. To my knowledge, in all the species that have been studied, the larva molts into a first nymph and molts to the second nymph without feeding. Hoogstraal *et al.*, (1970) studied the life cycle of *O. muesebecki* in the laboratory and found that it took 43 to 85 days to complete one tick generation. They postulated that the life cycle could be completed in a single bird-nesting season. The length of the total life cycle will undoubtedly vary depending on the climatic conditions of the study area and breeding habits of the bird hosts.

IV. VIRUSES FROM IXODID TICKS

To date 19 viruses of five serogroups have been isolated from ticks of the genus *Ixodes* (Table I). As the number of viruses increases, a pattern seems to be emerging. Group B and Uukuniemi agents have been isolated from a restricted portion of the total distribution of *I. uriae*. The possibility of further isolations of some of these viruses is heralded by the finding of serological evidence for the existence of a group B virus complex in marine birds and *I. uriae* in eastern North America (Main *et al.*, 1976b).

The most frequently isolated viruses are members of the Sakhalin and Kemorovo serogroups. Also viruses belonging to both the Sakhalin serogroup (Taggert) and Kemorovo serogroup (Nuggett) have been isolated from the southern distribution of *I. uriae*.

An interesting situation exists in a bird refuge known as Three Arch Rocks off the Oregon coast in the western United States. We have isolated numerous virus strains from *I. uriae*, which included all of the four serogroups known from this species. An *Ornithodoros* tick, currently designated as *O.* sp. near *denmarki*, is also present in large numbers on these rocks and we have tested hundreds of ticks and have found no viruses. These ticks appear to be in similar ecological niches, however; the vector potential of these two species is apparently quite different or there are microenvironmental factors that are not readily apparent and that effectively separate these species.

O. sp. near *denmarki* has also been collected from murre nesting sites on the mainland at nearby Oceanside, Oregon, and Farallon virus has been isolated from these ticks. In this instance, none of the viruses of the four serogroups associated with *I. uriae* has been recovered.

This separation of viruses in two families of ticks that are found in close association is not too surprising since Kaiser (1966a,b) found it could even occur between two closely related species of the genus *Argas*. His studies demonstrated that, although *Argas arboreus* and *A. persicus* occur at adjacent bird colonies, only the former can harbor and transmit Quaranfil virus. In that instance no sense could be made of the situation until *A. arboreus* was recognized as a morphologically and biologically distinct species (Kaiser *et al.*, 1964). This points out the importance of proper taxonomic studies for sorting out virus-tick relationships.

Some other interesting trends appear when comparing the antigenic similarity of these viruses. The viruses isolated from eastern Russia and western North America are quite similar antigenically. For example, the Tyuleniy isolates from Tyuleniy Island are identical to those from three Arch Rocks

TABLE 1
Ixodid Seabird Ticks Harboring Viruses[a]

Virus	*Serogroup*	*Region*	*Authority*
		I. (Ceratixodes) uriae	
Tick-borne Enc.	B	USSR (N)	Berkleshova *et al.*, 1969
Tyuleniy	B	USSR (E,N), USA (NW)	Lvov *et al.*, 1971b; Gaidamovich *et al.*, 1971; Clifford *et al.*, 1971
Oceanside	UUK	USA (NW)	Thomas *et al.*, 1973; Yunker, 1975
Zaliv Terpenya	UUK	USSR (E,N)	Lvov *et al.*, 1973a,b
Sakhalin	SAK	USSR (E)	Lvov *et al.*, 1972
Tillamook	SAK	USA (NW)	Thomas *et al.*, 1973; Yunker, 1975
Avalon	SAK	Canada (NE)	Main *et al.*, 1976a
Clo Mor	SAK	Scotland (N)	Main *et al.*, 1976a
Taggert	SAK	Australia (Macquarie Island)	Doherty *et al.*, 1975
Okhotsky	KEM	USSR (E)	Lvov *et al.*, 1973
Poovoot	KEM	USA (Alaska, St. Lawrence Island)	Yunker, 1975
Yaquina Head	KEM	USA (NW)	Yunker *et al.*, 1973
Great Island	KEM	Canada (NE)	Main *et al.*, 1973

Bauline	KEM	Canada (NE)	Main *et al*., 1973
Cape Wrath	KEM	Scotland (N)	Main *et al*., 1976c
Nugget	KEM	Australia (Macquarie Island)	Doherty *et al*., 1975
Paramushir	UNG	USSR (E)	Lvov *et al*., 1975
		I. (Scaphixodes) signatus	
Zaliv Terpenya	UUK	USSR (E)	Lvov, 1973
Kachemak Bay	SAK	USA (S. Central Alaska)	Yunker, 1975
Kenai	KEM	USA (S. Central Alaska)	Ritter and Feltz, 1974 (as Grest Island) (*fide* Yunker, 1975)
Okhotsky	KEM	USSR (E)	Lvov *et al*., 1973c
		A. loculosum	
Aride	UNG	Seychelles	Hoogstraal *et al*., 1976b

[a] *Abbreviations from Berge, T. O. (ed.) 1975. "International Catalog of Arboviruses." DHEW Publ. No. (CDC) 75-8301, USGPO, Washington, D. C.*

in Oregon; Oceanside virus is nearly indistinguishable from Uukuniemi virus and Zaliv Terpenya; Sakhalin virus is very similar to our Tillamook isolate, and finally Yaquina Head, Poovoot, and Okhotsky viruses are antigenically quite similar. Conversely, the Kemorovo and Sakhalin serogroups isolated from eastern Canada, Scotland, and Australia are not closely related to one another and are different from viruses in these groups from the western U. S. and eastern USSR. This may be due to the continuous distribution of the bird hosts for *I. uriae* between western U. S. and eastern USSR and the relative isolation of bird populations in eastern Canada, Scotland, and Australia.

Main *et al.*, (1976b) have suggested that the antigenic identity of the serogroups may be maintained by the isolation of the primary bird hosts. This is likely true for *I. uriae* since its primary hosts (murres in the north and penguins in the south) do not migrate great distances. Adult murres may move temporarily following the formation of ice packs but most of the movement in this species is confined to young birds (Tuck, 1960).

The dispersal of *I. uriae* and its associated viruses probably depends on the occasional movement of young birds and secondary hosts such as gulls, terns, and kittiwakes. Karpovich and Pilipas (1972) have demonstrated that kittiwakes from the Murmansk region moved during the winter to coastal areas in the North Sea and the littoral of Newfoundland and the southwest region of Greenland. Even more significant is their observation that the birds sometimes migrate over 2000 km and there is an exchange of young birds among neighboring colonies. These observations indicate how an exchange of group B viruses from the Murmansk region, where tick-borne encephalitis (TBE) has been reported in *I. uriae* and seabirds (Berkleshova *et al.*, 1969) (including murres and kittiwakes), to the east coast of North America could take place.

Of the four serogroups of viruses that have been associated with *I. uriae* throughout the world, only Sakhalin has been found exclusively in this species of tick and its primary host. The other three serogroups have been isolated from both argasid as well as ixodid ticks, and land birds and/or mammals (Hoogstraal, 1973). Further, until Oceanside virus from the U. S. and Zaliv Terpenya from the USSR were isolated from seabird ticks, viruses of the Uukuniemi serogroup were known only from *I. ricinus* and several species of land birds in Finland, western Ukraine, Czechoslovakia, and Egypt.

The finding of antigenically related or the same viruses in land bird ticks and seabird ticks is a theme that is repeated over and over again as these studies progress.

V. VIRUSES FROM ARGASID TICKS

There are 13 viruses from five serogroups known from soft ticks of the *O. capensis* group (Table II). For the most part they belong to different serogroups from those viruses isolated from *Ixodes* species. The two exceptions, Baku virus from the USSR and Huacho virus from Peru, are both Kemorovo serogroup. However, it should be pointed out that they are in a different antigenic complex, the Chenuda subgroup (Casals, 1971).

A serogroup that seems to be intimately associated with these ticks is Hughes. Although Hughes virus itself has only been found in *O. denmarki* from the Caribbean region, other viruses in this serogroup are scattered throughout the world. The most notable example is Soldado virus, which has been found in three of the species of the *O. capensis* group from Ethiopia, the Seychelles, Hawaii, Trinidad, Wales, and Ireland.

The distribution of viruses of two serogroups (Quaranfil and Nyamanini) that have been isolated from ticks of the *O. capensis* complex suggests a past or current association of the seabird tick-hosts between Africa and the mid-Pacific. Johnston Atoll virus isolated from *O. capensis* from the mid-pacific is closely related to Quaranfil, which was originally isolated in Egypt from *A. arboreus*, a parasite of herons. Johnston Atoll virus has also been isolated from *O. capensis* in Australia and New Zealand. The occurrence of Johnston Atoll virus in the mid-Pacific and Australia is difficult to explain since Amerson (1968) indicates little or no communication between the birds nesting in these two regions. Perhaps they could be carried on some of the other seabirds that migrate through the area, although Amerson (1968) indicates these transients seldom stop on these mid-Pacific Islands. Another possibility is that these viruses may have been introduced into the mid-Pacific and Australian regions from Africa by separate routes at different times.

A second possible Africa mid-Pacific connection was suggested by our isolation of a virus from *O. capensis* on Midway Island, which is antigenically similar to Nyamanini virus originally isolated from egrets in South Africa and later from *A. arboreus* and an egret in Egypt (Taylor *et al.*, 1966). Here again we see evidence of viruses that are found in ticks intimately associated with land birds having closely related counterparts in ticks found exclusively on seabirds.

One of our more interesting recent findings is the isolation of a virus (designated Aransas Bay; Yunker *et al.*, in manuscript) related to Upolu from *O. capensis* ticks collected from abandoned Brown Pelican nests on the southeastern coast of Texas. Upolu virus was originally isolated from this same tick species in association with tern colonies at Upolu Cay,

TABLE II
Argasid Seabird Ticks Harboring Viruses[a]

Virus	*Serogroup*	*Region*	*Authority*
		O. capensis group	
		O. (Alectorobius) capensis	
Soldado	HUG	Ethiopia, Seychelles, USA (SW)	Hoogstraal, 1973; Converse *et al.*, 1975; Keirans *et al.* (in press)
Johnston Atoll	QRF	Mid-Pacific Atoll, Australia, New Zealand	Clifford *et al.*, 1968; Doherty *et al.*, 1969; Queensland Int. Ann. Rep. 9/73
Midway	NYM	Midway I., Green Kure Atoll	Clifford *et al*, (unpublished)
Baku	KEM	Azerbaijan SSR	Lvov *et al.*, 1971a
Upolu	UPO	Australia	Doherty *et al.*, 1969
Aransas Bay	UPO	USA (SW)	Yunker *et al.* (in manuscript)
		O. (A.) denmarki	
Johnston Atoll	QRF	USA, Hawaii	Keirans *et al.* (in press)
Midway	NYM	USA, Hawaii	Keirans *et al.* (in press)
Hughes	HUG	(USA) (SE), Venezuela (Isla Aves) Trinidad	Hughes *et al.*, 1964; Keirans *et al.* (in press); Aitken *et al.*, 1968
Soldado	HUG	Trinidad, Hawaii	Jonkers *et al.*, 1973; Keirans *et al.* (in press)

Raza	HUG	Mexico (Raza I., San Lorenzo I.)	Clifford *et al.*, 1968; Keirans *et al.* (in press)
		O. (A.) sp. near denmarki	
Farallon	HUG	USA (NW) (SW)	Radovsky *et al.*, 1967; Keirans *et al.* (in press)
		O. (A.) maritimus	
Soldado	HUG	Wales, Ireland	Converse *et al.*, 1976; Keirans *et al.*, 1976
		O. (A.) amblus	
Punta Salinas	HUG	Peru	Johnson and Casals, 1972
Huacho	KEM	Peru	Johnson and Casals, 1972
		O. (A.) muesebecki	
Zirqa	HUG	Arabian Gulf	Hoogstraal *et al.*, 1970

[a] *A Kemerovo group agent (Mono Lake) was isolated from Argas cooleyi associated with gull nests at Mono Lake, California. It is not included here because the primary host for A. cooleyi is cliff swallows.*

Great Barrier Reef, Australia. It is anticipated that future studies will locate other foci of these agents that may give us indications of how this unusual distribution took place.

One possible reason for the extensive distribution of viruses associated with *O. capensis* is the fact that many of its primary hosts, such as terns migrate for great distances. This is in contrast to the rather limited movements of the primary hosts for *I. uriae*, i.e., murres and penguins. Also, the warmer climate in the areas where the *O. capensis* complex ticks would be deposited by migrating birds would tend to favor tick survival.

VI SUMMARY AND CONCLUSIONS

Viruses have been isolated from both argasid and ixodid ticks associated with seabirds in most regions of the world. In the colder areas, both in the north and south, the main vector of these agents is *I. uriae*, while ticks of the *O. capensis* complex replace *I. uriae* in this capacity in the more temperate regions of the world.

All of these ticks are closely aligned with the nesting habits of their bird hosts throughout their range and have evolved life cycle patterns that are compatible with the climatic conditions of the particular regions they inhabit.

Viruses of nine different serogroups have been isolated from seabird ticks. Group B., UUkuniemi, Sakhalin, and Kemorovo group viruses have been recovered from *I. uriae* and *I. signatus*, while Hughes, Nyamanini, Quaranfil, Kemorovo (Chenuda subgroup), and Upolu have been isolated from *O. capensis* complex ticks. Aride virus, which is ungrouped, has been found in *A. loculosum*.

The occurrence of these agents in widely separated populations of nesting birds has allowed sufficient isolation for the viruses to attain enough antigenci variation so they can usually be recognized as discrete viruses. This is especially true of the viruses isolated from ticks of the *O. capensis* complex. An exception is the viruses isolated from *I. uriae* in eastern Russia, which are antigenically similar to viruses in the same serogroup isolated along the western coast of North America. This similarity indicates that there may still be some exchange of genetic material resulting from the almost contiguous bird populations in these regions, or more likely, the separation of the ticks and breeding bird populations in these areas has occurred more recently than in some other regions.

Since the major portion of the tick's life cycle is closely allied to the nesting season of the bird host, the distribution of the ticks and their viruses must occur and have occurred by transport on nonbreeding young birds or on the adults when and if they migrate after breeding. Also the possibility that transient migrant birds may transport ticks must be considered.

The tremendous numbers of seabirds in confined areas during nesting and the vast number of ticks present during this time offer an effective means for the circulation of these viruses. In some instances, they can also be a threat to man if he intrudes in these areas during the nesting period. The illness of petroleum workers on Zirqa Island in the Seychelles (Hoogstraal *et al.*, 1970) and guano workers on islands near Peru (Hoogstraal *et al.*, 1976a) following the bites of *O. muesebecki* and *O. amblus*, respectively, provides circumstantial evidence that the transfer of these tick-borne viruses to man may be a reality.

Further, some of the viruses isolated from the bird hosts during mass die-offs indicate that these mass accumulations of birds may offer the potential for exposure of man and domestic animals to other virus groups, e.g., a myxovirus, Influenza A has been isolated from sick and dead terns during an apizootic among these birds at Cape Province, South Africa (Becker, 1961).

As additional viruses are isolated, the tremendous gaps in our present knowledge of the distribution of these viruses will be filled. Also, improvements and modifications in serological tests will be helpful in studying antigenic relationships among these viruses. Further, our ongoing studies on the taxonomy of the tick vectors, utilizing SEM, will undoubtedly help us to explain some of the tick-virus distribution patterns that do not make sense at the present time.

References

Aitken, T. H. G., Jonkers, A. H., Tikasingh, E. S., and Worth, C. B. (1968). *J. Med. Ent. 5*, 501-503.

Amerson, B. A., Jr. (1968). *J. Med. Ent. 5*, 332-339.

Arthur, D. R. (1960). *Parasitology 50*, 199-226.

Becker, W. B. (1961). *J. Hyg. Camb. 64*, 309-320.

Berkleshova, A. Yu., Terskikh, I. I., Bychkova, E. N., and Smirnov, V. A. (1969). *5th Study of the Role of Migrating Birds in the Distribution of Arboviruses* (Abstracts), p. 100. Acad. Sci. USSR, Siberian Branch Novosibirsk.

Casals, J. (1971). *Intr. Lect. Proc. Symp., Int. Symp. Tick-Borne Arboviruses (Excluding Group B), Smolenice, September 9-12, 1969*, pp. 13-20.

Clifford, C. M., Thomas, L. A., Hughes, L. E., Kohls, G. M., and Philip, C. B. (1968). *Am. J. Trop. Med. Hyg. 17*, 881-885.

Clifford, C. M., Yunker, E. E., Thomas, L. A., Easton, E. R., and Corwin, D. (1971). *Am. J. Trop. Med. Hyg. 20*, 461-468.

Converse, J. D., Hoogstraal, H., Moussa, M. I., Feare, C. S., and Kaiser, M. N. (1975). *Am. J. Trop. Hyg. 24,* 3-14.

Converse, J. D., Hoogstraal, H., Moussa, M. I., and Evans, D. E. (1976). *Acta Virol. 20,* 243-246.

Denmark, H. A., and Clifford, C. M. (1962). *Fl. Entomol. 45,* 139-142.

Doherty, R. L., Whitehead, R. H., Wetters, E. J., and Johnston, H. N. (1969). *Aust. J. Sci. 31,* 363-364.

Doherty, R. L., Carley, J. G., Murray, M. D., Main, Jr., A. J., Kay, B. H., and Domrow, R. (1975). *Am. J. Trop. Med. Hyg. 24,* 521-526.

Dumbleton, L. J. (1953). *Cape Exped. Ser. Bull. 14,* 1-28.

Flint, V. B., and Kostyrko, I. N. (1967). *Zool. Zh. 46,* 1253-1256.

Gaidamovich, S. Ya., Mlelnkova, E. E., Berkleshova, A. Yu., Terskikh, I. I., and Obukhova, V. R. (1971). Tezisy Dokl. Vop. Med. Virus, Inst. Virus imeni Ivanovsky, D. I., *Akad. Med. Nauk. SSSR, 2,* 68-69.

Hoogstraal, H. (1973). *In* "Viruses and Invertebrates" (A. J. Gibbs, ed.), Chapter 18, pp. 349-390. North-Holland Publ. Co., Amsterdam and London.

Hoogstraal, H., Oliver, R. M., and Guirgis, S. S. (1970). *Ann. Ent. Soc. Am. 63,* 1762-1768.

Hoogstraal, H., Clifford, C. M., Keirans, J. E., Kaiser, M. N., and Evans, D. E. (1976a). *J. Parasitol. 62,* 799-810.

Hoogstraal, H., Wassef, H. Y., Converse, J. D., Keirans, J. E., Clifford, C. M., and Feare, C. J. (1976b). *Ann. Ent. Soc. Am. 69,* 3-14.

Hughes, L. E., Clifford, C. M., Thomas, L. A., Denmark, H. A., and Philip, C. B. (1964). *Am. J. Trop. Med. Hyg. 13,* 118-122.

Johnson, H. N., and Casals, J. (1972). *5th Symp. Study of the Role of Migrating Birds in the Distribution of Arboviruses (Abstracts),* pp. 200-204. Acad. Sci. USSR, Siberian Branch, Novosibirsk.

Jonkers, A. H., Casals, J., Aitken, T. H. G., and Spence, L. (1973). *J. Med. Ent. 10,* 517-519.

Kaiser, M. N. (1966a). *Am. J. Trop. Med. Hyg. 15,* 964-975.

Kaiser, M. N. (1966b). *Am. J. Trop. Med. Hyg. 15,* 976-985.

Kaiser, M. N., Hoogstraal, H., and Kohls, G. M. (1964). *Ann. Ent. Soc. Am. 57,* 60-69.

Karpovich, V. N. (1973). *Parazitologiya Leningrad 7,* 128-134.

Karpovich, V. N., and Pilipas, N. I. (1972). *Mater. Simp. Itogi 6. Simp. Izuch. Virus Ekol. Svyazan. Ptits., Omsk, December, 1971,* pp. 91-94.

Keirans, J. E., Yunker, C. E., Clifford, C. M., Thomas, L. A., Walton, G. A., and Kelly, T. C. (1976). *Experientia 32,* 453-454.

Keirans, J. E., Yunker, C. E., and Clifford, C. M. (1977). *J. Med. Ent.* (in press).

Kitaoka, S., and Suzuki, H. (1973). *Nat. Inst. Anim. Health Quart. 13,* 142-148.
Lvov, D. K. (1973). *Sborn. Trud. Ekol. Virus 1,* 5-13.
Lvov, D. K., Gromashevsky, V. L., Sidorova, G. A., Tsyrkin, Yu. M., Chervonsky, V. I., Gostinshchikova, G. V., and Aristova, V. A. (1971a). *Virusol. 4,* 434-437.
Lvov, D. K., Timofeeva, A. A., Chervonski, V. I., Gromashevski, V. L., Klisenko, G. A., Gostinshchikova, G. V., and Kostyrko, I. N. (1971b). *Am. J. Trop. Med. Hyg. 20,* 456-460.
Lvov, D. K., Timofeeva, A. A., Gromashevski, V. L., Chervonsky, V. I., Gromov, A. I., Tsyrkin, Yu. M., Pogrebenko, A. G., and Kostyrko, I. N. (1972). *Arch. Ges. Virusforsch. 38,* 133-138.
Lvov, D. K., Smirnov, V. A., Gromashevsky, V. L., Veselovskaya, O. V., and Laurova, N. A. (1973a). *Med. Parazit. Moskva 42,* 728-730.
Lvov, D. K., Timopheeva, A. A., Gromashevski, V. L., Gostinshchikova, G. V., Veselovskaya, O. V., Chervonski, V. I., Fomina, K. B., Gromov, A. I., Pogrebenko, A. G., and Zhezmer, V. Yu. (1973b). *Arch. Ges. Virusforsch. 41,* 165-169.
Lvov, D. K., Timopheeva, A. A., Gromashevski, V. L., Tsyrkin, Yu. M., Veselovskaya, O. V., Gostinshchikova, G. V., Khutoretskaya, N. V., Pogrebenko, A. G., Aristova, V. A., Sazonov, A. A., Chervonski, V. I., Sidorova, G. A., Fomina, K. B., and Zhezmer, V. Yu. (1973c). *Arch. Ges. Virusforsch. 41,* 160-164.
Lvov, D. K., Timopheeva, A. A., Smirnov, V. A., Gromashevsky, V. L., Sidorova, G. A., Nikiforov, L. P., Sazonov, A. A., Andreev, A. P., Skvortzova, T. M., Beresina, L. K., and Aristova, V. A. (1972). *Med. Biol. 53,* 325-330.
Main, A. J., Downs, W. G., Shope, R. E., and Wallis, R. C. (1973). *J. Med. Ent. 10,* 229-235.
Main, A. J., Downs, W. G., Shope, R. E., and Wallis, R. C. (1976a). *J. Med. Ent. 13,* 309-315.
Main, A. J., Downs, W. G., Shope, R. E., and Wallis, R. C. (1976b). *J. Wildl. Dis. 12,* 182-194.
Main, A. J., Shope, R. E., and Wallis, R. C. (1976c). *J. Med. Ent. 13,* 304-308.
Murray, M. D., and Vestjens, W. J. M. (1967). *Aust. J. Zool. 15,* 715-725.
Radovsky, F. J., Stiller, D., Johnson, H. N., and Clifford, C.M. (1967). *J. Parasitol. 53,* 890-892.
Ritter, D. G., and Feltz, E. T. (1974). *Can. J. Microbiol. 20,* 1359-1366.
Roberts, F. H. S. (1970). "Australian Ticks," p. 254. Commonwealth Science and Industrial Research Organization, Australia.
Taylor, R. M., Hurlbut, H. S., Work, T. H., Kingston, J. R., and Hoogstraal, H. (1966). *Am. J. Trop. Med. Hyg. 15,* 76-86.

Thomas, L. A., Clifford, C. M., Yunker, C. E., Keirans, J. E., Patzer, E. R., Monk, G. E., and Easton, E. R. (1973). *J. Med. Ent. 10,* 165-168.
Tuck, L. M. (1960). "The Murres," *Can. Wildl. Ser. 1,* Ottawa, p. 260.
Yunker, C. E. (1975). *Med. Biol. 53,* 302-311.
Yunker, C. E., Clifford, C. M., Keirans, J. E., Thomas, L. A., and Cory, J. (1973). *J. Med. Ent. 10,* 264-269.
Yunker, C. E., Clifford, C. M., Keirans, J. E., and Thomas, L. A. (in manuscript).
Zumpt, F. (1952). *Aust. Nat. Antarctic Res. Exped. Ser. B. 1,* 12-20.

Chapter 7

ARBOVIRUSES IN ITALY

P. Verani, M. Balducci, and M. C. Lopes

I. INTRODUCTION

The presence of arboviruses in Italy was first documented during World War II in epidemiological studies on the American troops by Sabin (Sabin *et al.*, 1944): two viruses of the Phlebotomus fever group were isolated, Sandfly Sicilian and Sandfly Neapolitan. No further conclusive studies on the activity of arboviruses in Italy were carried out until 1965 when an investigation on arboviruses in selected Italian regions was initiated in this Department. The aim of these studies was to provide information on the existence and circultaion of arboviruses and their possible significance to human health.

Italy extends approximately between latitudes 36 and 47°N. It is almost completely surrounded by sea and bordered in the north by mountains. There is a north-south central mountain range (Appennines) with highest altitudes of over 2600 m. Consequently, climatic conditions vary considerably within Italy: from mountainous, to continental, and to coastal. The southern extremity of Italy is greatly influenced by its proximity to North Africa, sometimes producing subtropical climatic conditions. Furthermore, the formation of an abundant and varied arthropod population, the presence of domestic and small wild

ISBN 0-12-429765-X

animals in fairly large numbers in certain parts of the country, and a usually very temperate climate could maintain natural foci of arboviruses. The periodic migration of several species of birds between Europe and Africa across Italy can also contribute to the introduction and dissemination of arboviruses in the country.

The studies have been performed in selected localities chosen throughout Italy. The names and the description of the areas investigated are given in Table I. Throughout these studies, serologic methods were used and attempts were also made to isolate virus from blood-sucking arthropods, man, wild mammals, and birds. The present report describes the data accumulated during this research.

II. ISOLATION FROM VECTORS

A. Mosquito-Borne Viruses

During the summers and autums of 1966 to 1969, mainly in September and October, 76,000 female mosquitoes were collected in Central and Northern Italy and processed for virus isolation. The dominant mosquito species were *Culex pipiens, Aedes vexans and A. caspius*. *Aedes* mosquitoes were caught mainly on the vegetation, while most *C. pipiens* mosquitoes were collected inside stables. In addition, hibernating *C. pipiens* were caught at the end of autumn inside natural shelters and a few specimens were collected by animal traps. Virus isolation was achieved only from *Aedes* mosquitoes collected in northern Italy. No virus was recovered from nearly 30,000 *Aedes* mosquitoes from central Italy as well as from all *C. pipiens* processed up to now.

1. Tahyna Virus. Tahyna virus of the California group was isolated from two pools of *Aedes* mosquitoes (about 90% *A. caspius* and 10% *A. vexans)* collected in September 1967 in northern Italy along the branches of the Isonzo river delta (Balducci *et al.*, 1968).

In a previous study by Bardos and Sefcovicova (1961) neutralizing antibodies to Tahyna virus were detected in human sera (up to 9%). Following virus isolation, a serological survey by mouse neutralization test was performed using the Tahyna strain isolated in Italy (ISS.Z.186) and it was restricted to the inhabitants, mainly fishermen, living in the vicinity of the areas where the vector used to breed. A high prevalence of antibodies (70%) was found in this population. It was similar to those recorded in other Tahyna foci in Europe (Bardos *et al.*,

1962; Hannoun *et al.*, 1969; Brummer-Korvenkontio, 1973). Additional seroepidemiological surveys were carried out in other Italian regions. In Table II are presented the accumulated data on the results obtained. Low percentages of neutralizing antibodies were found in Parma (northern Italy) and in Sardinia (Sanna *et al.*, 1967). Two positive sera (3.3%) were found by hemagglutination inhibition (HI) test among residents in the province of Naples (Lopes *et al.*, 1970) but no positives in Calabria (Verani *et al.*, 1971) and in Sicily (Albanese *et al.*, 1976).

The data available on virus isolation and the results of serologic surveys suggest that Tahyna virus is not widespread in Italy (foci are probably present only in the northeastern areas), although high infection rates in humans have been repeatedly reported in countries of central and northern Europe (Henderson and Coleman, 1971).

2. *Other Viruses.* Antibodies against other mosquito-borne viruses have also been found in Italy, but their significance and specificity need to be clarified. Reactors to western equine encephalomyelitis (WEE) were found in domestic animals in central Italy (Latina province): 24% of bovine sera, 32% of goat sera, and 7% of sheep sera had antibodies to WEE by hemagglutination inhibition, but not by neutralization test (Verani *et al.*, 1967). In another central Italian region (Toscana) 28% of positive reactions to WEE were found among chicken sera (Verani *et al.*, 1978). Similarly, antibodies to West Nile virus were detected in human sera: 18% from northern Italy (Balducci *et al.*, 1967a), 8% from central Italy (Verani *et al.*, 1967), and 23% from a northwestern region (Sanna *et al.*, 1967). Furthermore, low incidence of HI antibodies to WEE (1.2%) and West Nile (10%) viruses was found in sera of migratory birds collected in northern Italy during their fall migration (Balducci *et al.*, 1973).

These results suggest that at least a virus of group A and a virus related to West Nile probably circulate in Italy. The isolation of Sindbis and West Nile viruses has been obtained from birds trapped in Czechoslovakia (Labuda *et al.*, 1974; Gresikova *et al.*, 1975a). Therefore, it is possible that these viruses could be introduced by birds in Italy.

B. Tick-Borne Viruses

Several tick species have been reported to be present in Italy (Starkoff, 1958). Their distribution, seasonal activity, and abundance vary greatly in different regions, according to climatic and ecologic conditions. During the present studies, a total of 32,567 Ixodid ticks of nine species were processed

TABLE I
Localities Where Arbovirus Studies Were Performed

Region	*Area (province)*
Northern Italy	
Friuli	Gorizia
Central Italy	
Toscana	Florence
	Siena
	Grosseto
Lazio	Rome
	Latina (Fondi)
Southern Italy	
Campania	Naples
Calabria	Cosenza

Description of area
Flat plain and surrounding hills Sea level Cultivated fields
Plain and surrounding hills Altitude 50 to 600 m Bushy pastures and deciduous wood
Appennines mountains Altitude about 600 m Grazing and cultivated hills
Seashore and surrounding hills Altitude from 0 to 400 m Deciduous wood
Littoral Mediterranean woodland Sea level National park
Plain and surrounding hills Sea Level Cultivated and grazing fields
Town and surroundings Sea level Cultivated fields
Flat plains and surrounding mountains Altitude from 0 to 1200 m Cultivated fields, bushy pastures, woods

TABLE II
Incidence of Human Antibodies to Tahyna Virus in Different Regions of Italy

Region	*Number of sera*		
	Positive/tested	*% positive*	*Reference*
Northern Italy			
Friuli	15/21	70	Unpublished data
Parma	8/600	1.3	Sanna *et al.*, 1967
Southern Italy			
Campania	2/60	3.3	Lopes *et al.*, 1970
Calabria	0/90	0	Verani *et al.*, 1971
Sicily (Eastern)	0/63	0	Castro *et al.*, 1976
Sicily (Western)	0/260	0	Albanese *et al.*, 1971a
Sardinia	3/183	1.6	Sanna *et al.*, 1967

for virus isolation (Table III). The ticks were collected during the years 1966 to 1973 either by handpicking from grazing animals or by blanket-dragging on the ground. The majority of ticks were collected in two regions: 18,538 in northeastern (Friuli) and 12,064 in central Italy (Latina). In the latter region where mostly adult ticks were found, a periodical tick collection was regularly performed throughout the whole year during 1967 and 1973. Fair numbers of ticks were also collected in two other central Italian localities and in southern Italy. The species most commonly found were *Ixodes ricinus, Haemaphysalis punctata, Rhipicephalus bursa, and Hyalomma marginatum.* In Friuli, ticks were collected only on vegetation and *I. ricinus,* mostly nymphs, was the only species found.

1. Bhanja Virus. Nine strains of Bhanja virus, an ungrouped tick-associated virus, were isolated from 2625 nonengorged adults of *H. punctata* collected from September to December 1967 in the Fondi area (central Italy) (Verani *et al.,* 1970a). A tenth strain was isolated from 1836 ticks of the same species collected six years later in the same locality (Table IV). The minimum infection rate was 1:290 in 1967, but only 1:1836 in 1973; considering all *Haemaphysalis* processed, the infection rate was 1:531. *H. punctata* was the species most commonly found in the Fondi area, whereas it was scarce in the other regions where collection of ticks has been attempted. The peak of activity of *H. punctata* in this area is between September and November (Saccà *et al.,* 1969), when all strains of Bhanja virus were isolated both in 1967 and in 1973. The vector *H. punctata* is primarily a parasite of large domestic animals and only few specimens were collected on wild mammals.

This was the first isolation of Bhanja virus outside tropical areas. The prototype strain was isolated in India in 1954 (Shah and Work, 1969) and strains of Bhanja virus were subsequently isolated from *Haemaphysalis* species in other European countries, such as Yugoslavia (Vesenjak-Hirjan, 1975, personal communication) and Bulgaria (Rosicky, 1977, personal communication).

The isolation of Bhanja virus prompted a large scale serological survey in Italy, both with respect to its geographical distribution and its significance for man, domestic mammals, birds, and wild rodents (Verani *et al.,* 1970b). The accumulated data on the results of the survey by hemagglutination inhibition test with Bhanja virus are presented in Table V. Antibodies against Bhanja virus were found in every surveyed region. The highest prevalence with high titers (up to 1:1280) was found in ovine species, mainly in goats: reactors were rare (2.3%) only in Campania, one of the most densely inhabited regions of Italy, where wild life has almost disappeared. The percentage of human

TABLE III

Species and Number of Ticks Processed for the Isolation of Virus in the Years 1966-1973

Locality and year	*Species*	*Total (ticks)*	*Number of isolations*
Northern Italy			
Friuli	*Ixodes ricinus*	18,538	6[a]
(1966-1970)			
(1972-1973)			
Central Italy			
Toscana (Siena)	*Ixodes ricinus*	218	1[a]
(1972-1973)	*Dermacentor marginatus*	9	
	Rhipicephalus	22	
	Hyalomma marginatum	124	
	Total	373	
Lazio (Rome)	*Ixodes ricinus*	30	
(1971-1973)	*Haemaphysalis punctata*	305	
	Haemaphysalis concinna	141	
	Rhipicephalus	51	
	Hyalomma marginatum	14	
	Total	541	

Lazio (Latina)	*Ixodes ricinus*	1,675	
(1966-1968)	*Haemaphysalis punctata*	5,310	10[b]
(1972-1973)	*Haemaphysalis otophila*	5	
	Dermacentor marginatus	26	
	Rhipicephalus sanguineus	545	
	Rhipicephalus bursa	2,342	
	Hyalomma marginatum	1,944	
	Boophilus calcaratus	217	
	Total	12,064	
Southern Italy			
Calabria	*Haemaphysalis punctata*	10	
(1969-1971)	*Dermacentor marginatus*	404	
	Rhipicephalus bursa	640	
	Rhipicephalus sanguineus	6	
	Hyalomma marginatum	3	
	Total	1063	
Total		32,579	

[a] *Tribec virus.* [b] *Bhanja virus.*

TABLE IV
Bhanja Virus from H. punctata *Ticks Collected in the Fondi area, 1967-1973*

	1967		*1968*		*1973*	
Monthly tick collection period	*Number of ticks*	*BHA positive/ number of pools*	*Number of ticks*	*BHA positive/ number of pools*	*Number of ticks*	*BHA positive/ number of pools*
January-April	-	-	476	0/12	-	-
September	673	4/26	373	0/9	-	-
October	646	1/19	-	-	290	0/10
November	851	4/20	-	-	850	1/26
December	455	0/11	-	-	496	0/8
Total	2625	9/76	849	0/21	.636	1/44

TABLE V
Results of Hemagglutination Inhibition Tests with Bhanja Antigen against Sera from Man and Other Vertebrates (1965-1975)

Origin of sera		*Hemagglutination inhibition*	
Species	*Locality*	*Positive/tested*	*% positive*
Man	Friuli	0/40	0
	Toscana	0/85	0
	Lazio	1/40	2.5
	Campania	4/92	4.3
	Calabria	5/370	1.3
	Total	10/627	1.6
Cattle	Friuli	0/29	0
	Toscana	4/50	8
	Lazio	3/56	5.3
	Campania	0/20	0
	Total	7/155	4.5
Goat	Lazio	44/61	72
	Campania	2/85	2.3
	Calabria	131/210	62
	Total	177/356	49.4
Sheep	Friuli	23/48	48
	Toscana	57/116	49
	Lazio	7/32	22
	Total	87/196	44
Wild rodents	Friuli	0/15	0
	Lazio	4/72	5.5
	Total	4/87	4.6
Migratory birds	Friuli	12/635	1.9

sera reacting with Bhanja virus was low (1.6%) and the titers were never greater than 1:40. The few reactors (4.5%) found among cattle breeders were bled in central Italy (Toscana and Lazio). Among wild rodents (mostly *Apodemus* sp.) a low percentage of positive sera was found (5.5%), suggesting an insignificant role of these mammals in the circulation of Bhanja virus. The incidence of antibodies in 635 migratory birds was similarly low (1.9%). In further surveys performed in Sicily, HI antibodies against Bhanja virus were found mainly in ovine species (Albanese *et al.*, 1971b; Castro *et al.*, 1976).

On the basis of the serological results, Bhanja virus seems to be widespread in Italy, goats and sheep being excellent indicators of its activity. The failure to isolate the virus in other parts of the country may be due to the small numbers of *Haemaphysalis* ticks collected and processed for virus isolation. No viremia was detected in sentinel goats introduced in a focus of Bhanja virus, although the animals bled at weekly intervals showed a seroconversion during the same period. Similarly, no virus was isolated from the blood of experimentally infected young goats, suggesting that these animals probably do not play a part in the cycle of Bhanja virus in nature (Verani *et al.*, 1971). Furthermore, there was no evidence of any disease caused by Bhanja virus in goats, in agreement with the data reported in India.

Clinical cases due to Bhanja virus are not known and only low titers of HI antibodies were so far found in a few human sera. Four laboratory workers handling Bhanja virus since 1967 in this laboratory developed specific antibodies, without any associated illness. However, a case of a mild disease has been reported recently in a laboratory worker (Calisher and Goodpasture, 1975). The pathogenicity of Bhanja virus for suckling mice and for monkeys was studied (Balducci *et al.*, 1970). After intrathalamic inoculation, the virus multiplied in the central nervous system of monkeys and migrated to the spinal cord. The pathologic picture, characterized by lesions in the neurons of the cerebral cortex, basal ganglia, and Purkinje cells, could suggest a possible neurotropic potentiality of this virus.

2. Tribec Virus. Seven strains of Tribec virus (Kemerovo group) were isolated from *I. ricinus* (unpublished studies from this laboratory). Six strains were derived from 3130 specimens, mostly nymphs, collected in May 1972 in northeastern Italy (Gorizia province) and one from 218 specimens collected in central Italy (Siena province). Five strains, including the one from Siena province, were isolated from females, one from nymphs, and one from a mixed pool of males and nymphs. The mean minimum infection rate of adults and nymphs together was 1:521 in northern Italy and 1:218 in central Italy.

The prototype strain, ISS.Ir.560, was identified in a one-way complement fixation (CF) test to be related to Tribec virus (M. Gresikova, personal communication, 1974). To further characterize this isolate, it was tested for its sensitivity to lipid solvents and was shown to be very sensitive to both ether (there was a decrease of 4.7 logs in infectivity) and sodium deoxycholate (SDC) (a decrease of over 4.0 logs). The sensitivity to ether and SDC was an exception from the pattern described for Tribec and other Kemerovo group viruses by Borden *et al.*, (1971), who reported a relative stability of Tribec virus (at the fourteenth mouse passage) to lipid solvents. Tests for the sensitivity of the ISS.Ir.560 strain of Tribec virus to ether and SDC were repeated at the fourth and fourteenth passages. The solvent sensitivity was retained also at the highest passage level, with a drop of 2.7 logs in infectivity after ether treatment and of 5.1 logs after SDC treatment. In this respect ISS.Ir.560 behaves as an atypical strain of Tribec virus.

The prototype Tribec virus has been previously isolated in Czechoslovakia (Gresikova *et al.*, 1965) from the same vector, *I. ricinus*. It may have caused a single case of human infection (Gresikova *et al.*, 1966) and antibodies have been reported in a few human sera (Gresikova *et al.*, 1965). Antibodies were also prevalent in domestic animals (Ernek and Kozuch, 1970; Saikku and Brummer-Kirvenkontio, 1975). No studies on the distribution of Tribec virus in man and animals in Italy have yet been carried out; only a few sera from patients with acute central nervous system diseases had been tested by CF and none had antibodies.

3. Thogoto Virus. Two virus strains identified as Thogoto virus (Thogoto group) have been isolated from *R. bursa* collected in 1969 in Sicily (Albanese *et al.*, 1972). The occurrence of antibodies to this virus in sera of cattle and sheep from different zones of Western Sicily has been reported by the same authors (Albanese *et al.*, 1971b). This was the first isolation of Thogoto virus in Europe; the prototype strain was isolated in Kenya in 1964 (Haig *et al.*, 1965).

Serologic studies for the presence of Thogoto virus in Italy outside Sicily have not yet been done. Virus isolation from *Rhipicephalus* species was attempted in other Italian regions: no virus was obtained from 3606 specimens processed (Table III).

4. Tick-Borne Encephalitis (TBE) Virus. Evidence for tick-borne encephalitis virus activity was obtained in a serologic survey in central Italy (Latina province) in 1966. In a limited study it was found that 10 out of 25 goat sera had hemagglutinating inhibition and neutralizing antibodies to TBE virus (Verani *et al.*, 1967). The initial survey was extended by investigating serial bleedings of locally grazing goats, which gave evidence of infection with TBE in a narrow area of the surveyed region (Balducci *et al.*, 1976b). In goats and other domestic animals grazing in the neighboring localities no antibodies were found. In a search for the suspected vector, *I. ricinus*, during the following year, a limited number of specimens of this tick species was collected in the area and no virus was isolated. The most numerous tick species in the area was *Haemaphysalis,* from which Bhanja virus was isolated.

C. Phlebotomus-Transmitted Viruses

Among the 20 presently recognized antigenic types of Phlebotomus fever group of arboviruses in the world, two serotypes, Sandfly Sicilian and Sandfly Neapolitan, were isolated in Europe. They were obtained by Sabin *et al.*, (1944) during World War II from the blood of patients during outbreaks of febrile illness among American troops in Southern Italy. The population of Phlebotomine flies was drastically reduced folfowing intensive insecticide treatment for the eradication of malaria. Nevertheless, surveys for antibodies to Phlebotomus fever viruses gave evidence for the persistence of natural cycles of these viruses in Italy.

In the summer of 1971 three virus strains were isolated from 329 sandflies, mostly nonengorged females of *Phlebotomus perniciosus,* collected in cattle and rabbit stables in Toscana (Central Italy) (unpublished studies from this laboratory). No virus was recovered from 1130 specimens collected in 1974 in a neighboring locality. The three viruses isolated in 1971 were shown to be closely related or identical by CF test. They were readily isolated in suckling mice without blind passages. By the third brain-to-brain passage, the time of death after inoculation was shortened to 3 to 4 days. The titer was 6.5 logs LD_{50}/0.01 ml. The prototype strain (ISS.Phl.3) was demonstrated to be an antigenically related member of the Phlebotomus fever group (J. Casals, personal communication, 1973). It reacted in CF, hemagglutination inhibition tests with neutralization antisera to Sandfly Neapolitan, although it seems to be somewhat distinct from the prototype Sandfly Neapolitan. Since some antigenic differences were established, ISS.PHI.3 can be considered as an SFN-like virus.

To further characterize this virus, its pathogenicity for laboratory animals was determined. The virus was lethal for suckling and weanling mice by intracranial and intraperitoneal inoculations. Guinea pigs, rabbits, and monkeys developed a neurological disease on intracranial inoculation, which was lethal for some of them. After intraperitoneal inoculation, no symptoms of disease were observed in these species, although the animals developed CF antibodies with low titers. These data revealed a marked difference in pathogenicity between the SFN-like virus and the prototype SFN, which is known to be nonpathogenic for the majority of laboratory animals (Berger 1975).

HI antibodies to Sandfly Sicilian and Sandfly Neapolitan viruses were found throughout Italy with an incidence varying from about 2 to 20% (Verani *et al.*, 1971; Balducci *et al.*, 1967a; Lopes *et al.*, 1970; Albanese *et al.*, 1971a; Castro *et al.*,

1976) (Table VI). The analysis of the distribution of positive sera by age groups seemed to indicate an age dependency, although a few reactors were found also among young people. A limited survey performed by HI test on the inhabitants of coastal areas (Grosseto province) near the locality where the SFN-like virus was isolated revealed that 1.8% of sera were positive to SFS, 9.1% to SFN, and 7.2% to SFN-like virus (Verani *et al.*, 1978). Reactors to SFN-like virus had no antibodies to SFN virus and vice versa. This finding confirmed the antigenic difference between the two viruses and further stressed the importance of using both viruses in antibody surveys. The incidence of antibodies was similarly low with SFN and SFN-like viruses, regardless of the fact that the latter was isolated in the region.

The serologic data and the lack of isolation of the virus from sandflies collected three years later in the same region could suggest that foci of the SFN-like virus are discrete and possibly widely separated.

Because of the mildness of the disease, laboratory data on the incidence of "phlebotomus fever" in human population are lacking. Nevertheless, seroepidemiological results and the recovery of three strains of an SFN-like virus from sandflies emphasize the endemicity of the infection in Italy. Furthermore, the importance of the newly isolated SFN-like virus to human health is still to be determined.

III. ISOLATION FROM VERTEBRATES (MIGRATORY BIRDS)

In order to get information on the possibility of new viruses being carried into the country, a study on migratory birds was started in 1967. The field study was restricted to the area of the eastern Alps (Gorizia province) where mountain passes are low and act to funnel birds migrating southward from northern and eastern Europe. From the frequencies of recaptures of banded birds by a local ornithological center it was evident that bird migration in the study area was prevalently from eastern Europe.

Two strains of Bahig virus and two strains of Matruh virus were isolated from the blood of 717 birds (Balducci *et al.*, 1973) captured by nets in the autumns of 1968 and 1969. Each virus was isolated in both years and both from Chaffinch *(Fringilla coelebs)* and brambling finch *F. montifringilla)*. Bahig and Matruh viruses are members of the Tete group of the Bunyamwera supergroup. The Tete group is composed of four viruses, which were all isolated originally from blood and organs of birds captured in Africa. Bahig and Matruh viruses

TABLE VI

Incidence of Antibodies to Phlebotomus-*Transmitted Viruses in Man in Different Regions of Italy*

Region	*SFS*		*SFN*		*SFN-like (ISS.Ph 1.3)*[a]	
	Positive/tested	*%*	*Positive/tested*	*%*	*positive/tested*	*%*
Northern Italy						
Friuli	27/166	16.2	27/166	16.2	-	
Central Italy						
Toscana	1/55	1.8	5/55	9.1	4/55	7.2
Lazio	42/205	20.5	20/205	9.7	-	
Southern Italy						
Campania	3/92	3.2	8/92	8.7	-	
Calabria	0/90	-	3/90	3.3	-	
Sicily (Eastern)[b]	0/63	-	9/63	14.2	-	
Sicily (Western)[c]	10/260	3.8	19/260	7.3		

[a] *Only sera from the region of virus isolation were tested.*

[b] *Castro* et al., *(1976)*.

Albanese et al., *(1971a)*.

were first isolated from autumn migrants in Egypt, suggesting that infections originated somewhere in eastern Europe or western Asia (Watson *et al.*, 1972). The isolation of the two viruses from birds near the northeastern Italian border seemed to confirm this hypothesis.

Until recently, there was no evidence that viruses of the Tete group circulate in Italy. In 1974, however, Bahig virus was isolated from larvae and nymphs reared from an engorged female of *H. marginatum* taken from a race horse near Naples (Converse *et al.*, 1974). Because data on the origin of this horse are lacking, the presence of "foci" of Bahig virus in Italy needs further field investigations.

IV. CLINICAL CASES OF TBE INFECTION

The possible relationship between some of the arboviruses circulating in Italy and human diseases has focused our attention. Several cases of acute disease of the central nervous system, causing often severe encephalitis or meningoencephalitis, are reported every year in different Italian regions, but diagnosis of a viral etiology is supported only by clinical observations. In order to study the possible role of arboviruses in the etiology of these diseases since 1975, sera from patients with acute CNS illnesses hospitalized at the Department of Neurology of the University and at the Infectious Diseases Hospital, Florence (Toscana) were examined. During 1975, sera from 25 patients hospitalized mainly during the period between May and September were received. Sera from three patients reacted with TBE antigen by the HI test at titers of 1:320 to 1:640. Treatment with 2-mercaptoethanol (2-ME) lowered the HI titers of the acute samples in two out of the three cases, indicating the presence of specific IgM antibodies (Amaducci *et al.*, 1976). The only serum sample available from the third positive patient did not show a reduction in the titer after 2-ME treatment, possibly because it was taken nearly three weeks after the onset of the disease. The combination of HI test and 2-ME treatment is the suggested method for the diagnosis of Tick-borne encephalitis (Kunz and Hofmann, 1971; Gresikova *et al.*, 1975b). Neutralizing antibodies were present in the acute serum samples of the two serologically confirmed cases of TBE infection. None of the 26 sera obtained from the same hospitals during 1976 was positive.

These results, although limited in significance due to the small number of positive cases, indicate that TBE virus could be implicated as responsible for human encephalitis in Italy even if it probably does not represent a great health hazard in this country.

V. CONCLUSIONS

The state of the art of arbovirus research in Italy has been described as a result of more than 10 years of work. Table VII summarizes the up-to-date account of arboviruses studied in Italy. Viruses have been isolated from every type of arthropods examined, i.e., mosquitoes, ticks, and sandflies. Some viruses, like Tahyna and Tribec, are closely related to agents circulating in central Europe, whereas other viruses, like the Phlebotomus-transmitted, Bhanja, and Thogoto, are related to agents typical of tropical and subtropical areas. As for the latter viruses, the strains isolated in Italy appear to represent the northernmost isolation. The observed distribution of arboviruses was to be expected, in view of the peculiar climatologic and ecologic features in Italy.

The isolation of Bahig and Matruh viruses from migratory birds provides additional confirming evidence of the role of birds as disseminators of arboviruses. It also emphasizes the possibility of new agents being thereby introduced in Italy.

Although at least eight different viruses have been detected in nature, their degree of association with human disease is still poorly understood, due mainly to difficulties in performing adequate clinical and serological surveys. On the other hand, clinical cases due to TBE virus have been reported. It follows that experimental, as well as clinical, studies on the epidemiology of arboviruses should be continued so that a final assessment of their medical importance will be made possible.

TABLE VII
Arboviruses in Italy

Group	*Virus*	*Year*	*Isolation source*	*Antibodies*		*Human disease*	*Reference*
				Man	*Mammals*		
California	Tahyna	1967	*Aedes* sp.	+	-	?	Balducci *et al.*, 1968
Phlebotomus fever	Sandfly Sicilian	1943	man	+	-	+	Sabin *et al.*, 1944
	Sandfly Neapolitan	1944	man	+	-	+	Sabin *et al.*, 1944
	ISS.Phl.3(SFN-like)	1971	Phlebotomus perniciosus	+	-	?	Unpublished data
Tete	Bahig	1968,1969	bird	-	-	-	Balducci *et al.*, 1973
	Matruh	1968,1969	bird	-	-	-	Balducci *et al.*, 1973
Kemerovo	Tribec	1972	*Ixodes ricinus*	-	-	-	Unpublished data
Thogoto	Thogoto	1969	*Rhipicephalus bursa*	-	+	?	Albanese *et al.*, 1972
Ungrouped	Bhanja	1967,1973	*Haemaphysalis punctata*	+	+	?	Verani *et al.*, 1970a
B	TBE	-	-	+	+	+	Amaducci *et al.*, 1976
B	WN	-	-	+	-	?	Balducci *et al.*, 1967a
A	?	-	-	-	+	-	Verani *et al.*, 1967

References

1. Albanese, M., Di Cuonzo, G., Randazzo, G., and Tringali, G. (1971a). *Boll. Soc. It. Biol. Sperim. 47*, 436-438.
2. Albanese, M., Di Cuonzo, G., Randazzo, G., Srihongse, S., and Tringali, G. (1971b). *Ann. Sclavo 13*, 641-647.
3. Albanese, M., Bruno-Smiraglia, C., Di Cuonzo, G., Lavagnino, A., and Srihongse, S. (1972). *Acta Virol. 16*, 267.
4. Amaducci, L., Arnetoli, G., Inzitari, D., Balducci, M., Verani, P., and Lopes, M. C. (1976). *Riv. Patol. Nervosa Mentale 97*, 77-80.
5. Balducci, M., Verani, P., Lopes, M. C., and Gregorig, B. (1967a). *Am. J. Trop. Med. Hyg. 16*, 211-215.
6. Balducci, M., Verani, P., Saccà, G., and Lopes, M. C. (1967b). *Ann. Ist. Super. Sanita 3*, 705-719.
7. Balducci, M., Verani, P., Lopes, M. C., Saccà, G., and Gragorig, B. (1968). *Acta Virol. 12*, 457-459.
8. Balducci, M., Verani, P., Lopes, M. C., and Nardi, F. (1970). *Acta Virol. 14*, 237-243.
9. Balducci, M., Verani, P., Lopes, M. C., and Gregorig, B. (1973). *Ann. Microbiol. (Inst. Pasteur) 124B*, 231-237.
10. Bardos, V., and Sefcovicova, L. (1961). *J. Hyg. Epidemiol. Microbiol. Immunol. 5*, 501-504.
11. Bardos, V., Adamcova, J., Sefcovicova, L., and Cervenka, J. (1962). *Ceskoslov. Epidemiol. Mikrobiol. Immunol. (Praha) 11*, 238-241. Berge, T. O. (ed.) (1975). "Catalogue of Arboviruses," 2nd ed. DHEW Publ. 75-8301, Atlanta.
12. Borden, E. C., Shope, R. E., and Murphy, F. A. (1971). *J. Gen. Virol. 13*, 261-271.
13. Brummer-Kovenkontio, M. (1973). *Am. J. Trop. Med. Hyg. 22*, 654-661.
14. Calisher, C. H., and Goodpasture, H. C. (1975). *Am. J. Trop. Med. Hyg. 24*, 1040-1042.
15. Converse, J. D., Hoogstraal, H., Moussa, M. I., Stek, M., and Kaiser, M. N. (1974). *Arch. Ges. Virusforsch. 46*, 29-35.
16. Ernek, E., and Kozuch, O. (1970). *Folia Parasit. (Praha) 17*, 331-335.
17. Gresikova, M., Nosek, J., Kozuch, O., Ernek, E., and Lichard, M. (1965). *Acta Virol. 9*, 83-88.
18. Gresikova, M., Rajcani, J., and Hruzik, J. (1966). *Acta Virol. 10*, 420-424.
19. Gresikova, M., Sekeyova, M., and Prazniakova, E. (1975a). *Acta Virol. 19*, 162-164.
20. Gresikova, M., Sekeyova, M., Stupalova, S., and Necas, S. (1975b). *Intervirology 5*, 57-61.
21. Haig, D. A., Woodall, J. P., and Danskin, D. (1965). *J. Gen. Microbiol. 38*, 389.

22. Hannoun, C., Panthier, R., and Corniou, R. (1969). *In* "Arboviruses of the California Comples and Bunyamwera Group." Publ. House, Slovak Acad. Sci., Bratislava.
23. Henderson, B. E., and Coleman, P. H. (1971). *Progr. Med. Virol. 13,* 405-461.
24. Kunz, C., and Hofmann, H. (1971). *Zent. Bakt. Hyg. I Orig. A 218,* 273-279.
25. Labuda, M., Kozuch, O., and Gresikova, M. (1974). *Acta Virol. 18,* 429-433.
26. Lopes, M. C., Verani, P., Balducci, M., and Da Villa, G. (1970). *Ann. Ist. Super. Sanita 6,* 121-124.
27. Sabin, A. B., Philip, C. B., and Paul, J. R. (1944) *J. Am. Med. Assoc. 125,*603-606, and 693-699.
28. Saccà, G., Mastrilli, M. L., Balducci, M., Verani, P., and Lopes, M. C. (1969). *Ann. Ist. Super. Sanita 5,* 21-28.
29. Saikku, P., and Brummer-Korvenkontio, M. (1975). *Med. Biol. 53,* 317-320.
30. Sanna, A., Fadda, G., Chezzi, C., Murruzzu, I., Bagella, G., and Gelli, G. P. (1967). *Ig. Mod. 60,* 739-749.
31. Shah, K. V., and Work, T. H. (1969). *Ind. J. Med. Res. 57,* 793-798.
32. Starkoff, O. (1958). "Ixodoidea d'Italia--Studio Monografico," Ed. Il Pensiero Scientifico, Roma.
33. Verani, P., Balducci, M., Lopes, M. C., Alemanno, A., and Saccà, G. (1967). *Am. J. Trop. Med. Hyg. 16,* 203-210.
34. Verani, P., Balducci, M., Lopes, M. C., and Saccà, G. (1970a). *AM. J. Trop. Med. Hyg. 19,* 103-105.
35. Verani, P., Balducci, M., and Lopes, M. C. (1970b). *Folia Parasit. (Praha) 17,* 367-374.
36. Verani, P., Lopes, M. C., Balducci, M., Serra, F., and Crivaro, G. (1971). *J. Hyg. Epidemiol. Microbiol. Immunol. 15,* 405-416.
37. Verani, P., Lopes, M. C., Balducci, M., Quercioli, A., and Bernardini, E. (1978). *Ann. Sclavo 19,* (in press).
38. Watson, G. E., Shope, R. E., and Kaiser, M. N. (1972). *Proc. 5th Symp. Study of the Role of Migrating Birds in the Distribution of Arboviruses.* pp. 176-180. Publ. House "Nauka", Siberian branch, Novosibirsk.

Chapter 8

TRANSMISSION OF GROUP C ARBOVIRUSES (BUNYAVIRIDAE)

John P. Woodall

I. INTRODUCTION

Group C of the arboviruses was established by Casals and Whitman (1961) and originally consisted of five serologically related viruses isolated at Belem, Brazil, by Causey *et al.*, (1961). These were Caraparu (CAR), Apeu, Marituba (MTB), Murutucu (MUR), and Oriboca (ORI). Subsequently six more viruses have been added: Itaqui (ITQ) (Shope *et al.*, 1961), Madrid (MAD) and Ossa (De Rodaniche *et al.*, 1964), Nepuyo (NEP) (Spence *et al.*, 1966), Restan (RES) (Jonkers *et al.*, 1967), and Gumbo Limbo (GL) (Henderson *et al.*, 1969). The 11 viruses fall into three complexes, as shown in Table I. (The classification and abbreviations used here for the group C viruses follow the International Catalogue of Arboviruses, Berge, 1975). Structural studies have shown that these viruses belong to the family Bunyaviridae.

All 11 agents have been isolated from mosquitoes; all except RES have been isolated from wild vertebrates and all except GL from humans with febrile disease (Berge, 1975; for NEP from man see Scherer *et al.*, 1976). They occur in the Neotropical zoo-

ISBN 0-12-429765-X

geographic region, where NEP has the widest distribution. This virus has been isolated in Brazil, Trinidad, Panama, Honduras, Guatemala, and Mexico. The fact that NEP is the only group C virus that has been recovered from flying vertebrates (*Artibeus* bats, Calisher *et al.*, 1971), may be relevant. Seven of the viruses coexist at Belem, Brazil, in a forested area near the mouth of the Amazon river.

II. EVIDENCE FOR MOSQUITO TRANSMISSION

All the group C arboviruses have been isolated from mosquitoes (Table I), and all but RES and GL have also been isolated from sentinel animals accessible only to flying arthropods (Berge, 1975; for Ossa from sentinels see Srihongse *et al.*, 1976). Six of them have been transmitted to laboratory mice or hamsters by naturally infected mosquitoes, all *Culex* species. Three have been transmitted experimentally between laboratory rodents by colonized *Aedes aegypti*, and five have multiplied in mosquitoes through at least five salivary gland passages in laboratory-reared *A. aegypti, C. quinquefasciatus,* or *Anopheles quadrimaculatus* (Table II). Thus multiplication in or transmission by mosquitoes has been demonstrated for all the group C viruses except ITQ and GL.

CAR and ORI have been recovered from the widest range of mosquito species (Table I). These are among the three strains most frequently recovered from sentinel and wild animals (632 CAR, 295 ITQ, and 128 ORI from sentinel mice at Belem). However, most of the mosquito isolations of group C viruses have been from *Culex*, particularly those of the subgenus *Melanoconion*. ORI has come mainly from *C. portesi*, and CAR from that species and *C. vomerifer*.

Wild-caught mosquitoes were not held until all blood meals had been digested before they were tested for virus. Therefore some of the isolations could represent infected gut contents--a possible explanation of the occasional recoveries from non-*Culex* species. Although six of the viruses multiplied in *A. aegypti*, only three have been recovered from *Aedes* species in nature (Table I), and none from *Anopheles*, despite the testing of almost 368,000 *Aedes* and 66,000 *Anopheles* (as compared to 238,000 *Culex*) in Trinidad (Aitken *et al.*, 1969), and comparable numbers at Belem. Intense transmission by *Culex* vectors may lead to such high levels of infection in vertebrate hosts that other mosquito genera are more likely to become infected. The *Aedes* isolates of GL and ORI may represent this type of spill-over of virus, since large numbers of these strains were recovered from *Culex*. But the two isolates of Apeu from *Aedes*,

TABLE I
Isolations of Group C Arboviruses from Mosquitoes[a]

Complex:	Caraparu				Marituba					Oriboca	
Type:	Caraparu		Apeu	MAD	Marituba			Nepuyo		ORI	ITQ
Subtype:	CAR	Ossa			MTB	MUR	RES	NEP	GL		
Aedes spp.			2[b]						2[c]	4[c,d]	
Culex (Culex) spp.	2[e]										
(Eubonnea)											
accelerans	3							1			
amazonensis	2										
(Melanoconion) spp.	4[f]	1[g]	1		2	1		1[h]	61	2[f]	5[f]
aikenii complex	2		4		2	4				6	1
portesi	22				1	5	2			46	3
vomerifer	14	6		5		1					10
Culex spp. *indet.*	5					2	1	5		6	1
Coquillettidia spp.						1[i]				3[i,j]	
Psorophora ferox										1	
Sabethini	3[k]					1				2	
Mixed spp.	1				2					1	1

[a]*Including transmissions by bite of wild-caught specimens to laboratory animals. Sources: Berge (1975), Belem Virus Laboratory (unpublished), Scherer et al., (1969, 1975).*
[b]*A. arborealis, A. septemstriatus.* [c]*A. taeniorhynchus.*
[d]*A. arborealis, A. argyrothorax, A. serratus.*
[e]*C. coronator, C. nigripalpus.*
[f]*Including some from C. spissipes (2 CAR, 1 ORI, 1 ITQ).*
[g]*C. taeniopus.* [h]*C. iolambdis.* [i]*Coq. venezuelensis.*
[j]*Coq. arribalzagai, Mansonia sp.*
[k]*Limatus durhamii, Wyeomyia medioalbipes, indet.*

TABLE II
Direct Evidence for Mosquito Transmission of Group C Arboviruses

Virus	*Transmission by naturally infected mosquitoes*	*Transmission from laboratory mouse-mosquito-mouse*	*Serial passage[a] in*		
			A. aegypti	*C. quinque-fasciatus*	*An. quadri-maculatus*
Caraparu	*Culex* spp.[b]	*A. aegypti*[c]	+[d]	+	×
Ossa	*C. taeniopus*, *C. vomerifer*[e]				
Apeu			+	+	×
Madrid	*C. vomerifer*[e]				
Marituba	*C. portesi*[b]		+	×	+
Murutucu			+	ND	×
Restan		*A. aegypti*[f]			
Nepuyo	*Culex* spp.[g]		×	×	ND
Oriboca	*C. portesi*[h]	*A. aegypti*[a]	ND	×	+

[a]*Whitman (1975).* [b]*Belem Virus Laboratory (unpublished).*
[c]*Metselaar (1966).*
[d]*+, five successful passages; ×, failure after 2-3 passages; ND, not done.*
[e]*Galindo and Srihongse (1967).* [f]*Jonkers et al., (1967).*
[g]*A. Toda, cited in Woodall (1967).* [h]*Toda and Shope (1965).*

compared with only five from *Culex*, suggest a more important role for *Aedes* in the transmission of that virus. The relatively large numbers of non-*Culex* isolates of ORI may indicate that this virus is also transmitted by mosquitoes of other genera.

III. VERTEBRATE HOST PREFERENCES

Culex (Melanoconion) mosquitoes are forest-dwelling species of nocturnal habits. Thus they would be expected to feed on nocturnal forest animals, such as rodents. Comparative bait studies in Trinidad (Aitken *et al*., 1968) and Belem have characterized certain mosquito species as primarily rodent-feeders or bird-feeders. Comparative data on these mosquitoes, their abundance, and the number of isolations of CAR and ORI made from them are presented in Table III. The minimum field infection rate/10^4 mosquitoes tested (MFIR) ranges from a maximum of 8.1 for the rodent-feeding *C. portesi* and CAR in Trinidad to less than 0.1 for most of the bird-feeding species. (The CAR rate for *C. vomerifer* is suspect since it is based on a single isolate.) The highest rate for ORI is 2.8 for the *C. aikenii* complex. This group and *C. spissipes*, which feed indiscriminately on rodents and birds, are intermediate in their rates for CAR. The MFIRs of the bird-feeding species are negligible in spite of the much greater numbers tested in most cases. There have been 37 isolations of CAR and 8 of ORI from rodents in Belem and Trinidad, but none from birds. Rates for CAR hemagglutination-inhibiting (HI) antibody in rodents at both places run up to 65%, depending on the season (Jonkers *et al* ., 1968b), whereas few of the thousands of bird sera tested at each locality had any group C antibody. CAR is the most frequently isolated virus at Belem and is common elsewhere, yet it has never been recovered from a marsupial. However, the mosquito species that carry it are attracted to marsupial bait and are found biting at canopy level as well as on the ground. This strongly suggests that marsupials are insusceptible to infection with CAR virus.

Anophelines are not much attracted to rodent bait in Belem or Trinidad; therefore they probably do not feed significantly on wild rodents, and this would explain why group C viruses have not been isolated from this abundant subfamily.

Table III shows that 28 of 29 isolations of CAR from Bush Bush forest (BBF) came from *Culex* mosquitoes; 25 of them were from species known to feed primarily on rodents in BBF (Aitken *et al*., 1968), and 19 of these were from *C. portesi*. That is, 66% of the mosquito isolates of CAR at BBF were recovered from a single rodent-feeding species that represented only 19%

TABLE III
Infection Rates of Caraparu and Oriboca Viruses in Mosquitoes According to Their Preferred Hosts

	No. tested		Caraparu				Oriboca			
	Belem	Trinidad	Belem		Trinidad		Belem		Trinidad	
Species	1960-69[a]	1959-63[b]	No.[a]	MFIR[c]	No.[d]	MFIR	No.	MFIR	No.	MFIR
Rodent feeders										
Culex accelerans[e]	*f*	12,781			3	2.3				
amazonensis[e]	*f*	24,393			2	0.8				
portesi	59,098	23,476	2	0.3	19	8.1	8	1.4	2	0.9
vomerifer	16,188	301	10	6.2	1	(33.2)[g]				
Intermediate										
C. aikenii complex	21,290	320	2	0.9			6	2.8		
spissipes	14,530	6,551	1	(0.7)	1	(1.5)	1	(0.7)		
Bird feeders										
Aedes serratus	71,349	100,323					1	(0.1)		
C. nigripalpus	*f*	34,660			1	(0.1)				
Coq. venezuelensis	127,556	33,327					1	(0.1)		
Psorophora ferox	27,107	17,704					1	(0.4)		
W. medioalbipes[e]	*f*	114,819			1	(0.1)				
Other mosquitoes			8[h]		1[i]		6[j]		0	
Total isolates			23		29		2.4		2	

[a]*APEG forests, Belem Virus Laboratory (unpublished).*
[b]*Bush Bush forest, Jonkers et al., (1968a); 1964 totals excluded.*
[c]*Minimum field infection rate/10^4.* [d]*Aitken et al., (1969).*
[e]*These species also feed to a great extent on lizards (Aitken et al., 1968).*
[f]*These species not separated from others in same genus.*
[g]*Rates in parentheses based on single isolations only.*
[h]*Culex spp. indet. (7), mixed Sabethini.*
[i]*C. coronator.*
[j]*A. arborealis, A. argyrothorax, Culex spp. indet. (2), Mansonia arribalzagai, Sabethini.*

of the *Culex* catch population, with an MFIR of 8.1. (Catches for 1964 were excluded from the calculations because there was a rodent population crash in BBF in that year, and not a single isolation of CAR was made from any BBF source in 1964.)

C. nigripalpus, which had a comparable share (28%) of the *Culex* catch population, had an MFIR almost a hundredfold less, but it is primarily a bird feeder. The other rodent-feeding species made up 30% of the *Culex* catch but had fourfold to tenfold lower MFIRs than *C. portesi*. *C. vomerifer* was so uncommon in BBF that it probably is unimportant to CAR transmission there. These data confirm the conclusions of Jonkers *et al.*, (1968b) that in Bush Bush forest, CAR is transmitted principally by *C. portesi* and, to a lesser extent, by other rodent-feeding *Culex*.

The situation in the APEG* Reserve at Belem is broadly similar; 22 of 23 strains isolated since 1960 came from *Culex*, at least 13 of these from rodent-feeding species. However, only two strains were recovered from *C. portesi*, which made up about 13% of the *Culex* catch but had an MFIR of only 0.3. In contrast, *C. vomerifer*, with only 3.5% of the *Culex* catch, produced 42% of the CAR isolates and had an MFIR of 6.2, a rate comparable to that of *C. portesi* in BBF. Therefore in Belem, CAR is also transmitted principally by rodent-feeding *Culex (Melanoconion)*, but in this case the main vector appears to be *C. vomerifer*.

The Bush Bush and APEG Reserve forests share many ecological features, but a difference that might be expected to account for the discrepancy in *C. portesi* MFIRs is the difference in rodent faunas.

At Belem, 92% of the catch is almost equally divided between two species, *Proechimys guyannensis* and *Oryzomys capito* (Table IV). These animals were trapped, bled for virus isolation, marked, and released. *Proechimys* were recaptured an average of 12 times each, *Oryzomys* an average of 5 times, and they were bled each time. Therefore there was more than twice the opportunity to isolate virus from *Proechimys*, and if the MFIR is adjusted accordingly, the rates for the two species are comparable.

In BBF, 97% of the catch is made up of three species, with *Oryzomys laticeps* predominating. These animals were trapped and then killed, and the viscera were tested for virus, so there was only one opportunity to make an isolation. However, virus may be recoverable from organs over a longer period than from the blood. At Belem, 28% of *P. guyannensis*, 17% of *O.*

*APEG: Area de Pesquisas Ecologicas de Guama (Guama Ecological Research Area); thanks are due to successive directors of the Instituto de Pesquisas e Experimentação Agropecuarias do Norte for permission to work there.

TABLE IV
Caraparu Virus Infection Rates in Rodents

Species	Belem, Brazil[a]				Bush Bush, Trinidad[b]			
	Total	%	*Isol.*	*MFIR*[c]	*Total*	%	*Isol.*	*MFIR*
Proechimys[d]	1052	44	17	16.2	16	1	0	-
Oryzomys[e]	1144	48	7	6.1	577	41	6	10.4
Nectomys[f]	182	7	1	(5.5)[g]	32	2	1	(31.3)
Zygodontomys[h]	21	1	0	-	440	29	4	9.1
Heteromys[i]	0	0	0	-	389	27	1	(2.6)
Total	2399	100	25	10.4	1454	100	12	8.3

[a]*Isolations from blood, 1962-1970, Woodall (1967) and Belem Virus Laboratory (unpublished).*
[b]*Isolations from viscera, 1959-1963, Jonkers et al., (1968a,b), excluding 1964 totals.*
[c]*Minimum field infection rate/10^3.* [d]*P. guyannensis.*
[e]*O. capito (Belem), O. laticeps (Bush Bush).* [f]*N. squamipes.*
[g]*Rates in parentheses based on single isolates.*
[h]*Z. lasiurus (Belem), Z. brevicauda (Bush Bush).* [i]*H. anomalus.*

capito, and 43% of *Nectomys squamipes* sera had CAR HI antibody, so that those percentages of CAR virus isolation attempts from the respective species would have been useless. Similarly 24% of *O. laticeps*, 11% of *Zygodontomys brevicauda* and 4% of *Heteromys anomalus* sera from Trinidad had CAR HI antibody (Jonkers *et al.*, 1968b). Moreover, seven nonimmune *Heteromys* inoculated with CAR did not produce detectable viremia (Jonkers *et al.*, 1968b), so transmissions to this species may be usually dead-ended. When these facts are taken into account, there is no significant difference in the availability of CAR-viremic rodents in the Belem and the Trinidad forests.

The attraction of *C. portesi* to the different rodent species was compared with that of white mice in the field in Trinidad (Aitken *et al.*, 1968) and Belem. Taking the level for mice as 1.0, *Proechimys* had an attractiveness of 1.04, *O. laticeps* of 1.75, *Heteromys* of 0.37, and *Zygodontomys* of 0.17. Thus in Trinidad *C. portesi* was 10 times more attracted to *O. laticeps* than to *Zygodontomys*. Jonkers *et al.*, (1964) found no significant differences between *O. laticeps* and *Zygodontomys* in the titers of natural or experimental CAR viremias (approx. 5 dex LD_{50}/0.02 ml) or in the duration of experimental CAR viremias (3 days). Thus there are no factors that tend to compensate for the lack of attractiveness of *Zygodontomys* to *C. portesi*, and this mosquito presumably picked up most of its infections from *O. laticeps*.

For Belem rodent species, we have no information about titer or duration of CAR viremia, or the attraction of *C. portesi* to any species other than *Proechimys*, which has an attractiveness ratio to *O. laticeps* of 1.04:1.75. If we assume that *O. capito* is as attractive as *O. laticeps* , then in order to account for the lower MFIR of *C. portesi* at Belem on the basis of inadequate viremias, both *Proechimys* and *O. capito* would have to circulate average CAR titers too low to infect 90% of *C. portesi* feeding on them, and yet still infect *C. vomerifer* heavily. This is considered possible but unlikely.

IV. VECTOR SUSCEPTIBILITY

The above considerations lead to the conclusion that the critical difference between CAR infection of *C. portesi* in Belem and Trinidad lies not with the vertebrate host, but with the strain of virus or the susceptibility of the mosquito to infection. The two strains of CAR are distinguishable serologically (Shope, 1965), but are so closely related that they are considered varieties, a difference even smaller than that between subtypes. A marked biological difference in infectivity might conceivably exist without being reflected serologically. The alternative is that the Belem population of *C. portesi* has a level of susceptibility to CAR an order of magnitude less than the BBF population. Gubler and Rosen (1976) found geographical strains of *Aedes albopictus* that differed in susceptibility to dengue-2 virus by as much as 1.5 dex oral ID_{50}. Such a difference was not noted for ORI in the two locations (Table III), nor for the wide range of other arboviruses of many serological groups recovered from *C. portesi* in both Trinidad and Belem. But Tesh *et al.*, (1976) have shown that two lines of *A. albopictus* with a sixfold difference in susceptibility to oral infection with dengue-2 virus had similar infection rates when fed chikungunya virus. A similar situation might account for the difference in the susceptibility of the *C. portesi* populations in Trinidad and Belem to CAR but not to ORI. Since *C. portesi* is rather difficult to colonize in the laboratory, it will not be easy to confirm this experimentally.

V. VECTOR ABUNDANCE AND AGE

There were two study areas in forests at Belem, 3 km apart. The sentinel mouse isolation rate for CAR was consistently higher in the Utinga forest than in the Aura forest. There were also two sentinel mouse locations 300 m apart in the Aura forest, one in primary forest and the other in secondary growth. The CAR rate was higher in the second growth. The relative abundance of the same three mosquito species in these forests in 1967 is shown in Table V.

The 3.5 times greater CAR transmission rate in second growth can be accounted for by the presence there of around fourfold larger populations of all three species. But there must be some other explanation for the greater CAR transmission in Utinga forest, since populations of *C. portesi* and *C. vomerifer* were similar in both areas, the *C. aikenii* complex was less abundant in Utinga, and the total *Culex* catch there was, in fact, proportionately smaller.

We assume here that the transmission rate of a mosquito species is correlated with its infection rate, although the relation between these two parameters may vary from species to species. The *C. aikenii* complex at Belem consists of at least two closely related species. Table VI shows that this complex has a greater infection rate for Apeu, MTB, MUR, and ORI than the other two species shown, but has seven times less for CAR and tenfold less for ITQ than *C. vomerifer*.

TABLE V
Total Catches from Ground Level Causey Hoods for 365 Days (1967)

	Aura Forest[a]					
	1	*2*	*Ratio*	*(1+2)/2*	*Utinga*	*Ratio*
C. aikenii complex	67	257	1:3.8	162	41	1:0.3
C. portesi	119	423	1:3.8	271	298	1:1.1
C. vomerifer	108	683	1:6.3	396	335	1:0.9
Sentinel mouse isolation rate	1.9%	6.6%	1:3.5	3.8%	13.7%	1:3.6

[a]*Primary (1) and secondary (2) forest locations.*

TABLE VI

	No. inoc.	*Caraparu*		*Marituba*		*Oriboca*	
	1960-69	*CAR*	*APEU*	*MTB*	*MUR*	*ORI*	*ITQ*
C. aikenii complex	21,290	0.9[a]	0.9	0.9	1.4	2.8	(0.5)[b]
C. portesi	59,098	0.3	0	(0.2)	0.3	1.4	0.3
C. vomerifer	16,188	6.2	0	0	(0.6)	0	4.9

[a]*MFIR/10*^{4}. [b]*Rates in parentheses based on single isolates.*

However, if catch figures are proporational to population, *C. portesi* is almost three times more abundant than the *C. aikenii* complex, and four times more abundant than *C. vomerifer*. As a result, *C. portesi* emerges as equal in importance to the *C. aikenii* complex for CAR transmission at Belem, and more important for ORI. The rarity of ORI transmission in the BBF in Trinidad could be related to the virtual absence of the *C. aikenii* complex there (Table III).

Differences in age structure between mosquito species would affect the MFIR, since an older population would have had a greater opportunity to acquire infective blood meals. Davies *et al.* (1971) found that at Belem during May 1970, parous rates were high and comparable for all species studied (*C. portesi*, 76%; *C. vomerifer*, 88%; *C. aikenii* complex, referred to as Culex B19 and B19 complex, 81 to 86%).

VI. COMPETITION FOR HOSTS

R. E. Shope (1964, personal communication) noted that about one-third of the blood specimens from *Proechimys* tested for virus at Belem in 1964 were from animals with group C HI antibody, indicating the presence of neutralizing antibody. Therefore about one-third of the isolations of group C viruses from *Proechimys* should be expected to come from animals with preexisting heterologous group C antibody. In fact, in no case did *Proechimys* have viremia with such antibody. This finding suggested that heterologous group C antibody has a considerable deterring effect on the ability of the host to develop viremia, and therefore on transmission. Allen *et al.*, (1967) showed that rhesus monkeys infected with one of four different group C viruses produced either minimal viremias or none at all after subsequent challenge with a different member of the group.

Karabatsos and Henderson (1969) tested all the group C arboviruses by crossneutralization and determined that the closest relationships were between the pairs of ORI-ITQ, CAR-Ossa, and RES-MUR. Other slightly more distant pairings were between CAR-Apeu and MAD-Ossa, and between the Belem and Trinidad strains of CAR. Neither Ossa nor MAD has been recovered at Belem, possibly because these viruses have been shut out by severe pressure from the superabundant CAR. As measured by number of sentinel mouse isolations, ITQ has only half the transmission rate of CAR, and ORI only a quarter (see Section II).

VII. THE ARBOREAL NICHE

R. E. Shope (1964, personal communication) noted that on the basis of HI antibody rates, the specific hosts of MTB, ORI, and MUR are possibly canopy-dwelling vertebrates. T. H. G. Aitken (1967, personal communication) considered that MTB and Apeu could have arboreal cycles, since neither had ever been isolated from a rodent, both have been recovered from opossums, which spend time in the trees, and HI antibody rates for both viruses are higher in marsupials than in rodents. From 1954 through 1965 at Belem, these were the only two viruses in group C that were isolated more frequently from sentinel monkeys than from sentinel mice. Almost half the isolations of MTB had come from sentinel monkeys and mice exposed in the forest canopy, and from mosquito species that are found as much in the trees at ground level. Experimentally inoculated sloths *(Bradypus tridactylus)* circulated high titers of MTB (Belem Virus Laboratory, unpublished).

In 1967 the exposure of sentinel mice and the collection of mosquitoes at canopy level was begun in Belem on an intensive basis. Table VII summarizes the isolations of group C arboviruses from those two sources in the APEG Reserve from 1967 through 1970, according to level.

Only those isolations from ground level sentinel mice that were exposed in localities where canopy sentinels were also exposed have been included in Table VII. Since the ratio of ground level sentinels to canopy sentinels was 1.6:1, a proportionate adjustment of the tree level rates has been made in the last column. No attempt has been made to adjust for different numbers of mosquitoes caught at the two levels. Many more mosquitoes were taken at ground level than in the canopy, so that the preponderance of isolations from mosquitoes in ground level catches merely reflects a fairly even distribution of infected mosquitoes at all levels. Even Apeu virus had been isolated before 1967 from the ground-caught *C. aikenii* complex. Many other species of mosquito have been found to be infected with the same group C viruses at both levels.

Table VII confirms that transmission of Apeu and MTB usually occurs above ground level in Belem. The failure to recover these two types from wild rodents, in spite of their

TABLE VII

	Mosquitoes		*Sentinel mice*			
Source:					*Tree/ total*	*Adjusted*
Capture level:	*Ground*	*Tree*	*Ground*	*Tree*	*(%)*	*(%)*
Apeu	0	4	2	9	82	88
Marituba	4	1	8	13	62	72
Oriboca	10	3	28	32	53	65
Murutucu	4	1	15	8	35	46
Caraparu	8	0	123	57	32	43
Itaqui	3	3	52	20	28	38
Nepuyo	2	0	8	0	0	--
Total sentinel groups exposed:			3320	2067	38	50

occurrence in mosquito species at ground level and their transmission to sentinel mice and man on the ground, suggests that the wild rodent species found at Belem are not susceptible to these viruses and that the antibody found in them is heterologous.

The slight predominance of ORI sentinel mouse isolates at tree level is interesting, because that virus has been recovered eight times from rodents and only twice from marsupials (*Didelphis*) at Belem. Its HI antibody rates are strikingly different in rodents (*Proechimys*, 9%) and arboreal marsupials (*Marmosa cinerea*, 36%).

VIII. CONCLUSIONS AND SPECULATION

The foregoing data indicate that the group C arboviruses are maintained in nature by cycles involving small forest mammals, principally rodents, and nocturnal mosquitoes, principally *Culex* species of the subgenus *Melanoconion*. One type, Apeu, and one subtype, MTB, are preferentially transmitted in the forest canopy between arboreal marsupials, monkeys, and possibly sloths. The mosquito species differ in importance as vectors from place to place, depending on their relative abundance and their level of vector competence for each virus. Competence within a single species may vary from locality to locality, for one virus and not for another. This appears to be true for CAR and *C. portesi* in Belem and Trinidad. *C. vomerifer* and the *C. aikenii* complex are important to group C virus transmission in Belem, but almost absent from Trinidad.

The coexistence of two group C virus strains within the same complex in one place seems to depend on two alternative mechanisms. CAR and Apeu, MUR and MTB coexist because the second virus in each pair predominates in the forest canopy. Here the number of vertebrate hosts is more limited, and so their transmission is less intense than that of the other member of each pair. ORI and ITQ survive in competition because ORI is principally transmitted by *C. portesi*, and ITQ by *C. vomerifer*, the latter mosquito species possibly being resistant to infection by ORI. It is tempting to speculate that this adaptation to different vectors may have been responsible for the divergence of ORI and ITQ from a common ancestor, and even that when *C. vomerifer* feeds on ORI, it changes this type into ITQ before transmitting it. Competition between these two viruses in the vertebrate host may nevertheless be the reason their combined transmission rate is below that of CAR, and that the ORI rate is half that for ITQ. Not enough information is available on the remaining group C viruses to determine how they coexist with others in the same locality.

Acknowledgments

Previously unpublished data reported here from the period 1965 through 1970 were obtained under the direction of the author at the Belem Virus Laboratory of the Instituto Evandro Chagas, Belem, Para, Brazil, which operated with the support and under the auspices of the Brazilian Ministry of Health and The Rockefeller Foundation.

References

Aitken, T. H. G., Worth, C. B., and Tikasingh, E. S. (1968). *Am. J. Trop. Med. Hyg. 17,* 253-268.

Aitken, T. H. G., Spence, L., Jonkers, A. H., and Downs, W. G. (1969). *J. Med. Ent. 6,* 207-215.

Allen, W. P., Belman, S. G., and Borman, E. R. (1967). *Am. J. Trop. Med. Hyg. 16,* 106-110.

Berge, T. O. (ed.) (1975). "International Catalogue of Arboviruses Including Certain Other Viruses of Vertebrates." U. S. Department of Health, Education, and Welfare Publication No. (CDC) 75-8301.

Calisher, C. H., Chappell, W. A., Maness, K. S. C., Lord, R. D., and Sudia, W. D. (1971). *Am. J. Trop. Med. Hyg. 20,* 331-337.

Casals, J., and Whitman, L. (1961). *Am. J. Trop. Med. Hyg. 10,* 250-258.

Causey, O. R., Causey, C. E., Maroja, O. M., and Macedo, D. G. (1961). *Am. J. Trop. Med. Hyg. 10,* 227-249.

Davies, J. B., Corbet, P. S., Gillies, M. T., and McCrae, A. W. R. (1971). *Bull. Ent. Res. 61,* 125-132.

De Rodaniche, E., De Andrade, A. P., and Galindo, P. (1964). *Am. J. Trop. Med. Hyg. 13,* 839-843.

Galindo, P., and Srihongse, S. (1967). *Am. J. Trop. Med. Hyg. 16,* 525-530.

Gubler, D. J., and Rosen, L. (1976). *Am. J. Trop. Med. Hyg. 25,* 318-325.

Henderson, B. E., Calisher, C. H., Coleman, P. H., Fields, B. N., and Work, T. H. (1969). *Am. J. Epidemiol. 89,* 227-231.

Jonkers, A. H., Spence, L., Downs, W. G., and Worth, C. B. (1964). *Am. J. Trop. Med. Hyg. 13,* 742-746.

Jonkers, A. H., Metselaar, D., De Andrade, A. H. P., and Tikasingh, E. S. (1967). *Am. J. Trop. Med. Hyg. 16,* 74-78.

Jonkers, A. H., Spence, L., Downs, W. G., Aitken, T. H. G., and Tikasingh, E. S. (1968a). *Am. J. Trop. Med. Hyg. 17,* 276-284.

Jonkers, A. H., Spence, L., Downs, W. G., Aitken, T. H. G., and Worth, C. B. (1968b). *Am. J. Trop. Med. Hyg. 17,* 285-298.

Karabatsos, N., and Henderson, J. R. (1969). *Acta Virol.* 13, 544-548.

Metselaar, D. (1966). *Trop. Geogr. Med. 18,* 137-142.
Scherer, W. F., Zarate, M. L., and Dickerman, R. W. (1969). *Bol. Of. Sanit. Panam. 66,* 325-338.
Scherer, W. F., Madalengoitia, J., Flores, W., and Acosta, M. (1975). *Bull. PAHO 9,* 19-26.
Scherer, W. F., Dickerman, R. W., Ordonez, J. V., Seymour, C., Kramer, L. D., Jahrling, P. B., and Powers, C. D. (1976). *Am. J. Trop. Med. Hyg. 25,* 151-162.
Shope, R. E. (1965). *Ciencia Cultura 17,* 30-32.
Shope, R. E., Causey, C. E., and Causey, O. R. (1961). *Am. J. Trop. Med. Hyg. 10,* 264-265.
Spence, L., Anderson, C. R., Aitken, T. H. G., and Downs, W. G. (1966). *Am. J. Trop. Med. Hyg. 15,* 231-234.
Srihongse, S., Scherer, W. F., and Galindo, P. (1967). *Am. J. Trop. Med. Hyg. 16,* 519-524.
Tesh, R. B., Gubler, D. J., and Rosen, L. (1976). *Am. J. Trop. Med. Hyg. 25,* 326-335.
Toda, A., and Shope, R. E. (1965). *Nature 208,* 304.
Whitman, L. (1975). *In* "International Catalogue of Arboviruses" (Berge, T. O., ed.) U. S. Department of Health, Education, and Welfare, Publ. No. (CDC) 75-8301.
Woodall, J. P. (1967). *Atas Simpos. Biota Amazon. 6,* 1-63.

Chapter 9

TRANSOVARIAL TRANSMISSION OF CALIFORNIA ARBOVIRUS GROUP

Wayne H. Thompson

ISBN 0-12-429765-X

I. INTRODUCTION

Transovarial transmission of arboviruses in mosquitoes has so far been observed with most members of the California arbovirus group. Previously transovarial transmission had been defined with other arthropod-borne agents, including some tick-borne (Burgdorfer and Varma, 1967; Burgdorfer and Brinton, 1975; Plowright *et al.*, 1970) and *phlebotomus* transmitted pathogens (Tesh *et al.*, 1972).

Arbovirus maintenance and possible overwintering in mosquitoes continues to be of major long-term ecologic and public health interest (Chamberlain and Sudia, 1961; Reeves, 1974; Danielova, 1975. Rapidly developing interests and progress with studies of transovarial transmission of California arbovirus group arboviruses make it timely to briefly review recent work in Wisconsin and elsewhere. It is of special interest to better define distribution and rates of transovarial transmission and its role in the natural maintenance of California group arboviruses.

II. THE CALIFORNIA ARBOVIRUS GROUP

Following the first isolation of a California encephalitis group virus in California several thousand related isolates have been obtained from mosquitoes in many parts of the United States and throughout the northern temperate and subarctic zones, including Canada, Alaska, Europe, and Finland, as well as from Trinidad and Africa. Most of these arboviruses have demonstrated ability for endemic persistance in adverse climates.

Those so far sufficiently studied have been classified into about a dozen antigenically and morphologically related but distinct California group serotypes within the family *Bunyaviradae* (Berge, 1975).

Most have been found in certain geographic regions, often with several different viruses active within the same area. Natural transmission cycles usually involve *Aedes* mosquitoes, with primary vector as well as nonavian vertebrate hosts described for most of the better known California group arboviruses.

III. TRANSOVARIAL TRANSMISSION OF CALIFORNIA GROUP ARBOVIRUSES IN MOSQUITOES FIRST OBSERVED WITH LA CROSSE VIRUS IN WISCONSIN

Continuing association of La Crosse virus with human disease in the United States has stimulated intensive and widespread investigation of its epidemiology and natural cycle.

Following the first isolation of La Crosse virus (Thompson *et al.*, 1965) from the brain of a fatal case of encephalitis in a 4-year-old girl at La Crosse, many cases have been subsequently serologically diagnosed each year in Wisconsin (Thompson and Evans, 1965) and other midwestern and eastern states (Henderson and Coleman, 1971).

A. Endemic Activity Defined

The geographic distribution of cases of California encephalitis caused by La Crosse virus occurring year after year has been defining endemic areas. Seasonally, cases in children are first diagnosed about the last week of June, and continue with increased numbers throughout the summer until the time of frost in the fall months (Thompson and Inhorn, 1967), thus providing evidence of its mosquito-borne nature.

1. Distribution of Human Infection and Disease

Epidemologic studies indicated an association with exposure to suburban or rural areas, where hardwood deciduous forests remain, as in southwestern Wisconsin and in portions of Minnesota, Iowa, Indiana, Ohio and other states.

2. Infections in Vertebrates

Serologic studies of animals in forests found antibodies to La Crosse virus mainly in chipmunks (up to 60%) and tree squirrels in endemic areas.

Experimental studies in the laboratory have shown that La Crosse virus can produce brief viremias for 2 or 3 days, high enough to infect mosquitoes (Pantuwatana *et al.*, 1972). The chipmunk has therefore been described as the main vertebrate host and is considered as a summer-season enhancing agent for the virus cycle in our hardwood deciduous forests.

First studies of possible overwintering mechanisms for La Crosse virus were pursued in these semihibernating tree-dwellers.

In a prospective study in endemic forests, recaptured chipmunks and squirrels acquired antibodies in summer but not during winter seasons (Moulton and Thompson, 1971).

B. The Main Vector *Aedes triseriatus*

1. Highest Isolation Rates

Of the first eight isolates of La Crosse virus obtained from mosquitoes in Wisconsin during 1964-1968, five were from *A. triseriatus*. Thirteen additional isolates were obtained during 1969 (Thompson *et al.*, 1972).

La Crosse isolates from mosquitoes in other states (Sudia *et al.*, 1971) also incriminate *A. triseriatus* as the main vector. Besides having the highest isolation rates, it has demonstrated excellent transmission ability (Watts *et al.*, 1972), distribution related to endemic areas, and natural habitat closely associated with chipmunks and squirrels (Gauld *et al.*, 1974).

2. Oviposition and Overwintering Sites

The natural oviposition sites for *A. triseriatus* are basal tree-holes, which are present in about one of every 100 trees in our deciduous hardwood forests. They often hold up to a quart or more of rainwater, and frequently have small openings that retard evaporation (Hanson and Hanson, 1970). The water in these tree-holes usually contains decaying leaves and other organic material, which provides nutrient media for these larvae. Adult *A. triseriatus* are very susceptible to wind, drying, and freezing, but in developmental egg and larval forms they have a very protected place in basal tree-holes (Fig. 1).

During the summer season *A. triseriatus* deposits eggs mainly in these basal tree-holes, old tires, and other water-holding containers, where they overwinter in diapause. After eggs hatch early the following season, larvae develop slowly in relatively cool water in tree-holes, with first emergence of adults delayed until late spring, usually mid-June in Wisconsin. Males usually emerge from tree-holes before females and remain in the area to inseminate females, which also seek blood meals from chipmunks and other forest-dwelling mammals.

Fig. 1 Typical basal tree-hole oviposition site from which La Crosse virus emerges with A. triseriatus through transovarial transmission.

C. First Isolates from Larvae

Finding that *A. triseriatus* was the main vector and chipmunks and squirrels vertebrate hosts explained summer season maintenance of La Crosse virus, but the overwintering mechanism was still unknown.

A summary of the first reported isolates of La Crosse virus from larvae collected in southwestern Wisconsin, with a comparison of isolation rates from various sources, is presented in Table I. Details have been documented in University of Wisconsin PhD theses of the three graduate students who were mainly involved in these studies and in publications listed in the references.

La Crosse virus was first obtained from *A. triseriatus* larvae during August 1972 (Pantuwatana *et al.*, 1974) from a pool of 43 larvae collected from a basal tree-hole in endemic western Wisconsin on June 7, 1972. A second isolate was obtained from a pool of 40 larvae collected from the same tree-hole on September 2. Both isolates were obtained from suspensions of ground larvae inoculated into 1- and 2-day-old suckling mice.

Since these larvae could have emerged from overwintered diapause eggs, this isolate provided a clue for a possible overwintering mechanism. However, further studies were needed to prove that the virus isolated from these larvae originated from transovarial and overwintered sources.

TABLE I

Summary of Reported Isolates of La Crosse California Group Arbovirus from Field-Collected A. triseriatus Larvae and Adults Reared from Larvae from Southwestern Wisconsin

Isolates/number of larvae tested		*Isolation rate (%)*	*Source*	*Collection dates (references)*
1/43		(2.4)	A tree-hole with La Crosse	June 7, 1972
1/40			Same tree-hole	September 2, 1972
2/555		(0.4)	All tree-holes sampled	June-September 1972 (Pantuwanta *et al.*, 1974)
4/509	males and	(0.8)	2 tree-holes with La Crosse	April 20-June 7, 1973
4/1722	females	(0.2)	All 8 tree-holes sampled	
39/437	"	(8.9)	2 old tires with La Crosse	May 23, 1973 (Watts *et al.*, 1974, 1975)
10/510	males and	(2.0)	9 tree-holes with La Crosse	April 30-June 7, 1974
10/1698	females	(0.6)	All 64 tree-holes sampled	
		[5 other males and 6 females with La Crosse were obtained throughout the summer season from 5 of 12 tree-holes screen closed June 11]		(Beaty and Thompson, 1975)
39/437	Totals	(8.9)	2 tires with La Crosse	(The 55 isolates were obtained from 23 larvae, 23 males, 9 females)
16/1102		(1.5)	12 tree-holes with La Crosse	
16/3975		(0.4)	All 78 tree-holes sampled	

D. Transovarial Transmission in the Laboratory

During subsequent studies in the laboratory (Watts *et al.*, 1973) La Crosse virus was recovered from eggs (larvae and adults) deposited by colonized female *A. triseriatus* previously infected by virus in a blood meal. Virus was found in 39% of male and in 30% of female progeny. Most transovarially infected females transmitted virus by bite to mice.

La Crosse virus was transmitted within eggs that had been surface sterilized. Transovarial infection rates in our laboratory colonies have averaged 30 to 40% with rates variable from one strain to another.

Transovarial passage of La Crosse virus has also been demonstrated in the laboratory (Tesh and Gubler, 1975) with another species, *A. albopictus*, but with lower rates indicating less vector potential.

E. Demonstration of Overwintering

Overwintering of La Crosse virus in *A. triseriatus* was described in nature (Watts *et al.*, 1974), demonstrating that the eggs of this mosquito could be a natural reservoir for the survival of this arbovirus throughout the frigid winter season as in Wisconsin.

In larvae collected from tree-holes during early spring prior to seasonal emergence of adults, virus was isolated from two of eight tree-holes. Of four isolates obtained from these tree-holes, three were from larvae and one from a female reared from field-collected larvae.

The minimum field infection rates (MFIR), based on one infected larvae per pool, was 1/110 larvae (or 0.8%) from the two tree-holes with La Crosse and in 0.2% of the larvae collected from all eight tree-holes sampled.

Two old automobile tires collected May 23, 1973, from the back yard of a suburban La Crosse home contained water in which 39 (8.9%) of 437 *A. triseriatus* larvae contained virus. Isolates were obtained from 11.5% of larvae, 8.6% of females, and 6.4% of males reared from the larvae from these tires. Virus was transmitted by bite to a laboratory mouse from a transovarially infected female.

F. Frequency of Overwintering

The distribution of La Crosse virus overwintering in diapause *A. triseriatus* eggs was studied by following many more (64) tree-hole oviposition sites in four enzootic hardwood forests in suburban La Crosse during 1974. La Crosse antigen was observed in dissected mosquitoes by fluorescent antibody technique. La Crosse virus was found in larvae collected from each of four areas before seasonal emergence of adults, but in only 10 (0.6%) of 1698 individually processed adults reared from these larvae (Beaty and Thompson, 1975).

The overall isolation rates in larvae from tree-holes during the three years of studies in Wisconsin were 1.5% (16/1102) collected from 12 tree-holes with La Crosse virus and 0.4% of 3975 larvae from all 78 tree-holes sampled.

G. Summer-Long Emergence of Virus from Overwintered Eggs

Isolates of La Crosse virus were found in *A. triseriatus* emerging from screen enclosed tree-holes containing only overwintered ova throughout the summer months, demonstrating that these sites not only can serve as foci for overwintering, but also for continuing summer season emergence of virus in these endemic areas (Beaty and Thompson, 1975).

La Crosse virus infection rates in the five positive tree-holes in this area varied from 1/135 (0.7%) of the larvae collected all season long from one screen enclosed positive tree-hole to 8/48 (17%) in another.

H. Distribution of Virus in Developmental Stages of Vector

La Crosse antigen has been observed by fluorescent antibody technique in all stages from egg through first, second, third, and fourth instar larvae, pupae, and adults, in 95 (25%) of 387 larvae from a transovarially infected laboratory colony (Beaty and Thompson, 1976).

In larvae, most antigen was observed in the alimentary tract, then ganglia, malpighian tubules, muscle, and other tissues. Most tissues and organs of *A. triseriatus* are involved in maintaining La Crosse virus during transovarial transmission.

In pupae and young adults, antigen was also present in gonadal tissues and salivary gland tissues, indicating that females are infective through eggs or by bite to vertebrates, upon emergence as adults. The high level of fluorescence detected in gonads and accessory sex organs during late pupal and adult stages suggested the possibility of venereal transmission of La Crosse virus in *A. triseriatus*.

IV. ISOLATIONS OF LA CROSSE VIRUS FROM LARVAE IN OTHER STATES

Table II summarizes data published on isolations from early season collected *A. triseriatus* larvae by those in Ohio and Minnesota.

In Ohio, two isolations were obtained with a rate of 3.1% from larvae collected from several tree-holes with La Crosse. The rate was 1.1% in larvae collected from all the tree-holes on June 7, 1973, before seasonal emergence of adults in the area (Berry *et al.*, 1974).

In Minnesota, five isolates were obtained from eggs and larvae collected from a tree-hole on April 29 and May 16, 1974, a month before seasonal emergence of adults in the area. The overall isolation rate in larvae from all sources was 0.3% (Balfour *et al.*, 1975).

The overall isolation rate of La Crosse virus from *A. triseriatus* larvae collected in the Ohio and Minnesota studies was 1.4% (7/515) larvae collected from the three positive tree-holes and 0.4% of 1748 collected from all tree-holes studied, similar to the 1.5% and 0.4% rates observed in Wisconsin.

V. ISOLATIONS FROM LARVAE OF CALIFORNIA GROUP ARBOVIRUSES OTHER THAN LA CROSSE

Table III summarizes published reports of isolates from larvae of other California group arboviruses, including Keystone, Tahyna, and Snowshoe Hare viruses.

A. Keystone Virus in Maryland

Keystone virus has been isolated from first generation larvae of *A. atlanticus* collected in Maryland during August, 1973 after summer rains flooded developmental sites in forested areas (Le Duc *et al.*, 1975 a,b). Nine isolates were obtained, including one from 518 larvae tested, three from 1688 reared females processed, and five from 2040 reared males. Isolates were obtained from each of two sites sampled.

Other evidence that transovarial transmission of Keystone virus is occurring includes high infection rates in adult *A. atlanticus*, constant isolation rates when significant numbers of emerging adults occur, and isolations from the earliest adult *A. atlanticus* collected and tested during 1974.

TABLE II

Summary of Reported Isolates of La Crosse California Group Arbovirus from Field-Collected A. triseriatus *Eggs and Larvae in States Other Than Wisconsin*

Isolates/number of larvae (pools) tested	*Isolation rate (%)*	*Source*	*Area*	*Collection dates (references)*
2/65 (2)	(3.1)	Several tree-holes with La Crosse	Ohio	June 7, 1973
2/178 (7)	(1.1)	All sampled		(Berry *et al.*, 1974)
1/150 (4)	(0.7)	One tree-hole	Minnesota	April 29, 1974
4/300 (15)	(1.3)	Same tree-hole		May 16, 1974
5/1570 (67) eggs & larvae	(0.3)	Many tree-holes and old tires sampled		April-July 1974 (Balfour *et al.*, 1975)
7/515 (11) eggs & larvae	(1.4)	3 tree-holes with La Crosse	Totals	
7/1748 (64) eggs & larvae	(0.4)	All sources sampled		

TABLE III

Summary of Reported Isolates of California Group Arboviruses Other Than La Crosse from Field-Collected Larvae (or Adults Reared from Them)

Virus vector	*Isolates/number of larvae (pools) tested*	*Isolation rate (%)*	*Source of larvae*	*Area*	*Collection dates (references)*
Keystone	1/518 (21)	(0.2)	rain-pools in forest	Maryland	August 26-31, 1973
A. atlanticus	3/1688 (35) females 5/2040 (42) males				(Le Duc *et al.*, 1975)
Tahyna	1/320 (32)	(0.3)	river-bed water pools	Moravia	July 3-5, 1974
C annulata					(Bardos *et al.*, 1975)
Snowshoe Hare	1/240	(0.4)	ice-water pools	Yukon; Canada	May 16, 1974
Aedes species					(McLean *et al.*, 1975)
			water in an old well	Central Saskatch-ewan	June 10, 1975
A. implicatus	1/959 (41) all adults	(0.1)			May 28-July 11, 1975 (McLintock *et al.*, 1976)
Total	12/5765	(0.2)	all reported sampled		1973-1976

Studies indicate that *A. atlanticus* may not often take more than one blood meal, although it is believed that both transovarial transmission and infection of adult female mosquitoes from viremic vertebrates may be involved in the natural cycle of Keystone virus (Watts and Eldridge, 1975).

B. Tahyna Virus in Europe

Tahyna virus has been reported isolated (Bardos *et al.*, 1975) from field-collected larvae of *C. annulata* (SCHRK.) obtained during July, 1974 from puddles in a forested river bed in southern Moravia. This isolate was obtained from a collection of 320 first-generation larvae.

Although transovarial transmission in mosquito vectors has been considered as an overwintering mechanism for Tahyna and other California group arboviruses, several other possibilities including virus overwintering in hibernating mosquitoes, in blood-sucking ectoparasites, or in animals have also been considered (Danielova, 1975). Danielova observed that the seasonal appearance of Tahyna virus is often associated with the simultaneous appearance of *A. vexans* mosquitoes, suggesting transovarial transmission.

C. Snowshoe Hare Virus in Canada

Snowshoe Hare virus has been isolated from one of 240 *Aedes* species larvae collected during May 1974 from ice water pools before spring thaw in the Yukon (McLean *et al.*, 1975a,b), evidence that this virus is also likely overwintering by transovarial transfer. Snowshoe Hare virus has also been obtained from unengorged female *A. canadensis* and *A. communis* mosquitoes.

Minimum field infection rates in collections of adults have been highest with *A. canadensis*, although there has been considerable variation in infection rates between areas.

An isolate of Snowshoe Hare virus has also recently been reported from 959 larvae of *Aedes implicatus* collected from water in an old well in central Saskatchewan (McLintock *et al.*, 1976).

California encephalitis group arboviruses have been isolated from mosquitoes in many areas throughout subarctic Canada and Alaska, from several *Aedes* species and from *C. inornata*, which supports long-term replication of a California group virus during incubation at 13°C or lower temperature (McLean, 1975).

Virus transmission by bite in saliva by *Aedes* and *Culiseta* mosquitoes following infection on viremic vertebrates probably occurs infrequently due to their reluctance to take more than one blood meal during their lifetime.

These researchers suggest that maintenance of California encephalitis viruses in the frigid Yukon territory may be accomplished through lengthyy persistance of infectivity in adult mosquitoes as well as by possible transovarial transmission.

D. Studies with Others in Progress

Evidence has also been obtained by those in Iowa that Trivittatus virus may overwinter in the main vector *A. trivittatus*. Sentinel rabbits caged near Ames in central Iowa acquired antibodies to this virus between June 20 and 27 during the first seasonal peak of *A. trivittatus* in the area. Further studies of transovarial transmission are in progress there (Pinger *et al.*, 1975).

VI TRANSOVARIAL TRANSMISSION WIDESPREAD--WITH LOW RATES

The role of transovarial transmission in the maintenance and overwintering of La Crosse virus has been described. La Crosse virus simultaneously emerges from transovarially infected sources in many forested hillsides each spring. However, rates of La Crosse virus in transovarially infected larvae from all tree-holes so far average only 0.4% in endemic areas and only 1.5% in all collections from basal tree-holes in which virus has been found. Even in endemic regions, La Crosse virus is routinely isolated from larvae from only a portion of tree-holes sampled.

Isolates of many other California group arboviruses from early season field-collected larvae indicate that laboratory demonstration of transovarial transmission may soon also be demonstrated with them. In considering transovarial transmission with California group arboviruses, it so far appears that although there are endemic foci in many areas, overall rates in nature are low.

These low rates perhaps explain why most of the researchers involved in obtaining these important first isolates of various California group viruses from larvae did so only after careful delineation of endemic areas, identification of specific vectors and oviposition sites, and processing of adequate numbers to obtain infected larvae within those sampled.

VII. HORIZONTAL TRANSMISSION ALSO NECESSARY FOR MAINTENANCE

Evidence so far indicates that La Crosse virus cannot be maintained endemically without transmission other than transovarial. Filial infection rates are not considered high enough to persist by vertical transmission alone. Occasional horizontal transmission from infected *A. triseriatus* to noninfected females is therefore believed necessary to maintain the natural cycle.

A. Blood Meals on Viremic Vertebrates

Horizontal transmission of La Crosse to previously noninfected females through infectious vertebrate blood meals can be demonstrated in laboratory studies and frequently occurs in nature as evidenced by antibodies in vertebrates.

B. Male to Female Transmission

Venereal transfer of La Crosse virus directly from males from transovarially infected sources to female *A. triseriatus* during mating has recently been observed in our laboratory (Thompson and Beaty, 1977). Subsequent transfer of La Crosse virus has been observed from some of these venereally infected females through saliva to vertebrates and through transovarial transmission to eggs and progeny.

Although veneral transmission has not yet been demonstrated in nature, similar filial infection rates with La Crosse virus in male as well as female *A. triseriatus* emerging from transovarially infected sources provides an opportunity for male mosquitoes to participate in the natural cycle of an arbovirus.

VIII. CONCLUSIONS

Following studies describing transovarial transmission of La Crosse virus in *A. triseriatus*, isolates of at least three other California arbovirus group have been reported from early season collected mosquito larvae by others. These widespread isolates indicate that transovarial transmission will soon be demonstrated in the laboratory for many of these. Whether transovarial transmission in mosquitoes will eventually be demonstrated in all members of the California group, in Bunyaviruses in general, or with most arboviruses, is of considerable interest and importance.

Transovarial transmission provides a basic means for vertical transmission and overwintering of California group arboviruses most of which are distributed in our seasonally freezing northern temperate zone. However, low rates of transovarial transmission and filial infection in progeny still indicate a need for occasional supplemental horizontal transmission, via blood meals on viremic vertebrates or by other means (including possible venereal transmission direct from infected males to females) for continuing maintenance cycles in nature.

Data from present and future studies concerning these rates and means of transmission should eventually provide a better understanding of how these and other arboviruses persist endemically in our northern temperate and subarctic zones.

References

Balfour, H. H. Jr., Edelman, C. K., Cook, F. E., Barton, W. I., Buzicky, A. W., Siem, R. A., and Bauer, H. (1975). *J. Infect. Dis. 131,* 712-716.

Bardos, V., Ryba, J., and Hubalek, Z. (1975). *Acta Virol. 19,* 446.

Beaty, B. J., and Thompson, W. H. (1975). *Am. J. Trop. Med. Hyg. 24,* 685-691.

Beaty, B. J., and Thompson, W. H. (1976). *Am. J. Trop. Med. Hyg. 25,* 505-512.

Berge, T. O. (1975). "International Catalogue of Arboviruses Including Certain Other Viruses of Vertebrates." U.S. Department of Health, Education, and Welfare Publ. (CDC) 75-8301.

Berry, R. L., LaLone, B. J., Stegmiller, H. W., Parsona, M. A., and Bear, G. T. (1974). *Mosq. News 34,* 454-457.

Brummer-Korvenkontio, M., and Saikku, P. (1975). *Med. Biol. 53,* 279-281.

Brummer-Korvenkontio, M. Saikku, P., Korhonen, P., Ulmanen, I., Reunala, T., and Karvonen, J. (1973). *Am. J. Trop. Med. Hyg. 22,* 404-413.

Burgdorfer, W., and Brinton, L. P. (1975). *Ann. N.Y. Acad. Sci. 266,* 61-72.

Burgdorfer, W., and Varma, M. G. R. (1967). *Ann. Rev. Entomol. 12,* 347-376.

Chamberlain, R. W., and Sudia, W. D. (1961). *Ann. Rev. Entomol. 6,* 371-390.

Chernesky, M. A. (1968). *Can. J. Microbiol. 14,* 19-23.

Danielova, V. (1975). *Med. Biol. 53,* 282-287.

Fine, P. E. (1975). *Ann. N.Y. Acad. Sci. 266,* 173-194.

Gauld, L. W., Hanson, R. P., Thompson, W. H., and Sinha, S. K. (1974). *Am. J. Trop. Med. Hyg. 23,* 983-992.
Hanson, R. P., and Hanson, M. G. (1970). *Mosq. News 30,* 215-221.
Henderson, B. E., and Coleman, P. H. (1971). *Progr. Med. Virol. 13,* 401-461.
Jousset, F. X., and Plus, N. (1975). *Ann. Microbiol. 126,* 231-249.
Le Duc, J. W., Suyemoto, W., Eldridge, B. F., Russell, P. K., and Barr, A. H. (1975a). *Am. J. Trop. Med. Hyg. 24,* 124-126.
Le Duc, J. W., Burger, J. F., Eldridge, B. F., and Russell, P. K. (1975b). *Ann. N.Y. Acad. Sci. 266,* 144-151.
McLean, D. M. (1975). *Med. Biol. 53,* 264-270.
McLean, D. M., Bergman, S. K. A., Gould, A. P., Grass, P. N., Miller, M. A., and Spratt, E. E. (1975a). *Am. J. Trop. Med. Hyg. 24,* 676-683.
McLean, D. M., Gubash, S. M., Grass, P. N., Miller, M. A., Petric, M., and Walters, T. E. (1975b). *Can. J. Microbiol. 21,* 453-462.
McLean, D. M., Grass, P. N., Judd, B. D., and Wong, K. S. K. (1976). *Can. J. Microbiol. 22,* 1128-1136.
McLintock, J., Curry, P. S., Wagner, R. J., Leung, M. K., and Iversen, J. O. (1976). *Mosq. News 36,* 233-237.
Moulton, D. W., and Thompson, W. H. (1971). *Am. J. Trop. Med. Hyg. 20,* 474-482.
Murphy, F. A. (1975). *Ann. N.Y. Acad. Sci. 266,* 197-203.
Pantuwatana, S., Thompson, W. H., Watts, D. M., and Hanson, R. P. (1972). *Am. J. Trop. Med. Hyg. 21,* 476-481.
Pantuwatana, S., Thompson, W. H., Watts, D. M., Yuill, T. M., and Hanson, R. P. (1974). *Am. J. Trop. Med. Hyg. 23,* 246-250.
Pinger, R. R., Rowley, W. A., Wong, Y. W., and Dorsey, D. C. (1975). *Am. J. Trop. Med. Hyg. 24,* 1006-1009.
Plowright, W., Perry, C. T., and Peirce, M. A. (1970). *Res. Vet. Sci. 2,* 582-584.
Reeves, W. E. (1974). *Progr. Med. Virol. 17,* 1973-220.
Spielman, A. (1975). *Ann. N.Y. Acad. Sci. 266,* 115-124.
Sudia, W. D., Newhouse, V. F., Calisher, C. H., and Chamberlain, R. W. (1971). *Mosq. News 31,* 576-600.
Tesh, R. B., and Gubler, D. J. (1975). *Am. J. Trop. Med. Hyg. 24,* 876-880.
Tesh, R. B., Chaniotis, B. N., and Johnson, K. M. (1972). *Science 175,* 1477-1479.
Thompson, W. H., and Beaty, B. J. (1977). *Science 196,* 530-531.
Thompson, W. H., and Evans, A. S. (1965). *Am. J. Epidemiol. 81,* 230-244.
Thompson, W. H., and Inhorn, S. L. (1967). *Wis. Med. J. 66,* 250-253.
Thompson, W. H., Kalfayan, B., and Anslow, R. O. (1965). *Am. J. Epidemiol. 81,* 245-253.
Thompson, W. H., Anslow, R. O., Hanson, R. P., and DeFoliart, G. R. (1972). *Am. J. Trop. Med. Hyg. 21,* 90-96.

Watts, D. M., and Eldridge, B. F. (1975). *Med. Biol. 53*, 271-278.

Watts, D. M., Morris, C. D., Wright, R. E., DeFoliart, G. R., and Hanson, R. P. (1972). *J. Med. Entomol. 9*, 125-127.

Watts, D. M., Pantuwatana, S., DeFoliart, G. R., Yuill, T. M., and Thompson, W. H. (1973). *Science 182*, 1140, 1141.

Watts, D. M., Thompson, W. H., Yuill, T. M., DeFoliart, G. R., and Hanson, R. P. (1974). *Am. J. Trop. Med. Hyg. 23*, 694-700.

Watts, D. M., Pantuwatana, S., Yuill, T. M., DeFoliart, G. R., Thompson, W. H., and Hanson, R. P. (1975). *Ann. N.Y. Acad. Sci. 266*, 135-143.

Chapter 10

VECTOR COMPETENCE OF *CULEX TARSALIS* AND OTHER MOSQUITO SPECIES FOR WESTERN EQUINE ENCEPHALOMYELITIS VIRUS

James L. Hardy, William C. Reeves, James P. Bruen, and S. B. Presser

I. INTRODUCTION

The vector competence or the vector efficiency of mosquito species for arboviruses is controlled by extrinsic and intrinsic factors, both of which can be influenced by environmental factors (Hardy *et al.*, 1975). Extrinsic factors, many of which have been identified and studied, influence the exposure of mosquitoes to viremic and clinically susceptible hosts (Reeves, 1967). Less is known about intrinsic factors, which determine if the virus, once it has been ingested, can initiate infection and eventually spread to the salivary glands from which it can be transmitted by bite (Chamberlain and Sudia, 1961; Chamberlain, 1968).

The potential importance of intrinsic factors of vector competence in arboviral transmission cycles became increasingly evident during recent studies in California when different field and colonized populations of *Culex tarsalis* were shown to vary significantly in their susceptibility to infection after

ISBN 0-12-429765-X

ingestion of western equine encephalomyelitis (WEE) virus (Hardy *et al.*, 1976). Similar findings were reported when colonized strains of *Aedes albopictus* that originated from different geographical locations were fed on dengue and chikungunya viruses (Gubler and Rosen, 1976; Tesh *et al.*, 1976). Although it had been recognized for a long time that different mosquito species varied in their ability to become infected with a single arbovirus (Chamberlain, 1968), the above studies represented the first demonstrations that individuals of a single mosquito species varied in their susceptibility to infection with a single arbovirus. When examined carefully, both intraspecific and interspecific variations in susceptibility to infection after ingestion of an arbovirus were associated with mesenteronal barriers that could be circumvented by introduction of virus directly into the hemocoel (Merrill and TenBroeck, 1935; McLean, 1955; Hardy *et al.*, 1976; and Gubler and Rosen, 1976).

Following the initial studies on the vector competence of *C. tarsalis* for WEE virus in California (Hardy *et al.*, 1976), further studies were undertaken to confirm the initial observations by more extensive field studies, to select WEE viral resistant (WR) and susceptible (WS) strains of *C. tarsalis* and to determine if there were seasonal variations in vector competence. Several mosquito species other than *C. tarsalis* were evaluated incidentally for vector competence to WEE virus. The results of these studies are summarized in this report.

II. MATERIALS AND METHODS

A. Mosquito Collection and Processing

Field populations of *C. tarsalis* were collected during the summers of 1972 and 1973 as larvae and pupae from various sites in the Sacramento Valley, San Joaquin Valley, and Southern California. Each sample was identified by the name of a site near where the collection was made. Approximately 5000 pupae were required to provide sufficient adults for the series of tests on each sample, and so most collections were made from receding pools of water in roadside ditches, canals, sloughs, and irrigated pastures where pupae were produced in large numbers or had become concentrated. Thus, certain habitats where *C. tarsalis* commonly breeds, such as rice fields, were excluded because pupae were not in sufficient densities.

Larvae and pupae were transported in ice-cooled chests to the laboratory and segregated within 2 to 24 hours after collection. Fourth instar larvae were removed and tested for susceptibility to ethyl parathion according to the procedures described by Gillies and Womeldorf (1968). A sample of 50 to 100 newly emerged adult females was transferred to pint ice cream cartons with nylon netting on top and provided with a cotton pledget soaked with a 5 to 10% sucrose solution. After incubation for 10 days at 24°C they were examined for autogenous egg development (Spadoni *et al.*, 1974). The remaining adult females, which were not provided with a carbohydrate source, were evaluated for susceptibility to infection with WEE, St. Louis encephalitis, and Turlock viruses when they were 2 to 5 days old.

During 1975 and 1976, additional collections were made of adult female populations at the Poso West study site in Kern County. These collections were made by CO_2 baited lard-can traps, CO_2 baited CDC light traps, or both. Vector competence tests usually were done within 24 to 48 hours of collection.

B. Viral Susceptibility Tests

Only results of vector competence studies with WEE virus are reported here. The strain of WEE virus as well as the methods used for infection of mosquitoes by the intrathoracic inoculation and pledget feeding techniques, the conditions for extrinsic incubation of mosquitoes, and the procedures for testing individual mosquitoes for virus have been described elsewhere in detail (Hardy *et al.*, 1976).

Some field samples contained several species of mosquitoes in addition to *C. tarsalis*. When this occurred, adult females of all species were inoculated or fed on pledgets simultaneously and then the different species were identified and segregated at the time the specimens were frozen for viral tests.

Susceptibility profiles and infectious dose fifties (ID_{50}) were derived from results of tests on 10 to 20 individual females for each concentration of virus inoculated and ingested. The ID_{50} was calculated using the formula of Reed and Muench (1938).

C. Viral Transmission Tests

WEE viral transmission rates were determined for several samples of *C. tarsalis* collected at Poso West. Females were fed on pledgets containing virus in high concentrations, i.e., $10^{6.8}$ to $10^{7.6}$ plaque-forming units (PFU) per 0.2 ml. Fed mosquitoes were incubated for 13 to 14 days at 24°C and then each mosquito was exposed individually to a normal White Leghorn chick (4 to 8 days of age). Mosquitoes that refed were frozen at -70°C for viral tests. Blood samples for viremia determinations and serum samples for antibody determinations were obtained at 48 hours and 21 days postfeeding, respectively, from each chick that was fed upon by a mosquito. Viral titers in each infected mosquito and viremic blood were measured by plaquing serial tenfold dilutions in duck embryonic cell cultures (Hardy *et al.*, 1976). Antibody titers were determined by the hemagglutination-inhibition (HI) test (Clarke and Casals, 1958). The detection of viremias and HI antibodies (titer of 1:20 or greater) in chicks that survived for 21 days postfeeding and viremias in chicks that died before a serum sample could be obtained were considered evidence for viral transmission. There was a 100% correlation between the development of viremias and antibodies in chicks for which both tests were performed.

D. Genetic Selection

The procedures used for genetic selection of two hybrid strains of *C. tarsalis* that were highly refractory to WEE virus will be described elsewhere in detail (Hardy *et al.*, manuscript submitted for publication) and will not be repeated here. Briefly, the WR strains were derived from two or three parental strains (Fort Collins, Colorado; Knights Landing, California; and Chico, California) after over 20 generations of selection at high titers of viral challenge. Random brother-sister mating among progeny of refractory female parents was employed initially. However, problems arising partly from inbreeding necessitated a switch to the cyclical family mating scheme (Ward, 1963) during the later generations of selection. Viral challenges were provided by viremic chicks that were circulating $10^{7.0}$ to $10^{9.5}$ PFU of virus per 0.2 ml of blood. All females were allowed to oviposit in individual containers and after oviposition were tested individually for virus. Likewise, progeny of each female were reared as a separate group until the viral tests were completed on the parent, at which time the progeny of several WR females were sometimes pooled for the next generation of selection.

A WS strain was selected from the Knights Landing strain of *C. tarsalis*,which was the most susceptible strain available. In this case progeny were retained for further selection when the female parent became infected after feeding on chicks with viremia titers of $10^{3.0}$ to $10^{5.0}$ PFU per 0.2 ml. Progeny of five to ten susceptible female parents were pooled and allowed to mate under controlled conditions (i.e., one male plus four to six females) during each of ten generations of selection for a WS strain.

III. RESULTS AND DISCUSSION

A. Intraspecies Variations in Vector Competence

1. *Susceptibility to Infection*

a. Survey of field populations. Forty-five field collections of *C. tarsalis* were made from 27 sites in California during the summers of 1972 and 1973 and were evaluated for susceptibility to infection with WEE virus by pledget feeding (Tables I and II). The median ID_{50}s were $10^{4.7}$ and $10^{4.2}$ in 1972 and 1973, respectively. The $\log_{10}$ of the ID_{50}s obtained for different populations ranges from 3.3 to >5.8 in 1972 and from 3.3 to >5.7 in 1973. The greater than 100-fold variations in ID_{50}s obtained by pledget feeding were considered significant since the ID_{50} of a colonized strain of *C. tarsalis* had varied less than tenfold when tested five times over a 3-year period.

WEE viral susceptibility profiles obtained by pledget feeding for four selected populations each in 1972 and 1973 are graphically illustrated in Fig. 1. In addition to showing the range of susceptibilities between different populations, these profiles revealed that many field populations of *C. tarsalis* were quite heterogeneous or polymorphic in their susceptibility to infection following ingestion of WEE virus.

Twenty-one of 22 and 2 of 23 samples collected in 1972 and 1973, respectively, were evaluated for susceptibility by intrathoracic inoculation. The $\log_{10}$ of the ID_{50} obtained for different populations ranged from -0.8 to 0.8 with a median of 0.2. Since nearly this amount of variation was observed when the same colonized strain was tested repeatedly (Hardy *et el.*, 1976), these results indicated that field populations of *C. tarsalis* did not vary significantly in their susceptibility to infection with WEE virus by intrathoracic inoculation. When considered with the findings obtained by pledget feeding, these results suggested that resistance of *C. tarsalis* to WEE virus was associated with a mesenteronal barrier.

This study on the vector competence of a large number of field populations of *C. tarsalis* for WEE virus confirmed the observations reported earlier for a more limited sample (Hardy *et al.*, 1976).

b. Seasonal variations. We attempted to determine if there were seasonal variations in the susceptibility of *C. tarsalis* to infection following ingestion of WEE virus. When ID_{50}s obtained by pledget feeding were compared with dates of collection in 1972 and 1973, it was found that eight of ten populations collected through the first week in July had ID_{50}s of 4.8 or greater, whereas only seven of 35 populations collected there-

TABLE I

Comparison of WEE Viral Susceptibility, Ethyl Parathion Resistance, and Autogeny Rates in Field Populations of C. tarsalis *Collected in California during 1972 and 1973*

Mosquito source		*Date collected*	*WEE viral ID_{50} (log_{10})[a]*	*Ethyl parathion LC_{50} (ppm)*	*Autogeny rate (%)*
Sacramento Valley					
Butte County	Boeger's Pasture	8-30-72	3.3	--	22
		8-10-73	4.3	0.004	60
	Llano Seco Slough	9-11-73	4.0	0.004	70
Glenn County	Road P	7-5-72	>4.8	0.003	38
	Road 44	7-18-73	4.2	--	13
	Willows	8-16-73	5.3	0.004	67
	Glenn	8-21-73	4.2	0.004	80
Tehama County	Red Bluff	8-6-73	4.3	0.005	46
	Tehama-Colusa Canal	8-13-73	5.2	0.003	68
Yuba County	Bingham's Ditch	8-14-72	5.3	0.003	8
	Shintaffer's Pig Farm	8-14-72	4.6	0.002	93
		8-28-72	4.8	0.005	80
	N. Beale Road	8-28-72	4.4	0.002	78
	Ellis Road	7-26-73	3.9	--	74
	Marysville	9-13-73	4.4	0.004	44

San Joaquin Valley					
Kern County	Meadow Gold Farms	8-16-72	4.5	0.002	63
	Panama Lane	8-17-72	4.5	--	33
	Richfield Oil Sump	8-30-72	4.3	0.002	64
	N. Rim Ditch	6-20-73	3.7	0.025	86
		7-11-73	3.5	0.017	33
	Belridge Oil Well	6-21-73	>5.7	0.003	3
	Paloma Field	8-22-73	4.2	0.004	82
Stanislaus County	Freitas Pasture	7-21-72	5.0	--	22
	Oakdale	8-1-73	3.3	--	97
Tulare County	Correia Ranch	7-17-72	4.1	--	81
Southern California					
Los Angeles County	Flintkote Gravel Pit	7-8-72	>5.8	0.003	1
Riverside County	UCR Ponds	5-16-73	4.1	0.001	4

[a] *Derived by the Reed and Muench (1938) equation from data obtained for groups of 10-20 female mosquitoes that were fed on pledgets with varying tenfold concentrations of virus and tested individually for virus after an extrinsic incubation period of 10-14 days at 24°C.*

TABLE II
Comparison of WEE Viral Susceptibility, Ethyl Parathion Resistance, and Autogeny Rates in Field Populations of C. tarsalis *Collected Sequentially during the Summer from the Same Locality in California*

Mosquito source	Date collected	Infection with WEE virus: Estimated log_{10}(±0.3) PFU ingested							Ethyl parathion	
		0.9	1.9	2.9	3.9	4.9	5.9	log_{10} ID_{50}	LC_{50} (ppm)	Autogeny rate (%)
Kern County										
Poso Creek										
Upper Pasture	6-20-72	0[a]	0	0	4	40	72	5.6	0.008	44
" "	7-18-72	0	0	5	45	60	88	4.3	0.006	54
Lower Pasture	8-15-72	0	5	5	25	30	80	5.0	0.006	47
Upper Pasture	9-11-72	5	10	21	22	78	95	4.1	0.007	67
Upper Pasture	5-14-73	0	0	0	10	60	90	4.8	0.003	6
Lower Pasture	6-26-73	5	5	5	15	40	75	5.2	0.003	45
	7-23-73	0	5	20	65	74	95	3.6	0.004	37
	8-13-73	5	15	45	75	75	85	3.7	0.003	40
	9-11-73	0	13	38	30	64	85	4.1	0.003	24
Bakersfield										
Sewer Farm[b]										
Drainage Sump	5-8-72	0	0	10	20	50	50	5.1	0.019	70
	6-15-72	0	0	0	0	25	50	5.6	0.018	79
	7-17-72	-	-	20	10	50	84	4.7	0.017	84
	7-31-72	0	0	0	25	50	75	4.8	0.010	74
Irrig. Pasture	8-16-72	10	10	20	50	90	90	3.4	0.008	57
	9-12-72	0	10	35	45	40	70	4.4	0.012	90
Madera County										
Stoetzl Ranch	7-21-73	-	0	10	61	40	95	4.2	0.005	87
	8-1-73	0	0	30	37	55	70	4.6	--	87
	9-21-73	0	10	30	40	60	89	4.2	--	6

[a] *Expressed as percentage of 10-20 females tested individually and found to be infected after a 10-14 day extrinsic incubation period at 24°C.*
[b] *The two sites at the Sewer Farm where samples were collected were within 25 yards of each other.*

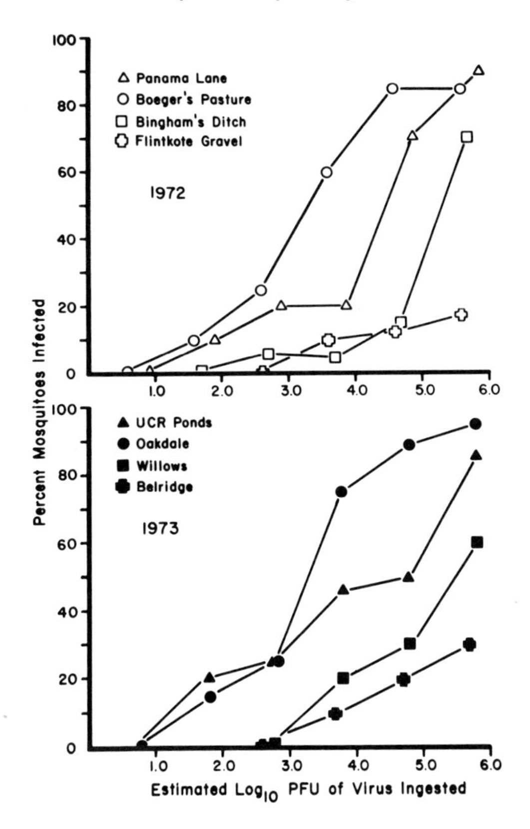

Fig. 1. Comparative WEE viral susceptibility profiles obtained by pledget feeding for field populations of C. tarsalis *collected in California during 1972 and 1973.*

after had similar ID_{50}s (Tables I and II). This suggested that late spring and early summer populations of *C. tarsalis* were less susceptible to infection than were subsequent populations. This conclusion seemed to be supported by susceptibility data obtained from tests on monthly collections of *C. tarsalis* made at Poso Creek study area in 1973 and possibly 1972 (Table II). Furthermore, the *C. tarsalis* population tended to become more heterogeneous for susceptibility as the summer progressed. However, collections in this area could not always be made at precisely the same pasture site. Thus, the observed variations could have reflected unknown factors associated with mosquitoes breeding in different microhabitats.

The preceding conclusion was supported by similar studies conducted in 1972 at the Bakersfield Sewer Farm (Table II). WEE viral susceptibility at this site was essentially the same for four collections made from the same drainage sump. A significant increase in susceptibility was observed, however, when the August collection was made from an irrigated pasture 25 yards from the above sump that received the tailwaters from the pasture.

The question of whether *C. tarsalis* vary seasonally in their susceptibility to infection with WEE virus cannot be answered at this time. It is possible that minor variations in environmental conditions at different study sites and in different years might explain the conflicting results that we obtained. Unfortunately, no climatological data were collected at the specific study sites.

c. Factors potentially associated with susceptibility. Limited data obtained in the studies in 1971 (Hardy *et al.*, 1976) suggested that there was no correlation between variations in WEE viral susceptibility of field populations of *C. tarsalis* and larval resistance to organophosphorus insecticides or autogeny rates in females. This initial conclusion was verified by more extensive data obtained from mosquito populations evaluated in 1972 and 1973 (Tables I and II). A lack of correlation was most evident when multiple samples were collected from the same site during the year (Table II).

The heterogeneity of many populations of *C. tarsalis* in their susceptibility to infection by ingestion of WEE virus (Fig. 1) suggested that susceptibility was an inherited trait. A study was undertaken to select for strains of *C. tarsalis* that were highly susceptible and highly refractory to infection following ingestion of WEE virus from viremic chicks. The susceptibility profiles following feeding on viremic chicks are shown in Fig. 2 for the WS and WR strains that were derived from 10 and more than 20 generations of selection respectively. The ID_{50}s were $10^{2.2}$ for the WS strain and $10^{7.4}$ for the WR strain. This represented a greater than 100,000-fold difference in susceptibility. Resistance was associated with a mesenteronal barrier since WEE virus multiplied in both strains when they were inoculated intrathoracically. Further, viral multiplication occurred in the mesenteron of WS but not WR females when virus was inoculated intrathoracically. This finding clearly indicated that the barrier to resistance and the pathway of viral spread was bidirectional in WR and WS females respectively. However, a low proportion (10 to 20%) of WR females became infected after ingestion of relatively small concentrations of virus (Fig. 2). Other studies suggested that these mosquitoes became infected by a nongenetically controlled mechanism since individuals with a phenotype for susceptibility had the genotype of a resistant individual (Hardy *et al.*, manuscript submitted for publication).

Although the mode of inheritance was not established, these results suggested that WEE viral susceptibility in *C. tarsalis*

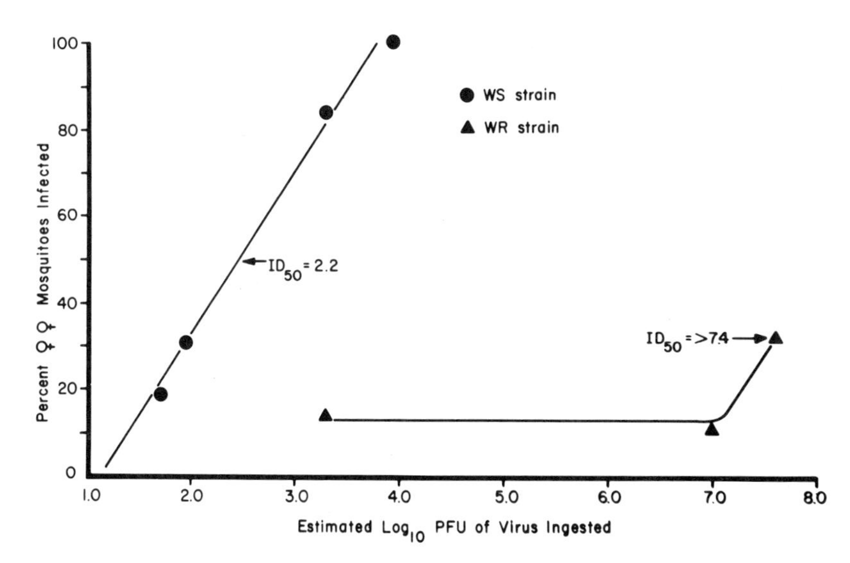

Fig. 2. Susceptibility profiles of C. tarsalis *strains selected for susceptibility (WS) and resistance (WR) to infection following ingestion of WEE virus from viremic chicks.*

was controlled at least in part by genetic mechanisms. A similar conclusion was reached in studies on the susceptibility of *Culicoides variipennis* to infection with bluetongue virus (Jones and Foster, 1974) and of *A. albopictus* to infection with dengue (Gubler and Rosen, 1976) and chikungunya (Tesh *et al.*, 1976) viruses. In addition, the results obtained in the genetic selection study with WEE virus help to explain the variations seen in the susceptibility of field populations of *C. tarsalis* to infection after ingestion of WEE virus. Further studies are required to determine what environmental factors, if any, influence selection for susceptibility or resistance and to ascertain whether the trait for viral resistance is of potential value for the genetic control of vector competence in field populations.

2. Transmission of Virus

In 1972, *C. tarsalis* from the Poso Creek area were fed on chicks having viremia titers of $10^{5.0}$ to $10^{7.0}$ PFU per 0.2 ml of blood. These mosquitoes were evaluated for their ability to transmit virus after 14 days extrinsic incubation. Only four of 18 (22%) infected mosquitoes transmitted virus. In contrast, a transmission rate of 100% was obtained with a colonized strain of *C. tarsalis* fed on chicks with similar levels of viremia. These findings suggested that some field populations of *C. tarsalis* might be poor transmitters of WEE virus. Subsequently, attempts were made to obtain transmission rates for mosquitoes from nine collections made from the Poso West study site in 1975 and 1976. These mosquitoes were infected by the pledget feeding technique. Refeedings on chicks for viral transmission were poor; however, the accumulated results indicated that only 26 to 29% of the infected females were capable of transmitting WEE virus (Table III). Furthermore, viral titers in nontransmitting females were 100-fold lower than in transmitting females. It is not known whether transmission rates would have been higher if the time of extrinsic incubation had been extended.

If these results represent the true transmission potential of the Poso West population, then it is obvious that the vector competence of field populations cannot be assessed by only doing tests for viral susceptibility. Also, these results may partially explain the rather consistent differences that have been observed between WEE viral infection and transmission rates for *C. tarsalis* collected in chick-baited traps (Reeves *et al.*, 1961). In this early study it was frequently observed that only one in four infected females transmitted virus when they fed on susceptible chicks. At that time it was assumed that nontransmitters had not completed the extrinsic incubation period. The present studies provide an alternative or additional explanation.

TABLE III
WEE Viral Transmission Rates Obtained with Field Populations of C. tarsalis, *Kern County, California, 1975 and 1976*

Adult mosquito source	*Fraction (%) of mosquitoes infected*[a]	*Fraction (%) of infected mosquitoes transmitting*[a]	*Mean viral titer/mosquito*	
			Trans-mitters	*nontrans-mitters*
Pupae	35/64 (55%)	9/35 (26%)	5.5[b]	3.2[b]
CO_2/ light traps	51/98 (52%)	15/51 (29%)	5.3	3.4

[a]*Determined 13 to 14 days after ingestion of $10^{5.0}$ to $10^{5.8}$ PFU of virus from infected pledgets.*
[b]*Expressed in log_{10}.*

B. Interspecies Variations in Vector Competence

As expected, some of the field collections contained several species of mosquitoes. When this occurred, it provided an opportunity to simultaneously evaluate the susceptibility of different mosquito species to infection with WEE virus. The results from one such collection from the Sacramento Valley which contained *Culex peus*, *Culex pipiens* and *C. tarsalis* are depicted in Fig. 3 along with results of a collection of *Aedes melanimon* that was made in the same locality and tested simultaneously. The susceptibility profiles of different mosquito species did not vary significantly by intrathoracic inoculation, but they did by pledget feeding. *C. peus* and *C. pipiens* were highly refractory to infection when fed on virus. Similar results were observed with numerous other populations of *C. peus* and *C. pipiens* that were collected in 1972 and 1973 throughout California. The results were verified when females from colonized strains of *C. peus* and *C. pipiens* were fed on viremic chicks. *C. pipiens* was also found to be an incompetent vector of WEE virus in earlier studies (Reeves and Hammon, 1962). *C. peus,* however, was thought to be an efficient vector of WEE virus.

The population of *A. melanimon* shown in Fig. 3 was more susceptible than the population of *C. tarsalis* to infection following ingestion of WEE virus. However, other populations

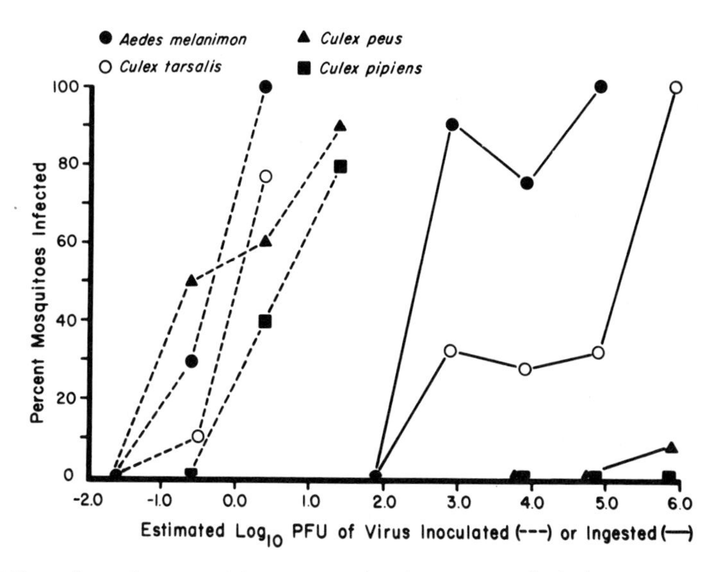

Fig. 3. Comparative WEE viral susceptibility profiles obtained by intrathoracic inoculation and pledget feeding with four mosquito species.

of *A. melanimon* were ten to 100-fold more resistant than was this population, thus suggesting that *A. melanimon* also varied in its susceptibility to WEE virus.

Acknowledgments

These studies were supported in part by funds from the Office of Naval Research and by Research Grant AI03028 from the National Institute of Allergy and Infectious Diseases.

The cooperation of the Managers of the Mosquito Abatement Districts where mosquitoes were collected is greatly appreciated and the technical assistance of Ms. B. Penney, Ms. S. Samimi, Mr. V. Geer, and Mr. G. Apperson is gratefully acknowledged. The authors are indebted to Dr. R. D. Sjogren conducting organophosphorus insecticide resistance tests on larval populations collected in 1972.

References

1. Chamberlain, R. W. (1968). *Curr. Top. Microbiol. Immunol. 42*,38-58.
2. Chamberlain, R. W., and Sudia, W. D. (1961). *Ann. Rev. Entomol. 6*,371-390.
3. Clarke, D. H., and Casals, J. (1958). *Am. J. Trop. Med. Hyg. 7*,561-573.
4. Gillies, P. A., and Womeldorf, D. J. (1968). *Vector Views 15*,45-50.
5. Gubler, D. J., and Rosen, L. (1976). *Am. J. Trop. Med. Hyg. 25*,318-325.
6. Hardy, J. L., Reeves, W. C., and Asman, S. M. (1975). *Proc. Calif. Mosq. Control Assoc. 42*,15-18.
7. Hardy, J. L., Reeves, W. C., and Sjogren, R. D. (1976). *Am. J. Epid. 103*,498-505.
8. Jones, R. H., and Foster, N. M. (1974). *J. Med. Entomol. 11*,316-323.
9. McLean, D. M. (1955). *Austral. J. Exp. Biol. 33*,53-66.
10. Merrill, M. H., and TenBroeck, C. (1935). *J. Exp. Med. 62*,687-695.
11. Reed, L. J., and Muench, H. (1938). *Am. J. Hyg. 25*,493-497.
12. Reeves, W. C. (1967). *Vector Views 14*,13-18.
13. Reeves, W. C., and Hammon, W. McD. (1962). "Epidemiology of the Arthropod-Borne Viral Encephalitidies in Kern County, California 1943-1952," Vol. 4. Univ. of Calif. Press, Berkeley.
14. Reeves, W. C., Bellamy, R. E., and Scrivani, R. P. (1961). *Am. J. Hyg. 73*,303-315.

15. Spadoni, R. D., Nelson, R. L., and Reeves, W. C. (1974). *Ann. Entomol. Soc. Am.* *67*,895-902.
16. Tesh, R. B., Gubler, D. J., and Rosen, L. (1976). *Am. J. Trop. Med. Hyg.* *25*,326-335.
17. Ward, R. N. (1963). *Exp. Parasitol.* *13*,328-341.

Chapter 11

ISOLATION OF EYACH VIRUS FROM IXODID TICKS

R. Ackermann, B. Rehse-Küpper, J. Casals,
E. Rehse, and V. Danielova

I. INTRODUCTION

Since the first isolation of louping ill virus in 1929, 21 different tick-borne viruses have been identified in Europe. As established so far, six can attack man. On account of the seriousness or special characteristics of the clinical symptoms, tick-borne encephalitis (TBE), louping ill, Crimean-Congo-hemorrhagic fever, and Kemerovo viruses are of particular significance. For Uukuniemi and Tribec viruses, only their immunogenic properties indicate their human pathogenicity.

In the Federal Republic of Germany, two tick-borne viruses, TBE and Tettnang, have been found. In addition, from a TBE focus in Baden-Württenmberg in 1972 Eyach virus has been isolated from *Ixodes ricinus* ticks. Because of its close relationship to Colorado tick fever (CTF) virus, Eyach virus is a further candidate for human pathogenicity in Europe.

ISBN 0-12-429765-X

II. ISOLATION OF EYACH VIRUS

Eyach virus was found accidentally during virus isolation experiments in a TBE focus 60 km southwest of Stuttgart, Baden-Württemberg, Federal Republic of Germany (Rehse-Küpper *et al.*, 1976). The virus was isolated from one pool of two male and two female *I. ricinus* ticks by intracerebral (IC) and subcutaneous (SC) inoculation into suckling mice. The virus is named after the place where it has been found, a small village in the Neckar valley. Eyach (EYA) virus was registered by the American Committee on Arthropod-borne Viruses on June 6, 1975 as a member of the CTF serogroup.

III. HOST RANGE AND PATHOGENICITY

A. Mice

Eyach virus is pathogenic only for suckling mice after IC, intraperitoneal (IP), and SC inoculation, but not for adult mice after IC and IP inoculation (mouse strain NMRI Han).

The first isolation of Eyach virus by IC inoculation of suckling mice and its adaptation to this host proceeded with moderate difficulties. In the first passage, 8 days after the inoculation, one mouse showed symptoms of disease; six out of 11 animals died between the ninth and seventeenth day after inoculation. Already in the second IC passage in suckling mice the lethality was 100%. From the tenth passage onward, the survival time was reduced to 6-8 days. The titer increased thereby from 4.0 log LD_{50}/0.02 ml in the seventh passage to 6.9 in the nineteenth passage.

Four days after IC inoculation of Eyach virus, suckling mice develop symptoms of encephalitis, very often lasting 4 days, characterized by reduced activity, cramped body posture, clumsy movements, lack of suckling, and finally, prior to exitus, laying lethargically on one side.

In contrast, after IP and SC inoculation the pathogenicity of Eyach virus for suckling mice is much lower. With a dose of 8.4×10^4 LD_{50} inoculated IP or SC, only about 50% of the mice die after 7-13 or 10-11 days, respectively.

B. Tissue Culture

Eyach virus can be propagated in cultures of secondary mouse embryo and Vero cells, as demonstrated by IC inoculation of the cell culture material into suckling mice after propagation over a period of 6 days. After nine continuous passages in Vero cells, lasting for 3-4 days each, the titer in suckling mice reached 4.8 log ID_{50}/0.02 ml (IC). In a plaque test with Vero cells, the tites was 4.1 log_{50}/0.1 ml and in tube test the titer was 4.3 log ID_{50}/0.1 ml, as measured by cytopathogenic effect.

Eyach virus could not be propagated in FL and HeLa cells. IC inoculation of the cell culture material (after propagation for 6 days) into suckling mice had no effect.

IV. PHYSICAL AND CHEMICAL PROPERTIES

Eyach virus passes through membrane filters of 200 nm (Sartorius) but not of 100 nm (Millipore) pore diameter. At 37°C in 0.03 *M* tris buffer, pH 7.5 containing 10% inactivated calf serum, half of the infectivity is inactivated within 30 to 40 minutes.

Under standard experimental conditions (Hammon and Sather, 1969), Eyach virus is resistant to treatment with ether and sodium deoxycholate but is sensitive to treatment with chloroform. Applying the method of Kolman (1970), the virus titer decreases by more than 4.4 log 10 units in a chloroform containing medium at room temperature within 30 minutes.

V. IDENTIFICATION OF SEROLOGIC METHODS

Serologic methods identified Eyach virus as being related to CTF virus. In complement fixation (CF) tests with 49 antigens, unrelated to the group B tick-borne viruses, Eyach virus showed cross reactivity with CTF antigen. This relationship was also demonstrated by neutralization tests (NT). However, while Eyach virus was neutralized by CTF antiserum, CTF virus was not neutralized by Eyach antiserum (Table I).

TABLE I
The Relationship between CTF and EYA Virus as Revealed by Complement Fixation (CF) and Virus Neutralization (NT) Tests

	CF test with antigen[a]		NT with virus[b]			
			EYA		CTF	
Serum	*EYA*	*CTF*	*log* LD_{50}	*NI*	*log* LD_{50}	*NI*
anti-EYA	128/128	32/64	2.6	3.3	6.4	0.3
anti-CTF	64/64	256/512	3.0	2.9	2.7	4.0
Normal			5.9		6.7	

[a]*Serum titer/antigen.*
[b]*Neutralization tests in suckling mice by IC inoculation.*

VI. TOPOLOGY AND ECOLOGY OF THE EYACH VIRUS ISOLATION AREA

Eyach virus was isolated in a region where TBE virus infections repeatedly occurred. Earlier serologic investigations revealed antibodies in 7.9% of the rural population (Ackermann *et al.*, 1968). Ticks were collected for virus isolation experiments in an area where two patients had been infected with TBE virus.

This region belongs to a moderate climate zone at 48°27' N latitude and 8°27' E longitude 374 m above sea level, situated southwest of Stuttgart, Baden-Württemberg, in the district of Freudenstadt. The landscape alongside the river Neckar is characterized by low hills, pastures, farmland, hardwood, and evergreen forests, where *I. ricinus* ticks occur abundantly. The ticks, from which Eyach virus was isolated, were collected from a trail in a hardwood forest, overgrown with grass and underbush.

Eyach virus was isolated in 1972, during TBE virus isolation experiments from *I. ricinus* ticks of that area. Attempts to isolate Eyach virus from ticks of the same focus were unsuccessful in later years (Table II). Nevertheless, in 1974 one and in 1976 two strains of TBE virus were isolated from ticks collected in the same general area (6 km^2).

TABLE II
Isolation of Eya and TBE Virus from I. Ricinus Ticks in the Vicinity of Eyach Village in 1972-1976

Year	*Sites*	*Imagines*	*Nymphs*	*Isolates*
1972	3	143	349	EYA (1)
1974	11	739	3348	TBE (1)
1976	5	247	870	TBE (2)

VII. ANTIBODY STUDIES IN HUMAN POPULATION

In order to obtain some initial insight into the eventual rate of occurrence of antibodies in the population in the vicinity of Eyach, neutralization tests were performed in suckling mice. The sera of 84 individuals (34 male and 50 female), aged between 15 and 81 years, were tested for antibodies to Eyach virus. All sera were negative with neutralization indices (NI) less than 0.9.

VIII. DISCUSSION AND CONSLUSIONS

The physical, chemical, and pathogenic properties of Eyach virus are similar to those of CTF virus. Particularly, the sensitivity to chloroform but resistance to ether and sodium deoxycholate are remarkable features of both CTF and Eyach virus. As demonstrated by CF tests, Eyach and CTF virus are related. However, the reciprocal NT with the two agents and their corresponding antisera indicates that they are distinct viruses.

Our present knowledge about the host range and the virus host cycle of the European relative of the CTF virus is still poor. Moreover, it is still unknown whether or not mice are regular members of the cycle and *I. ricinus* ticks, from which Eyach virus was isolated, are main vectors for this agent. Because of the discrepancies in the fauna of North America and Europe, different host cycles for CTF and Eyach virus might conceivably exist.

Another important question, as to whether Eyach virus is pathogenic for man is still open. Due to the limited number of human sera tested, our own serologic investigations cannot give a final answer. Yet, the results obtained with Vero cells might indicate that primates are just as susceptible to Eyach virus than to CTF.

The isolation of Eyach virus is another example for the existing relationships between certain viruses from different areas of the world. To our present knowledge both viruses, Eyach and CTF, occur in zones with temperate climate on the northern hemisphere in regions more than 8000 km apart. It remains to be established whether other viruses related to CTF also occur in the tropics or whether they are limited to temperate climates.

Eyach is the first virus related to CTF that is found outside North America. Only speculation can be formulated at the present time about the mechanism of its geographical distribution in the west-east direction.

References

Ackermann, R., Rehse-Küpper, B., Löser, R., and Scheid, W. (1968). *Deutsch. Med. Wschenschr. 93,* 1747-1754.

Hammon, W. McD., and Sather, G. E. (1969). *In* "Diagnostic Procedures for Viral and Rickettsial Infections" (E. H. Lenette and N. J. Schmidt, eds.), 4th ed., pp. 227-280. American Public Health Association, New York.

Kolman, J. (1970). *Acta Virol. 14,* 159-162.

Rehse-Küpper, B., Casals, J., Rehse, E., and Ackermann, R. (1976). *Acta Virol. 20,* 339-342.

Chapter 12

PERSISTENT TICK-BORNE ENCEPHALITIS INFECTION IN MAN AND MONKEYS: RELATION TO CHRONIC NEUROLOGIC DISEASE

David M. Asher

I. INTRODUCTION

We shall review chronic neurological syndromes that sometimes follow infection with tick-borne encephalitis (TBE) virus, mention their similarities to other brain diseases, and describe an experimental TBE virus infection of monkeys that may be a model for human diseases.

Chronic seizure disorder following encephalitis was described by Aleksei Kozhevnikov in Moscow at the end of the last century. In October of 1889, 20-year-old Michael L. was admitted to Kozhevnikov's neurology ward because of constant jerking of the right foot. At age 15 the patient had a severe febrile illness with loss of consciousness and convulsion; he seemed to make full recovery, but the following year began to have several generalized seizures each month. He also noted

ISBN 0-12-429765-X

shaking in his right foot while walking. The shaking became more severe, and by age 19 he was unable to walk at all. On examination the boy could not stand or lie flat, but sat with the right knee flexed, gripping the foot with one hand. There were constant clonic flexing and extending movements of the toes. Releasing the foot from the boy's grasp, touching the toes, or even reaching toward the foot as if to touch it, made the jerking movements worse. As the clonic jerks became more severe, first the right leg and then the whole body would jerk. The movements diminished during sleep and after a seizure. Michael L. failed to improve after bromide therapy and an operation to stretch the popliteal nerve and was discharged in January 1890. Kozhevnikov (1894) described this patient and three others with similar seizures as examples of a "particular type of cortical epilepsy," which he named *epilepsia corticalis sive partialis continua,* now usually called *epilepsia partialis continua* or Kozhevnikov's epilepsy. No tissue from these patients was available for diagnosis, but one possible etiology suggested by Kozhevnikov was encephalitis. (A similar patient without history of encephalitis was demonstrated at a meeting in Hanover, Germany by Bruns in 1894).

II. CHRONIC AND PROGRESSIVE SYNDROMES FOLLOWING TICK-BORNE ENCEPHALITIS

Thirty years later, Omorokov (1927a,b) summarized some 50 reported cases of Kozhevnikov's epilepsy with 52 of his own cases from Siberia, where the condition seemed to affect mostly children and young adults in rural villages. Illness typically began in spring or summer with high fever, delirium, often loss of consciousness, and convulsions; patients then improved, though there might be residual paralysis. During the convalescent period, or months later, the focal jerking of epilepsia partialis continua began. Thirty of Omorokov's patients had craniotomy for excision of epileptogenic cortex; the tissue usually showed chronic changes with loss of neurons and sclerosis, but there were also areas of fresh acute inflammation, with round cell nodules in parenchyma and perivascular cuffs.

In 1937 the tick-borne virus responsible for Russian spring-summer encephalitis was identified (Zilber *et al.*, 1938; Levkovich *et al.*, 1938; Chumakov, 1939). Soon cases of Kozhevnikov's epilepsy following typical Russian spring-summer encephalitis were recognized (Kanter, 1940; Golman, 1941). Chumakov *et al.* (1944) found that many patients with Kozhevnikov's epilepsy had serum neutralizing antibodies to the tick-borne virus, while only a few control patients did. So tick-borne encephalitis (TBE) seemed to be a probable cause of many cases of Kozhevnikov's epilepsy in the USSR. Of course, many other causes were also assigned to epilepsia partialis continua, for example, cysticercosis, syphilis, trauma, granuloma, and vascular accident or other focal CNS lesions (Omorokov, 1927a,b, 1938, 1941). In a recent review of 32 North American patients with epilepsia partialis continua at the Mayo Clinic, only five had probable encephalitis (Thomas *et al.*, 1977).

During the past 35 years there has been a flood of clinical reports by Soviet neurologists, describing various chronic disorders caused by TBE (Shapoval, 1945; Galant, 1946), and there is still no agreement on proper classification of these syndromes (Klyuchikov, 1965a). They seem to fall into two major groups:

A. Movement and Seizure Disorders

The movement and seizure disorders, collectively called "hyperkinesias" (Shapoval, 1945), include epilepsia partialis continua (most common), ordinary Jacksonian epilepsy frequently associated with epilepsia partialis continua (Kozhevnikov, 1894; Bruns, 1894; Omorokov, 1927a; Galant, 1946) and rarer conditions such as myoclonic epilepsies (Kanter, 1961; Strokina *et al.*, 1969), chorea (Galant, 1951), choreoathetosis (Strokina *et al.*, 1969), and other tremors and dystonias (Klyuchikov, 1965b; Baishtruk, 1975).

B. Paralytic Disorders

The paralytic disorders include a variety of polioencephalomyelitic syndromes. Such syndromes were recognized in the last century (Pervushin, 1899), although first attributed to preceding TBE much later (Pervushin, 1940, 1943; Shefer and Polikovsky, 1940; Golman, 1941). Patients with paralytic syndromes tend to be older than those with seizure and movement disorders (Klyuchikov, 1965b). Some Soviet authorities have concluded that amyotrophic lateral sclerosis is a common sequela of TBE (Kuimov and Dubov, 1958), though others have disagreed (Protas and Votyakov, 1970).

C. Other Syndromes

Mixed types of syndrome are frequent, often with mental retardation (Galant, 1946) or dementia (Gulyaeva, 1975). In all types of chronic disease, but especially in the polioencephalomyelitic syndromes, there may be progressively severe damage to the nervous system called "progredient" disease, which usually ends in death.

There is no agreement on the frequency of various chronic forms of TBE; estimates by different authorities have varied from 1 to 20% (Komandenko *et al.*, 1972) or more (Umansky, 1975a) of patients with acute TBE who get chronic disease. There may be a tendency for progressive disease to occur in patients with residual paralysis (Baishtruk, 1975), and in one series chronic disease was more common after meningoencephalitis than after milder aseptic meningitis (Vaneeva, 1969). However, progressive disease has also been reported after inapparent acute infection (Magazanik and Robinzon, 1966; Shubin *et al.*, 1975). Signs of progressive disease frequently appear during clinical recovery (Shapoval and Garishina, 1975) and usually occur within 6 months of acute infection (Magazanik and Robinzon, 1966; Strokina *et al.*, 1969). Outcome is not inevitably poor, and in a significant number of patients progression has stopped, allowing return to normal life (Shapoval, 1975; Shapoval and Garishina, 1975).

In a recent summary of TBE in some 3000 patients (Umansky, 1975a,b) all clinical types of both acute and chronic disease were found in generally similar incidence throughout the USSR, except that the severe paralytic disease was more common in the Far East and mild aseptic meningitis without paralysis in Byelorussia; small endemic foci of specific clinical forms of TBE occurred in many places. Chronic disease has followed biphasic milk-borne encephalitis as well as the more common tick-borne disease (Klyuchikov, 1961). Chronic disease has recently been observed after vaccination with a Langat-like strain of TBE virus previously thought to be avirulent (Magazanik and Shakinko, 1975; Umansky *et al.*, 1975).

Outside the USSR, chronic progressive disease after TBE has also occurred in Japan (Harada, 1970; Kono *et al.*, 1972; Ogawa *et al.*, 1973). One patient, described by Ogawa and colleagues (Fig. 1) was a 51-year old man who developed progressive confusion and difficulty in seeing, hearing, and walking. At age 30, while a prisoner of war in Siberia, he had encephalitis, which left him with flaccid paralysis of the right shoulder but otherwise well. Thirteen years later he began to have trouble hearing with his left ear, and during the following years other symptoms of progressive neurological disease including ringing and deafness in both ears, double vision, increasing difficulty in walking, changes in personality, forgetfulness, headaches, and impaired speech. On physical examination he was found to be disoriented with bilateral nerve deafness, visual loss, optic atrophy, dysphagia, and scanning speech. There were muscular fasciculations and spastic weakness in all extremities except the flaccid right shoulder, intention tremors in the upper extremities, and grossly ataxic gait. Tendon jerks were hyperactive with bilateral Babinski

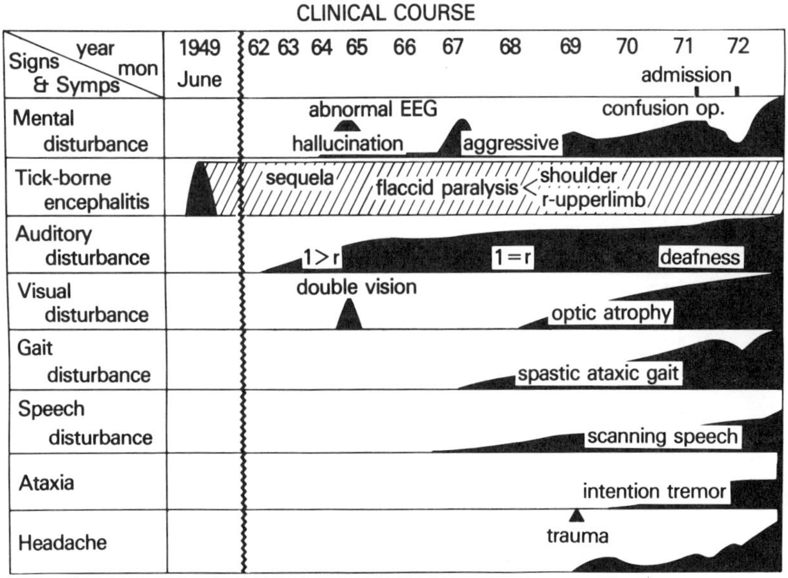

Fig. 1. Chronic illness beginning 13 years after TBE Reprinted with permission from M. Ogawa et al. *(1973)*, J. Neurol. Sci. 19,*363-373 (Elsevier Publ. Co., Amsterdam)*.

reflexes. The EEG showed diffuse slowing, and the pneumo-encephalogram showed atrophy of cerebrum and cerebellum. The CSF contained a few cells and increased globulin. Hemagglutination-inhibiting antibodies to TBE virus were present in serum at titer of 1:2560 and in CSF at 1:1280, suggesting active synthesis within CNS. (Antibody titers to Japanese encephalitis virus were only 1:80 in both serum and spinal fluid, and measles antibody titer was 1:128 in serum and not detected in spinal fluid, ruling out subacute sclerosing panencephalitis and demonstrating that the blood brain barrier was intact (Vandvik and Norrby, 1973; Norrby *et al.*, 1974.) A brain biopsy, performed 22 years after the acute encephalitis and at least 8 years after onset of progressive brain disease, showed cerebral atrophy with lymphocyte and plasma cell infiltrates in parenchyma and around vessels. In short, this patient had a slowly progressive panencephalitis, clinically and histologically similar to progredient paralytic cases of chronic TBE in the USSR. A very similar Japanese patient, also a former prisoner in Siberia, was described earlier by Harada (1970).

III. EVIDENCE THAT CHRONIC SYNDROMES RESULT FROM PERSISTENT INFECTION WITH TICK-BORNE ENCEPHALITIS VIRUS

The evidence that chronic syndromes are due to actual persistent infection with TBE virus is incomplete. Histological changes of fresh active inflammation, present in the brain long after onset of illness (Omorokov, 1927a; Magazanik and Robinzon, 1966; Ogawa *et al.*, 1973), suggest an active aggressive process, as does antibody to TBE virus secreted within the CNS (Ogawa *et al.*, 1973). At least two laboratories have reported isolating TBE virus from patients with Kozhevnikov's epilepsy long after the original acute illness, Chumakov and co-workers (1944) from two brain specimens 9 and 13 months, and Kraminskaia and co-workers (1969) from spinal fluid 3.5 months after onset of seizures. Other successful virus isolations from such patients may also have been made but not published (Ilyenko, 1961; Asher, 1971). Most attempts to isolate virus from patients with chronic post-TBE syndromes have apparently been unsuccessful.

IV. OTHER CAUSES OF CHRONIC ENCEPHALITIS

Chronic encephalitis occurs after other acute viral infections (Aguilar, 1959). Measles (Payne *et al.*, 1969; Horta-Barbosa *et al.*, 1969) and rubella viruses (Townsend *et al.*, 1975; Weil *et al.*, 1975) both cause panencephalitis, and the viruses of Western encephalitis (Platou, 1940; Noran and Baker, 1943, 1945; Noran, 1944; Bruyn and Lenette, 1953; Herzon *et al.*, 1957) and Japanese encephalitis (Kimoto *et al.*, 1968) have rarely been reported to do so as well. Rasmussen and colleagues in Montreal (Rasmussen *et al.*, 1958; Aguilar and Rasmussen, 1969; Rasmussen and McCann, 1968) reported a number of North American children with chronic focal seizures, hemiparesis, and mental retardation in whom fresh inflammatory changes were found in the cerebral cortex (Figs. 2 and 3). Although none of those children had a history of acute encephalitis, their illnesses showed similarities to the syndromes that follow TBE. Similar cases have also been reported from India (Gupta *et al.*, 1974). A recent study of a child with clinical and histological findings not unlike those of Rasmussen's patients revealed enterovirus-like particles in brain tissue (Friedman *et al.*, 1977).

We have obtained tissues from 13 such patients (Asher, 1975); no virus was isolated in a variety of assay systems and no particles were demonstrated by electron microscopy.

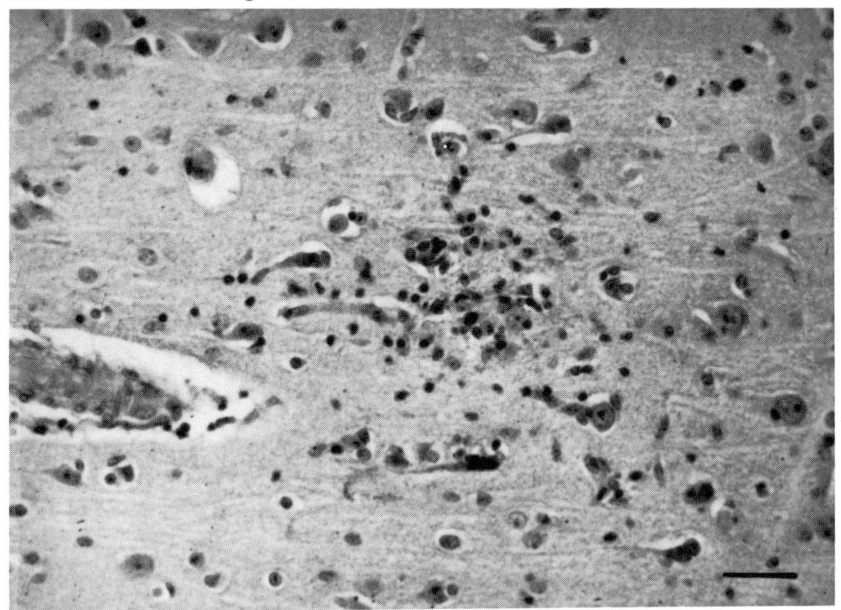

Fig. 2. Mild inflammation in the frontal cortex of a child with chronic epilepsy. Microglial infiltrate; hematoxylin and eosin stain,___,approximately 70 µm. Courtesy of T. Rasmussen and S. Carpenter, Montreal Neurological Institute.

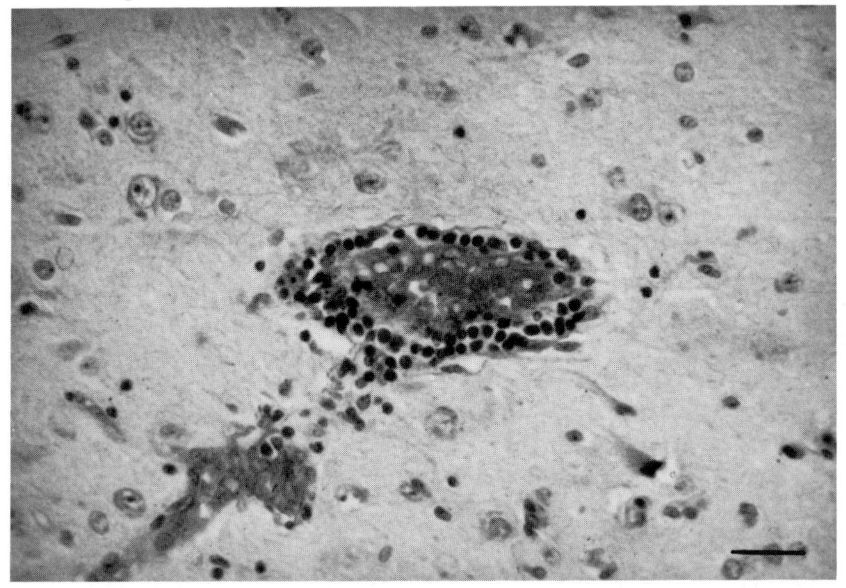

Fig. 3. Mild inflammation in the frontal cortex of a child with chronic epilepsy. Cuffing of a blood vessel with lymphocytes; hematoxylin and eosin stain;___,approximately 70 µm. Courtesy of T. Rasmussen and S. Carpenter, Montreal Neurological Institute.

V. EXPERIMENTAL MODELS OF PERSISTENT INFECTION WITH TICK-BORNE ENCEPHALITIS VIRUS

A. Persistence of Flaviviruses in Cell Cultures

TBE virus, like others in the flavivirus group, easily establishes persistent infections in a variety of mammalian cell cultures (Andzhaparidze and Bogomolova, 1961; Mayer, 1962; Deryabin *et al.*, 1977); one such line has been persistently infected for 17 years (Andzhaparidze *et al.*, 1977). Langat and Kyasanur Forest disease viruses (Illavia and Webb, 1969) and Powassan virus (Nemo, G., unpublished data) behave similarly.

B. Persistence of Flaviviruses in Animals

Many flaviviruses also persist for long periods in vertebrate and invertibrate animals (Reeves, 1961; Casals, 1973). Modoc virus (Johnson, 1970; Davis and Hardy, 1974), St. Louis encephalitis virus (Slavin, 1943), Kyasanur Forest disease virus (Price, 1966; Goverdhan and Anderson, 1972), Langat virus (Stárek *et al.*, 1977; Larina *et al.*, 1977), and some strains of TBE virus (Borsuk *et al.*, 1975; Vorobieva *et al.*, 1975) have all been reported to cause persistent infections of rodents. TBE virus also appears to persist in bats (Nosek *et al.*, 1961) and birds (Ernek *et al.*, 1969; Brummer-Korvenkontio *et al.*, 1970). There is circumstantial evidence that louping-ill virus may cause natural persistent infection of sheep with late onset of encephalitis (Brotherston, 1965). However, the animal model most resembling chronic disease in man is probably that developed by Ilyenko and co-workers (1974) in Leningrad.

VI. MOVEMENT DISORDERS AFTER EXPERIMENTAL TICK-BORNE ENCEPHALITIS IN MONKEYS

As part of a search for strains of TBE virus avirulent for man, Ilyenko, Smorodintsev, Platonov, and others looked for a test to predict virulence (Smorodintsev *et al.*, 1969). One marker that seemed promising was the ability of a strain to elicit encephalitis in rhesus monkeys after intracerebral inoculation; strains from patients with inapparent or mild illness usually caused little or no clinical encephalitis, while strains isolated from fatal human cases usually killed monkeys. (Strains of TBE virus are not virulent for rhesus monkeys except by direct inoculation into the central nervous system

(Morris *et al.*, 1955). Many strains of TBE virus isolated in various regions of Europe and Asia were inoculated intracerebrally into rhesus monkeys; a few monkeys with mild or inapparent encephalitis later developed peculiar athetoid movement disorders (Ilyenko *et al.*, 1974). Of 58 strains isolated in Central Europe and various parts of the USSR west of the Urals before 1968, 53 were always lethal for monkeys and only five caused mild encephalitis, two of which left some survivors with athetosis. East of the Urals more strains of low virulence were isolated, 17 of 44; however, all strains from the Far East and Khabarovsk, where severe encephalitis was common, were lethal for monkeys. Three less virulent Asian strains left some surviving monkeys with athetosis. Two of the three strains, both from Siberia, are of special interest: one was Kraminskaia *et al.'s* isolate (1969) from the CSF of a girl with Kozhevnikov's epilepsy; the other was isolated from the blood of a construction worker with a nonparalytic febrile illness. This second strain, the Vasilchenko strain, was used in the experiments to be reviewed.

Figure 4 shows the kind of clinical illness often seen in a rhesus monkey inoculated intracerebrally with 10^8 mouse LD_{50} of the Vasilchenko strain of TBE virus. The morning temperature usually rose above 40°C on about day 6, peaked on day 9 or 10, and fell to normal by the end of 2 weeks. If there were to be other signs of encephalitis, they appeared shortly after the

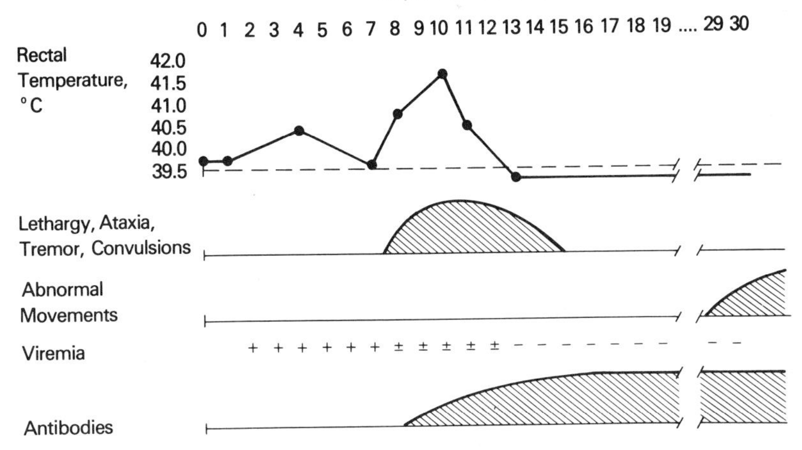

Fig. 4. Reprinted with permission from D. M. Asher (1975), Adv. Neurol. 10, *277-289 (Raven Press, New York).*

fever, and included lethargy, weakness, ataxia with intention tremors, and sometimes convulsions. These findings grew worse for several days and lasted into the third week of illness. Virus was usually isolated from blood until the sixth day, less frequently thereafter, and rarely after the twelfth day. Antibodies were detected in sera of some monkeys on day 6 and were almost always present by day 12. If death occurred, it was usually at the end of the second week postinoculation, 6 to 10 days after onset of illness. About a month after inoculation some of the survivors began to show repetitive movements of the extremities, most frequently rotatory movements of the forearm with flexion and extension of the fingers, less often movements of the legs, feet, face, and head. These movements resembled athetosis. If the movements had not appeared in 3 months after inoculation they did not appear at all.

Of 55 monkeys inoculated with the Vasilchenko strain in Leningrad before 1970 (Table I) 75% had evidence of acute encephalitis, and 37% of these (27% of all monkeys) either died or had to be destroyed. Eighty-five percent of survivors of encephalitis and almost 40% of animals that did not show encephalitis later developed athetosis. Of 19 monkeys inoculated in Bethesda the percentage with clinical signs of encephalitis (84%) was similar to that in Leningrad. But there was a marked difference in outcome; none died, only one appeared sick enough to sacrifice, and only a small percentage developed movement disorder. In animals infected in Bethesda, athetosis did not last more than a few months. In contrast, one of a larger num-

TABLE I
Clinical Illness in Rhesus Monkeys Inoculated Intracerebrally with TBE/Vasilchenko Virus[a]

Illness	*Leningrad*		*Bethesda*	
	No.	*%*	*No.*	*%*
Acute encephalitis	41/55	75	16/19	84
Deaths and sacrifices	15/55	27	1/19	5
Movement disorders				
Per total	27/55	49	5/19	26
Per survivors	27/41	67	5/18	28

[a]*Reprinted with permission from D. M. Asher (1975),* Adv. Neurol. 10, *277-289 (Raven Press, New York).*

ber of animals stricken in Leningrad apparently had persistent abnormal movements of both hands for more than two years.

In animals with movement disorder, the most striking histological abnormalities were in the cerebellum. Figure 4 shows the cerebellum of a typical rhesus monkey that had athetotic movements of the left forearm, 248 days after inoculation. There were patchy areas of thinned molecular layer, complete disappearance of Purkinje cells, and marked reduction of granule cells (Fig. 5). There were also nodules of inflammatory cells in the white matter (Fig. 6) and areas of meningeal infiltration (Fig. 7). Komandenko *et al.*, (1972) described similar mild inflammatory changes throughout the cerebral cortex, basal ganglia, thalamus, pons, medulla, cord, and nerves of other affected monkeys.

Of 11 monkeys inoculated with the Vasilchenko strain of tick-borne encephalitis in 1969, virus was recovered from the brains of seven (Asher, 1975); one was an animal with no neurological abnormalities 71 days after inoculation. After that

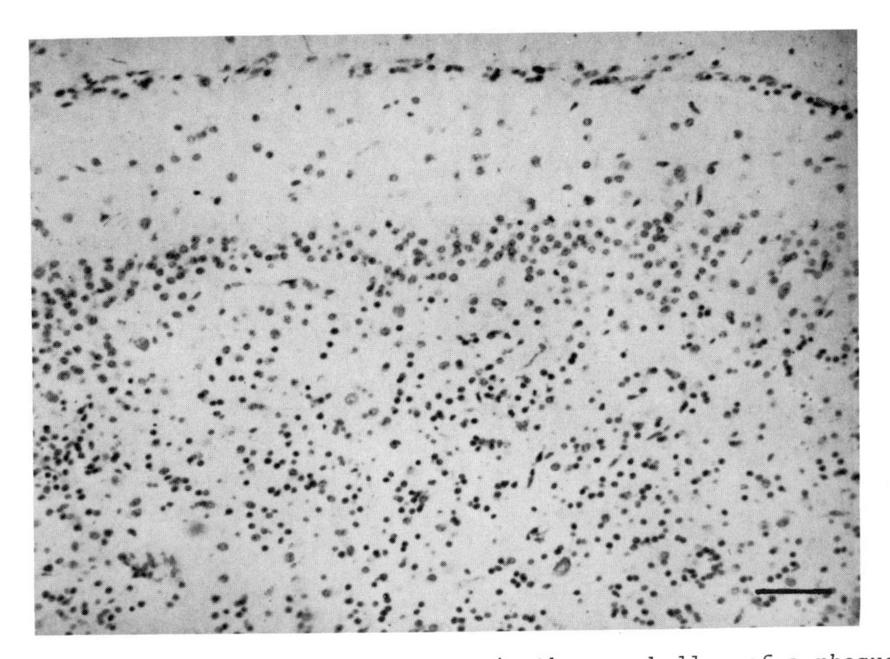

Fig. 5. Degenerative changes in the cerebellum of a rhesus monkey 248 days after intracerebral inoculation with TBE virus. Thin molecular layer, absent Purkinje cells, and severe reduction in the number of granule cells; Nissl stain; ___, approximately 100 μm. Courtesy of V. I. Ilyenko, V. G. Platonov, and O. A. Tsvetukhina, All-Union Research Institute of Influenza, Leningrad.

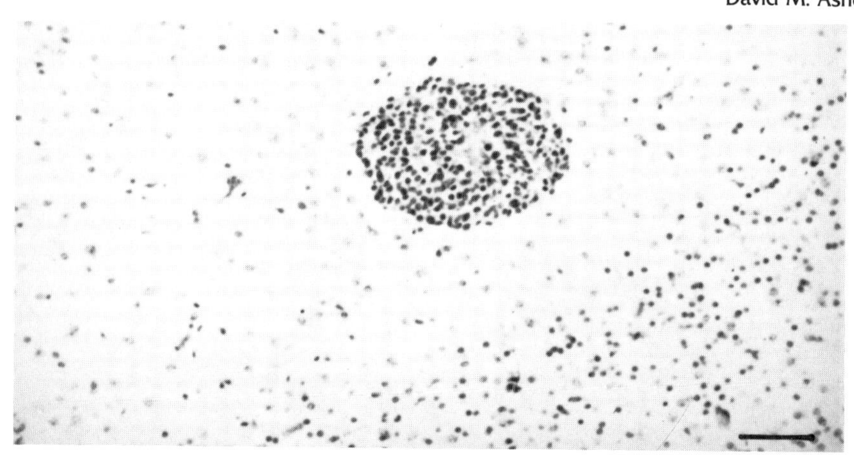

Fig. 6. Inflammation in the cerebellum of a rhesus monkey 248 days after intracerebral inoculation with TBE virus. Nodule of inflammatory cells; Nissl stain;——,approximately 100 μm. Courtesy of V. I. Ilyenko, V. G. Platonov, and O. A. Tsvetukhina, All-Union Research Institute of Influenza, Leningrad.

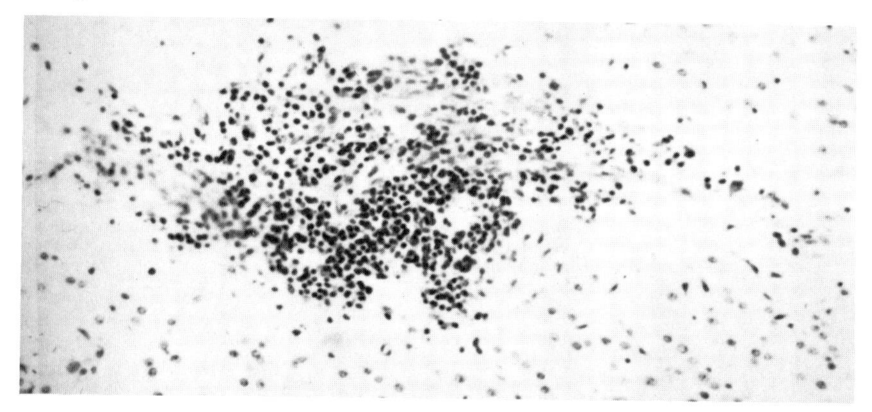

Fig. 7. Inflammation in the cerebellum of a rhesus monkey 248 days after intracerebral inoculation with TBE virus. Meningeal infiltrate of inflammatory cells; Nissl stain;——, approximately 100 μm. Courtesy of V. I. Ilyenko, V. G. Platonov, and O. A. Tsvetukhina, All-Union Research Institute of Influenza, Leningrad.

virus was isolated only from animals with athetosis. Up to 199 days postinoculation when the last animal was sacrificed, the brain of every monkey with athetosis yielded virus. No virus was recovered from spinal fluid or spleen. If the CNS yielded virus it did so from almost all areas tested--cerebral cortex, basal ganglia, thalamus, pons, medulla, and spinal cord. (Panov *et al.*, 1975, reported a more limited distribution of virus.) Indirect fluorescent antibody-staining showed antigen in the brain of one animal killed 78 days after inoculation; Four brains of subsequent animals were negative (O. A. Tsvetukhina *et al.*, unpublished data). In Bethesda, eight animals were sacrificed between 358 and 806 days after inoculation; virus was isolated from none of them, however all were tested only after abnormal movements had disappeared. No differences were found in titers of serum antibodies to TBE virus in animals with or without athetosis (Asher, 1975).

VII. SUMMARY

In humans, infections with TBE virus may be followed by chronic progressive movement disorders or paralytic diseases. These syndromes sometimes appear months or even years after the acute infection. There is suggestive evidence, not yet completely convincing, that such syndromes are associated with persistence of TBE virus in the nervous system. It seems likely that other viruses might also produce such syndromes. Experimental intracerebral inoculation with some strains of TBE virus elicits in monkeys a movement disorder that is delayed in onset and of long duration. This animal model merits more study.

References

1. Aguilar, M. J. (1959). *Am. J. Med. Sci. 238,* 354-362.
2. Aguilar, M. J., and Rasmussen, T. (1960). *Arch. Neurol. (Chicago) 2,* 663-676.
3. Andzhaparidze, O. G., and Bogomolova, N. N. (1961). *Vopr. Virusol. 3,* 343-347.
4. Andzhaparidze, O. G., Rozina, E. E., Bogomolova, N. N., and Boriksin, Yu. S. (1977). *Abstr. Int. Symp. Chronic Virus Infections, Smolenice, Czechoslovakia.*
5. Asher, D. M. (1971). *Proc. 13th Int. Congr. Pediatrics, III, 2,* 379-384.
6. Asher, D. M. (1975). *Adv. Neurol. 10,* 277-289.
7. Baishtruk, M. N. (1975). *Vopr. Med. Virusol.,* 381-382.
8. Borsuk, E. A., Levina, M. N., and Vorobieva, M. S. (1975). *Vopr. Med. Virusol.,* 270-271.
9. Brody, J. A. (1965). *In* "Slow, Latent, and Temperate Virus Infections" (D. C. Gajdusek, C. J. Gibbs, Jr., and M. Alpers, eds.), pp. 111-113. NINDB Monograph No. 2, USGPO, Washington, D. C.
10. Brotherston, J. G. (1965). *In* "Slow, Latent, and Temperate Virus Infections" (D. C. Gajdusek, C. J. Gibbs, Jr., and M. Alpers, eds.), p. 113. NINDB Monograph No. 2, USGPO, Washington, D. C.

11. Brummer-Korvenkontio, M., Saikku, P., and Korhonen, P. (1970). *Scand. J. Clin. Lab. Invest. 25 (Suppl. 113)*, 105.
12. Bruns, (1894). *Neurol. Centralblatt 13*, 388-392.
13. Bruyn, H. B., and Lennette, E. H. (1953). *Calif. Med. 79*, 362-366.
14. Casals, J. (1963). *Abstr. 9th Int. Congr. Trop. Med. Malaria, Athens, Greece*, p. 218.
15. Chumakov, M. P. (1939). *Zh. Mikrobiol. Epidemiol. Immunobiol.* (4), 5-14.
16. Chumakov, M. P., Vorobieva, N. N., and Bieliaieva, A. L. (1944). *Nevropatol. Psikhiat. 13*, 63-68.
17. Davis, J. W., and Hardy, J. L. (1974). *Infect. Immun. 10*, 328-334.
18. Deryabin, P. G., Loginova, N. V., Cherendnichenko, Yu. N., and Gavrilov, V. I. (1977). *Symp. Chronic Virus Infections, Smolenice, Czechoslovakia.*
19. Ernek, E., Kožuch, O., and Nosek, J. (1969). *Acta Virol. (Engl. Ed.) 13*, 303-308.
20. Friedman, H., Chien, L., and Parham, D. (1977). *Lancet 2*, 666.
21. Galant, I. B. (1946). *Pediatriya (Moscow) 6*, 36-40.
22. Galant, I. B. (1951). *Nevropatol. Psikhiat. 20*, 31-32.
23. Golman, S. V. (1941). *Nevropatol. Psikhiat. 4*, 36-47.
24. Goverdhan, M. D., and Anderson, C. R. (1972). *Indian J. Med. Res. 60*, 1002-1006.
25. Gulyaeva, S. E. (1975). *Vopr. Med. Virusol.* 386-387.
26. Gupta, P. C., Roy, S., and Tandon, P. N. (1974). *J. Neurol. Sci. 22*, 105-120.
27. Harada, K. (1970). *Psychiatr. Neurol. Japan 72*, 857-863.
28. Herzon, H., Shelton, J. T., and Bruyn, H. B. (1957). *Neurology 7*, 535-548.
29. Horta-Barbosa, L., Fuccillo, D. A., Sever, J. L., and Zeman, W. (1969). *Nature (London) 221*, 974-975.
30. Illavia, S. J., and Webb, H. E. (1969). *Br. Med. J. 1*, 94-95.
31. Ilyenko, V. I. (1961). Doctoral Dissertation, Institute of Experimental Medicine, Leningrad.
32. Ilyenko, V. I., Platonov, V. G., and Smorodintsev, A. A. (1974). *Vopr. Virusol. 4*, 414-418.
33. Ilyenko, V. I., Platanov, V. G., and Komandenko, N. I. (1975a). *Vopr. Med. Virusol.* 296-297.
34. Ilyenko, V. I., Smorodintsev, A. A., Panov, A. G., Platonov, V. G., Osetrov, B. A., and Prozorova, I. N. (1975b). *Vopr. Med. Virusol.* 295-296.
35. Johnson, H. D. (1970). *Am. J. Trop. Med. Hyg. 19*, 537-539.
36. Kanter, V. M. (1940). *Nevropatol. Psikhiat. 9*, 28-30.
37. Kanter, V. M. (1961). *Zh. Nevropatol. Psikhiatr. im S.S. Korsakova 1*, 48-51.

38. Kimoto, T., Yamada, T., Ueba, N., Kunita, N., Kanai, A., Yamagami, S., Nakajima, K., Akao, M., and Sugiyama, S. (1968). *Biken, J. 11,* 157-168.
39. Klyuchikov, V. N. (1961). *Klin. Med. (Moscow) 39,*59-62.
40. Klyuchikov, V. N. (1965a). Abstract of Doctoral Dissertation, 2nd Moscow State Medical Institute.
41. Klyuchikov, V. N. (1965b). *In* "Tick-borne Encephalitis," pp. 346-358. Belarus, Minsk, USSR.
42. Komandenko, N. I., Ilyenko, V. I., Platonov, V. G., and Panov, A. G. (1972). *Zh. Nevropatol. Psikhiatr im S.S. Korsakova 72,* 1000-1007.
43. Kono, R., Akao, Y., Matsunaga, Y., Ogawa, M., and Okubo, A. (1972). Presented at the Virology Section of the Japan-US Medical Program, Sapporo.
44. Kozhevnikov, A. (1894). *Med. Oboz. 42,* 97-118.
45. Kraminskaia, N. N., Meierova, R. A., and Zhivoliapina, R.R. (1969). *Dokl. Irkuts. Protivo-chumno. Inst. 8,* 189-192.
46. Kuimov, D. T., and Dubrov, A. V. (1958). *Zh. Nevropatol. Psikhiatr. im S.S. Korsakova 58,* 282-287.
47. Larina, G. I., Karpovic, L. G., Levkovic, E. N., and Ugbayeva, T. D. (1977). *Abst. Int. Symp. Chronic Virus Infections, Smolenice, Czechoslovakia.*
48. Levkovich, E. N., Shubladze, A. K., Tchumakoff, M. P., and Sololovieff, D. (1938). *Arkh. Biol. Nauk 52,* 162-183.
49. Magazanik, S. S., and Robinzon, I. A. (1966). *Vopr. Psikhiatr. Nevropatol. 12,* 38-47.
50. Magazanik, S. S., and Shakinko, L. A. (1975). *Vopr. Med. Virusol.,* 388-390.
51. Mayer, V. (1962). *Acta Virol. (Engl. Ed.) 6,* 317-326.
52. Morris, J. A., O'Connor, J. R., and Smadel, J. E. (1955). *Am. J. Hyg. 62,* 327-341.
53. Noran, H. H. (1944). *Am. J. Pathol. 20,* 259-267.
54. Noran, H. H., and Baker, A. B. (1943). *Arch. Neurol. Psychiatr. 49,* 398-413.
55. Noran, H. H., and Baker, A. B. (1945). *J. Neuropathol. Exp. Neurol. 4,* 269-276.
56. Norrby, E., Link, H., and Olsson, J.-E. (1974). *Arch. Neurol. (Chicago) 30,* 285-292.
57. Nosek, J., Gresiková, M., and Rehácek, J. (1961). *Acta Virol. (Engl. Ed.) 5,* 112-116.
58. Ogawa, M., Okubo, H., Tsuji, Y., Yasui, N., and Someda, K. (1973). *J. Neurol. Sci. 19,* 363-373.
59. Omorokov, L. I. (1927a). *Zh. Nevropatol. Psikhiatr. 20,* 13-24.
60. Omorokov, L. I. (1927b). *Z. Gesamte. Neurol. Psychiatr. 107,* 487-496.
61. Omorokov, L. I. (1938). *Nevropatol. Psikhiat. 10,* 26-37.
62. Omorokov, L. I. (1951). *Nevropatol. Psikhiat. 20,* 23-28.

63. Panov, A. G., Ilyenko, V., and Komandenko, N. I. (1975). *Vopr. Med. Virusol.*, 393-394.
64. Payne, F. E., Baublis, J. V., and Itabashi, H. H. (1969). *N. Engl. J. Med. 281*, 585-589.
65. Pervushin, V. P. (1899). Trudi VII Pirogovskogo S'ezda Vrachei, 647-653. Cited by Pervushin (1940) and Ilyenko (1961).
66. Pervushin, V. P. (1940). *Sov. Med.* (3), 3-7.
67. Pervushin, V. P. (1943). *Klin. Med. (Moscow) 21*, 17-25.
68. Platou, R. V. (1940). *Am. J. Dis. Child. 60*, 1155-1169.
69. Price, W. H. (1966). *Virology 29*, 679-681.
70. Protas, I. I., and Votyakov, V. I. (1970). *Zh. Nevropatol. Psikhiatr. im S.S. Korsakova 70*, 1124-1129.
71. Rasmussen, T., and McCann, W. (1968). *Trans. Am. Neurol. Assoc. 93*, 89-94.
72. Rasmussen, T., Olszewski, J., and Lloyd-Smith, D. (1958). *Neurology 8*, 435-445.
73. Reeves, W. C. (1961). *Progr. Med. Virol. 3*, 59-78.
74. Shapoval, A. N. (1945). *Nevropatol. Psikhiatr. 14*, 59-61.
75. Shapoval, A. N. (1975). *Vopr. Med. Virusol.*, 402-403.
76. Shapoval, A. N., and Garishina, M. F. (1975). *Vopr. Med. Virusol.*, 405-406.
77. Shefer, D. G., and Polikovsky, M. G. (1940). *Sov. Psikhonevrol.* 5-6, 29-38.
78. Shubin, N. V., Karpov, S. P., Terentiev, V. F., Dremov, D. P. and Erofeev, V. S. (1975). *Vopr. Med. Virusol.*, 407-408.
79. Slavin, H. B. (1943). *J. Bacteriol. 46*, 113-116.
80. Smorodintsev, A. A., Dubov, A. V., Ilyenko, V. I., and Platonov, V. G. (1969). *J. Hyg. 67*, 13-20.
81. Stárek, M., Kubistová, K., and Jirásek, A. (1977). *Abstr. Int. Symp. Chronic Virus Infections, Smolenice, Czechoslovakia.*
82. Strokina, T. I., Gurari, R. M., and Tatarinova, L. G. (1969). *Trudi Vladivostok. Inst. Epidemiol. Mikrobiol. 4*, 73-79.
83. Thomas, J. E., Reagan, T. J., and Klass, D. W. (1977). *Arch. Neurol. (Chicago) 34*, 266-275.
84. Townsend, J. J., Baringer, J. R., Wolinsky, J. S., Malamud, N., Mednick, J. P., Panitch, H. S., Scott, R.A.T., Oshiro, L. S., and Cremer, N. E. (1975). *N. Engl. J. Med. 292*, 990-993.
85. Umansky, K. G. (1975a). *Vopr. Med. Virusol.*, 399-400.
86. Umansky, K. G. (1975b). *Vopr. Med. Virusol.*, 400-401.
87. Umansky, K. G., Shefer, D. G., Magazanik, S. S., and Shapoval, A. N. (1975). *Vopr. Med. Virusol.*, 401-402.
88. Vandvik, E., and Norrby, E. (1973). *Proc. Nat. Acad. Sci. U.S.A. 70*, 1060-1063.

89. Vaneeva, G. G. (1969). *Pediatriya (Moscow) 48*, 46-48.
90. Vorobieva, M. S., Gavrilov, V. I., Dzagurov, S. G., Levenbuk, I. S., Robinzon, I. A., Ladyzhenskaya, I. P., Borsuk, E. A., and Chigirinsky, A. E. (1975). *Vopr. Med. Virusol.*, 58-59.
91. Weil, M. L., Itabashi, H. H., Cremer, N. E., Oshiro, L. S., Lennette, E. H., and Carnay, L. (1975). *N. Engl. J. Med. 292*, 994-998.
92. Zilber, L. A., Levkovich, E. N., Shubladze, A. K., Chumakov, M. P., Solviev, V. D., *et al.* (1938). Cited by Levkovich, E. N., Pogodina, V. V., Zasukhina, G. D., and Karpovich, L. G. "Viruses of the Tick-Borne Encephalitis Complex." Meditsina, Leningrad, 1967.

Chapter 13

NEPHROPATHIA EPIDEMICA IN FINLAND

N. Oker-Blom, C.-H. von Bonsdorff, M. Brummer-Korvenkontio,
T. Hovi, K. Penttinen, P. Saikku,
A. Vaheri, and J. Lähdevirta

I. INTRODUCTION

The clinical entity presently known as nephropathia epidemica (NE) was described in Sweden in 1934 by Zetterholm (1934) and Myhrman (1934) as a new disease in Scandinavia. Since then cases of the disease have also been reported from Finland, Norway, and Denmark (Lähdevirta, 1971; Nyström, 1977). The first documented cases in Finland occurred among the Finnish and German troops in North Finland during World War II (Stuhlfaut, 1943; Hortling, 1946) and a few sporadic cases were described since (Kuhlbäck *et al.*, 1964). In 1962 Lähdevirta started an intensive survey on NE. The first of the series (Lähdevirta, 1971) comprised a detailed study of 380 cases, followed by that of 444 additional cases (Lähdevirta and Elo, 1975). On the basis of these data, the current picture of the clinical features of the disease as well as that of its epidemiology have been formed. From the early 1960s, attempts to isolate and identify the causative agent have been

ISBN 0-12-429765-X

made at the Department of Virology, University of Helsinki, and the progress made in this field is given in this report. It also includes the description of an unusual family outbreak of NE and a comparison of the features of the Scandinavian form of NE to the more severe similar disease known as hemorrhagic fever with renal syndrome (HFRS), prevalent in the USSR and Eastern Asia.

II. CLINICAL FEATURES

NE is an acute systemic disease of, most probably, infectious origin with an "incubation time" of 3-6 weeks. The disease begins with an acute onset of high fever. Other common symptoms and signs during the first week are headache, backache, abdominal pain, nausea, vomiting, somnolence, and facial flush. Clinical and laboratory examinations show signs of a multisystemic infection: evidence for carditis, hepatitis, and meningoencephalitis has been obtained. There is also a tendency of low platelet counts, hemorrhagic manifestations, and slight immunological disturbances. The clinical picture is, however, dominated by the renal affection, which begins on the third to fifth day after the onset of fever. The first symptoms include tenderness over the kidneys, proteinuria, and oliguria. In most cases microscopical hematuria, leukocyturia, and azotemia develop. The oliguria phase is followed by polyuria and hypostenuria, and subsequent gradual recovery of renal functions.

Histopathological examination of kidney biopsies has revealed an acute, tubular, and interstitial hemorrhagic nephritis with the interstitial extravasation being the most prominent finding. A mild glomerulitis is also often seen. The typical course of the disease allows confirmation of the diagnosis by clinical findings alone. It should be noted, however, that the present clinical picture is based on the hospitalized cases, and possible "subclinical infections" cannot be excluded. As a whole, the disease has many features in common with the hemorrhagic fever with renal syndrome (HFRS) (Table I). The main difference is in the severity of the disease. Compared to HFRS, the hemorrhagic symptoms in NE are much less pronounced and the mortality is markedly lower.

TABLE I
Comparison between the Clinical Pattern of NE and of Hemorrhagic Fever with Renal Syndrome[a]

	Nephropathia epidemica (%)	*Hemorrhagic fever with renal syndrome (%)*
General symptoms		
Fever	100	100
Nausea	78	82
Headache	90	86
Backache	82	78
Enlarged lymph nodes	15	38
Petechial rash	12	32
Pharyngeal infection	67	55
Enlargement of liver	9	0
Enlargement of spleen	0	7
Nephrological findings		
Proteinuria	100	96
Hypostenuria	100	88
Polyuria	97	92
Azotemia	86	94
Hematuria		
Gross	3	85
Microscopic (5 RBC per field)	74	
Anuria	8	4
Oliguria	54	63
Mortality	0.2-0.5	5-10

[a]*According to Lähdevirta (1971).*

III. EPIDEMIOLOGY

A. Occurrence of NE in Finland

The disease has a pronounced seasonal incidence, with about 80% of the cases occurring between September and March. There seems to be a correlation between the cooling of the temperature and the appearance of new cases of the disease after the postulated incubation time. However, this correlation holds only until a stable snow cover is established (Fig. 1).

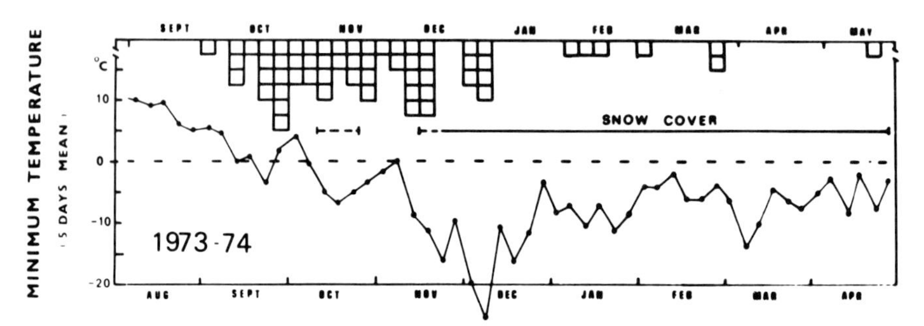

Fig. 1. The seasonal distribution of NE in Finland as exemplified by the epidemic of 1973-1974. Each square in the diagram corresponds to one patient and the location of the square corresponds to the day of the onset of the disease. Observe that the upper time scale has been shifted by 25 days (approximate incubation period). Note that most of the cases are concentrated to the period when the temperature falls and before the permanent snow cover is formed. Each year there are also a few cases during summer.

The disease has a pronounced regional occurrence (Fig. 2). The highest morbidity due to the disease is in the lake district in the central-eastern part of the country and in areas with abundant forests. About 80% of the cases occur among the rural population, affecting males of 15-45 years of age (Fig. 3).

B. The Role of the Voles

The incidence of NE shows a periodicity of 3 to 4 years. This periodicity coincides with the fluctuation of the population densities of voles (Fig. 4). Several investigators have already suggested that small rodents are the natural reservoir of NE, as well as of HFRS (Casals *et al.*, 1966; Lähdevirta, 1971; Trenscéni and Keleti, 1971; Nyström, 1977). It may be pointed out that during the year 1942, when the largest known outbreak of NE in Finland occurred, the voles were exceptionally abundant (Kalela, 1949). Since NE occurs anywhere in Finland and since the *Apodemus* species are encountered only in the southern parts of the country, these mice might be excluded as vectors. There are three vole species that occur throughout Finland: the bank vole (*Cletrionomys glareolus)*, the field vole (*Microtus agrestris)* , and the water vole (*Arvicola terrestris)*. Of these, the bank vole seems to be the best candidate. Both NE and the bank voles avoid open fields and occur mostly in regions of forests and bush vegetation. The latter has considerably increased in the epidemic area due to

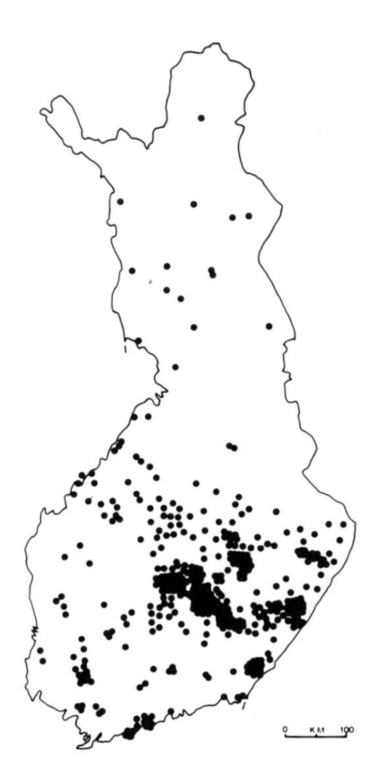

Fig. 2. The geographical distribution of 413 NE cases registered between 1966 and 1973. The spots correspond to the residence of each patient, which is not necessarily the site of contraction of the disease. Note that the southwest of Finland, which has the maximal population density, has far less cases than the lake district area (according to Lähdevirta and Elo, 1975).

the withdrawal of agricultural lands from production. This may explain the apparent increase of NE cases during the last years. In the fall, when the temperature starts to drop, bank voles often move from the forests to barns. The sporadic cases of NE observed during the summer are for the most part restricted to urban dwellers who spend their vacation in summer cottages in the country. During years of abundant occurrence of bank voles their average age increases. Under conditions of high population density, the bank vole, in contrast to other voles, shows increased climbing activity. These factors may favor the spreading of the infectious agent.

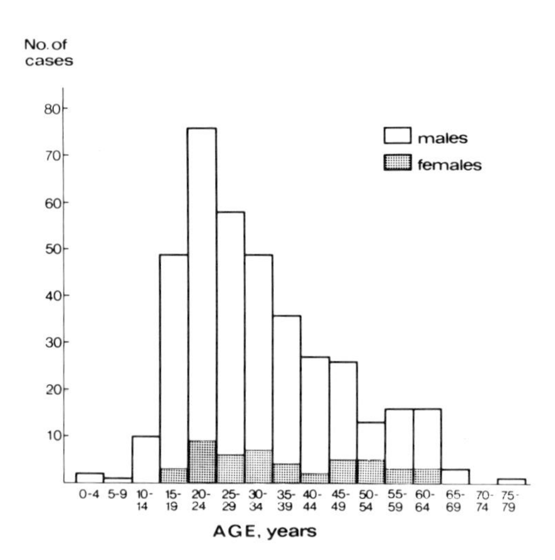

Fig. 3. The age and sex distribution of 383 NE patients registered between May 1966 and April 1973 (according to Lähdevirta and Elo, 1975).

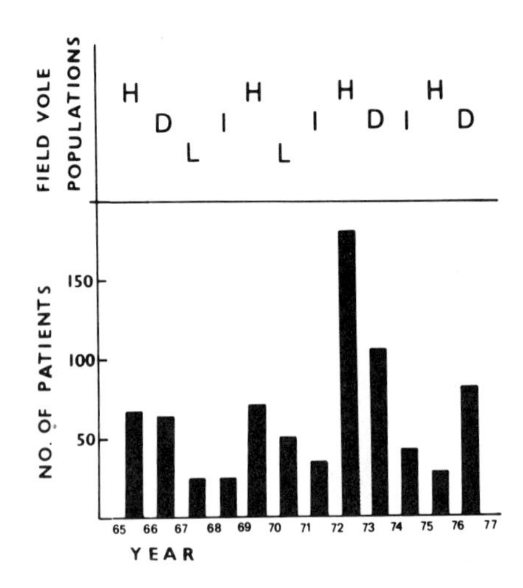

Fig. 4. Annual occurrence of NE cases in Finland as compared to the population density of voles. The field vole has been used as an indicator species but the determinations are also true for the bank vole (Myllymäki et al., 1978). The NE incidence appears to be high when the vole populations show a high density or a year after that. H, high; D, decreasing; I, increasing; L, low density of vole populations.

More direct evidence pointing to the role of the bank vole as transmitter of NE was obtained in the winter, 1976-1977. At the Department of Virology, University of Helsinki, three persons participated in the dissection of bank voles from the epidemic area. Two of them contracted NE 29 and 30 days later. Similar evidence was obtained from an unusual family outbreak of the disease (Fig. 5). In a six-membered family living in Helsinki, the youngest son kept a terrarium with bank voles collected from the epidemic area. About 2 months later five members of the family contracted NE (Fig. 5). Four of these were hospitalized and the diagnosis confirmed. The 13-year-old vole collector had a febrile disease with abdominal pain lasting for 5 days but the NE diagnosis remained unconfirmed.

These findings rather strongly support the idea that at least the bank vole *(C. glareolus)* functions as a reservoir of NE in Finland.

IV. ISOLATION ATTEMPTS

A. Background

Since the early 1960s numerous attempts to isolate the causative agent of NE have failed. The efforts for isolating the agent were resumed during the epidemic season 1976-1977 and concentrated in two different directions.

Previous isolation attempts and epidemiological observations indicated that patient material(s) obtained after the onset of the disease probably had very low, if any, infectivity.

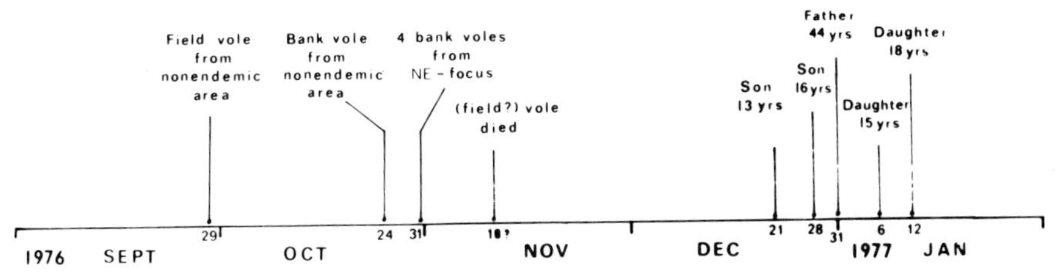

Fig. 5. Chronological presentation of an unusual family outbreak of NE. All the patients of the six-membered family were hospitalized with the exception of the 13-year old son for whom the diagnosis remained tentative.

This conclusion was mainly based on the fact that there were no known cases of disease contracted by hospital or laboratory personnel handling patients and/or samples. We therefore searched for NE cases as early as possible during the disease. Fortunately, some samples from the above presented case of family outbreak were obtained, even before the onset of the disease. On the other hand, the observation made in the USSR that severe laboratory infections of NE were caused by rodents brought from the epidemic areas into the laboratory (Kulagin *et al.*, cited by Trenscéni and Keleti, 1971) suggested that this material could be highly infectious. Consequently, small rodents, mainly voles, were collected from the farms where recent cases of the disease had occurred. This material included the pet voles of the above-mentioned family.

B. Materials and Methods

The samples collected for the isolation are presented in Table II. Patient material was obtained in one case starting 1 week before the onset of the disease and in another case from the day of the onset of the disease. Both crude, unpurified buffy coat cells and cultured lymphocytes were used as isolation material. The urine samples consisted of both the sediment and a concentrate prepared by ultrafiltration followed by ultracentrifugation. Renal biopsies were obtained more than 1 week after the onset of the disease.

TABLE II
Material Used for Isolation Purposes

Samples from patients	*Samples from voles*
Serum and plasma	Collected urine
Blood cells	Organ samples
"buffy coat"	kidney
lymphocyte culture	lung
	spleen
Urine	heart
sediment	salivary glands
concentrate	brain
Renal biopsy	Pooled organ cultures
Liquor	
Feces	
Throat swab	

From the rodents, the urine was collected by a special device that allowed the harvest of larger volumes (1-3 ml). The organs pooled for cultures were kidney, lung, heart, salivary gland (including the submandibular lymph nodes), and spleen.

In addition to the isolation attempts, the direct demonstration of the presence of an agent and/or the antigen in the primary samples by electron microscopy or immunofluorescence was also attempted. The latter was performed either as a direct method using fluorescein-conjugated IgG-fraction of pooled convalescent sera from five NE-patients (3 to 8 weeks post-infection) or as an indirect method with several individual convalescent sera and labeled rabbit anti human γ-globulin.

As experimental animals, newborn and weanling mice, guinea pigs, and three species of voles (*C. glareolus, C. rufocanus* and *M. agrestis*) were used. Mice were inoculated intracerebrally and intraperitoneally, the other animals only intraperitoneally. The cells used for the isolation attempts are listed in Table III.

TABLE III
Cells Used in Isolation Attempts[a]

Main types	*Additional types*
Human amnion	Human glia
Human embryonic skin	Field vole kidney
Green monkey kidney	Bank vole kidney
Mouse macrophage	McCoy, X-ray irradiated (mouse)
Mouse embryo, total	Lymphoid line (bovine)
Mouse embryo viscera	Lymphoid line (porcine)
Chick embryo	
HeLa (human)	Types used in passages
Vero (green monkey)	Human embryonic kidney
Kidney line DK (dog)	PK 15 (porcine)
Salivary gland SVG line (dog)	HAK (hamster)
RK 13 (rabbit)	Pt K2 (*Potorous tridactylis)*
BHK21/WI-2 (hamster)	Singh's *Aedes albopictus* (mosquito)

[a]*Origin of established lines is given in parentheses. The other cell types represent adherent primary or early passage cell cultures from the tissues.*

All cells not being available at any given time, each sample was initially inoculated, before freeaing, to approximately ten different cell types. The cells were primarily examined for a possible development of CPE. The negative cultures were used for three to four blind passages, after which they were stored at -70°C. For immunofluorescence the cell samples used for passaging were fixed with cold acetone, air-dried, and stored at -20°C until the assay. Electron microscopy was performed on the supernatant and/or of cell samples lysed in distilled water whenever signs of CPE were present.

Additional methods used for identifying a possible agent were (1) interference using Semliki Forest virus, (2) labeling the cultures with radioactive adenine followed by sucrose gradient centrifugation of the supernatants, (3) immune adherence using human convalescent serum and red cells, (4) immunodiffusion with the lysed cells or antigen against human convalescent sera, and (5) immunoelectron microscopy, searching for possible material coated or aggregated by convalescent sera. These methods were used mainly in cases where an established CPE had been encountered.

C. Results

In two instances the possible presence of an NE antigen in the initial samples has been detected by immunofluorescence. A pooled organ culture of a bank vole collected from the epidemic area showed a weak but clear cytoplasmic fluorescence (by the indirect method) in a small percentage of the cells. The possibility of nonspecific staining cannot be totally excluded since cytoplasmic fluorescence of kidney cell cultures can often be observed, suggesting the presence of antitissue antibodies in the patient sera. It must be stressed, however, that the two laboratory infections mentioned above were most likely contracted during the preparation of these bank vole organ cultures. The other positive FA-finding is presented in Fig. 6. In a buffy coat sample taken one week before the onset of the disease, fixed with acetone and stained by the direct method, about 10% of all white blood cells showed clear granular cytoplasmic fluorescence (Fig. 6). At the day of the onset of the disease, about 1% of the cells were positive in the two observed cases. No samples taken after the onset of the disease were positive. It must be pointed out that nonspecific attachment of immunoglobulins to some fraction of the white blood cells, e.g., monocytes, cannot be totally excluded. Nevertheless, the following results lend further support for the viral nature of the causative agent of NE. Mononuclear leukocytes, separated from one of the FA-positive buffy coats and cultured for 5 days *in vitro*, were inoculated into newly prepared cultures of mouse

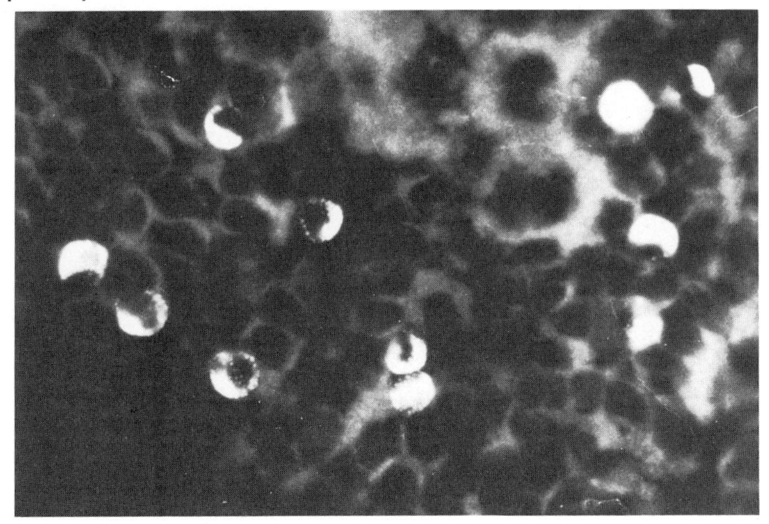

Fig. 6. Direct immunofluorescence of buffy coat cells of the 18-year old patient of the family epidemic (see Fig. 5). The sample was taken 1 week before the onset of the disease. Note the granular cytoplasmic fluorescence in several cells.

peritoneal macrophages. After 6 days at 37°C, these cultures showed an atypical CPE consisting of cell rounding and detachment from the glass surface. Supernatants of these cultures inoculated into fresh macrophages produced a similar CPE, which could be reproduced in eight sequential passages. These preliminary findings were confirmed on two other occasions when the cytopathogenic agent was isolated from macrophage cultures. Infectivity titers in these passages remained relatively low, only of the order of 10^4. A partial neutralization, by two logs, could be demonstrated using unheated convalescent serum from one of the patients, while the corresponding preinfection serum showed no effect on the CPE. Attempts to demonstrate antigens of the agent in cells by direct or indirect immunofluorescence have not yielded convincing results so far. Cultures of vole kidney cells were found to support growth of the agent isolated from the macrophages but a contaminating paramyxovirus in these cells invalidated this system for further studies. Recently the agent was also propagated in cultures of potoroo kidney (Pt K2) cells, where titers over 10^7 infectious units per ml were obtained 1 to 2 weeks after inoculation. An autointerference with no CPE in the dilutions up to 10^{-3} - 10^{-4} was seen in this system. Preliminary neutralization tests in these cells again suggested a relation of this agent to NE. Like in the macrophage system, attempts to demonstrate a NE-related antigen in the cells by immunofluorescence were unsuccessful.

V. CONCLUSIONS

Based on the study of nearly 1000 cases of NE in Finland and on the intensive virological and serological studies of the epidemic of NE and of an unusual family outbreak in 1976-1977, the following conclusions can be made:

(1) NE in Finland appears to represent a less severe form of the related disease found in the Far East. In particular, the hemorrhagic symptoms are less common.

(2) NE shows both a pronounced seasonal (onset after frost periods) and geographical incidence (lake districts in Finland).

(3) A role of the small rodents, and more specifically the bank vole *(C. glareolus)*, as a reservoir of the disease is supported by the analysis of the epidemiology of the disease, by the clustering of NE cases to the vole years, as well as by the features of recent laboratory and family outbreaks of NE.

(4) Extensive virological and immunological studies of the above outbreaks, including analysis of samples taken prior to the onset of NE, have been initiated. The results so far, while still preliminary, suggest that noncellular samples of NE patients (urine, plasma, cerebrospinal fluid, etc.) yield negative results, whereas cellular samples (buffy coat) from patients and from infected voles may yield positive findings with immunofluorescence or with specialized virus isolation techniques.

ACKNOWLEDGMENTS

This study was supported by a grant from Sigrid Jusélius Foundation.

References

1. Casals, J., Hoogstraal, H., Johnson, K. M., Shelekov, A., Wiebenga, N. H., and Work, T. H. (1966). *Am. J. Trop. Med. Hyg. 15*,751-764.
2. Hortling, H. (1946). *Nord. Med. 30*,1001-1004.
3. Kalela, O. (1949). *Ann. Zool. Soc. Zool. Bot. Fenn. Vanamo 13*(5), 1-90.
4. Kuhlbäck, B., Fortelius, P., and Tallgren, L. G. (1964). *Acta Pathol. Microbiol. Scand. 60*,323.
5. Lähdevirta, J. (1971). *Ann. Clin. Res. 3, Suppl 8*.

6. Lähdevirta, J., and Elo, O. (1975). *Suomen Lääkärilehti 30*, 677-682.
7. Myhrman, G. (1934). *Nord. Med. Tidsskr. 7*, 793-794.
8. Myllymäki, A. (1978). *EPPO Bull.* (in press).
9. Myllymäki, A., Christiansen, E., and Hansson, L. (1978). *EPPO Bull.* (in press).
10. Nyström, K. (1977). *Acta Med. Scand., Suppl 609.*
11. Stuhlfaut, K. (1943). *Deutsch. Med. Wochenschr. 69*, 474-477.
12. Trencséni, T., and Keleti, B. (1971). Clinical aspects and epidemiology of haemorrhagic fever with renal syndrome. *Akad. Kiado Budapest.*
13. Zetterholm, S. (1934). *Läkartidningen 31*, 425-429.

Note added in proof:

After the submission of this manuscript the demonstration of the antigen of the Korean haemorrhagic fever in vole tissues has been reported by Lee, H. W. and Johnson, K. M. (1978). J. Inf Dis. 137, 298-308.

Chapter 14

PHENETIC RELATIONSHIPS OF VIRUSES OF THE HUGHES SEROLOGICAL GROUP

C. E. Yunker, L. A. Thomas, and J. Cory

I. INTRODUCTION

The Hughes group is an assemblage of antigenically related but otherwise unclassified arboviruses found in ticks of the family Argasidae. All but one of these viruses are circulated among marine and shore birds by *Ornithodoros (Alectorobius)* species of the *capensis* complex (Hoogstraal, 1973). The exception is Sapphire II virus (Yunker *et al.*, 1972) of inland cliff swallow ticks, *Argas (Argas) cooleyi*. Viruses of this serogroup are widely distributed throughout the tropical and temperate regions of the world. Some have been associated with human disease (Hoogstraal *et al.*, 1970, 1976; Varma *et al.*, 1973) and catastrophic die-off of seabirds (Converse *et al.*, 1975; Feare, 1976).

ISBN 0-12-429765-X

The Rocky Mountain Laboratory (RML) program of study of tick-borne viruses associated with seabirds has recently been reviewed (Yunker, 1975) and Clifford, in a subsequent chapter of this volume, will list the vectors and distribution of tick-viruses of seabirds. The purpose of the present study is to obtain additional information on antigenic similarities (or extent of dissimilarities) of tick-borne viruses of the Hughes group in the hope of an understanding of their distribution and vector relationships.

A. History of Hughes Group Viruses

Hughes virus was first isolated 12 years ago from *Ornithodoros denmarki* ticks taken on Dry Tortugas, Florida, by workers at the RML (Hughes *et al.*, 1964). It has subsequently been recognized in Trinidad, West Indies (Aitken *et al.*, 1968) and Aves Island, of the Lesser Antilles (J. G. Keirans, C. E. Yunker, and C. M. Clifford, unpublished results).

A variant of Hughes virus, Farallon, was isolated by Dr. H. N. Johnson off the coast of northern California (Radovsky *et al.*, 1967). Shortly thereafter a Hughes-like virus from Raza Island in the Gulf of California, Mexico, was reported by Clifford and co-workers (1968). At that time the Raza isolate was thought to be identical to Hughes virus. These viruses and identification of the related Soldado virus (Jonkers *et al.*, 1973) prompted Casals (1970) to formally recognize the Hughes serological group, which grew to include Punta Salinas virus from coastal Peru (Johnson and Casals, 1972), Sapphire II virus from the inland U.S. (Yunker *et al.*,1972), and Zirqa virus from the Persian Gulf (Varma *et al.*, 1973).

Theiler and Downs (1973) reported Hughes group neutralizing antibodies in sera taken from migratory seabirds in South Georgia, Antarctica. A Hughes group agent later proving to be Soldado virus has been recorded from Oahu, Hawaii (Yunker, 1975; J. F. Keirans, C. E. Yunker, and C. M. Clifford, unpublished results). Soldado virus has been identified from various other parts of the world, including Texas, southern Ireland, northern Wales, Ethiopia, and the Seychelles Islands (Keirans *et al.*, 1976, unpublished results; Hoogstraal, 1973; Converse *et al.*, 1975, 1976).

B. Application of Serological Methods to the Study of Hughes Group Viruses

There has been no systematic serological study of viruses of this group. Upon primary isolation strains of these viruses have been identified by complement-fixation (CF) and neutralization (NT) procedures. Because no Hughes group virus is known to produce a hemagglutinin and none are reported to inhibit hemagglutination by other viruses, a valuable means of study is not available for these viruses. Agar-gel precipitin banding by one virus, Zirqa, has been observed (Williams and Moussa, 1972), which observation holds promise of an additional classifying device. As a whole, Hughes group agents are typical of the tick-borne viruses characterized as difficult to study and define by virtue of their relatively low pathogenicity and antigenicity (Yunker, 1975). Tissue culture techniques for the detection of a virus in this group were first utilized by Buckley (1964), who observed cytopathic effect (CPE) of Hughes virus upon Vero cells. These cells were found by Stim (1969) to form plaques in response to infection with both Hughes and Farallon viruses. These observations were confirmed by Buckley (1971) who also obtained plaques from Soldado virus in Vero cells in the absence of CPE. In our own experience the plaquing technique employing the Vero cell line is far superior to inoculation of the RML strain of Caesarian-derived suckling mice as a means of studying Hughes group agents. We have repeatedly isolated Hughes, Soldado, Farallon, Raza, and Sapphire II viruses in Vero cells from field-collected materials, often when the identical inoculum failed to cause illness or death, even in blind passage, in intracerebrally inoculated suckling mice (unpublished information). In addition, all seven Hughes group viruses may easily and economically be assayed and at least partially identified by plaquing and the plaque-reduction NT technique in Vero cells.

Immunofluorescence (IF) techniques have not been widely employed in the serological study of arboviruses (Buckley and Clarke, 1970; Hahon and Hankins, 1970) and evidently have not been used at all with Hughes group agents. These techniques are, in fact, quite useful in this regard and our findings are detailed below.

In all of these serological tests, CF, NT, and IF, we routinely determine the reciprocal reactions of available viruses of the serological group. This is necessary for differentiation of closely related strains. In this, heterologous and homologous endpoint titers of reactive antisera prepared from two or more viruses are compared. Accurate evaluation of these quantitative results is sometimes difficult and subject to errors of interpretation, as well as bias stemming from preexisting knowledge of host and locality of origin. Reduction of the values to simple titer ratios is only slightly more helpful. A method originally devised to simplify such comparisons among influenza serotypes has been put to use in arbovirus differentiation. The calculation, which results in an estimate of antigens shared among viruses (Calisher *et al.*, 1969), is detailed in Section II.

II. MATERIALS AND METHODS

A. Virus Strains

All seven viruses currently comprising the Hughes serological group were available for study. Of these, five came from North America or the Caribbean Sea, one from South America, and one from the Persian Gulf. These viruses and their strain designations are Hughes (Original, Dry Tortugas), Sapphire II (14), Raza (829), Punta Salinas (Ar888), Farallon (Ar846), Soldado (TRVL 52214), and Zirqa (Por 7866).

B. Serological Tests

CF tests were patterned after those of Clifford *et al.*, (1971). Tissue culture (plaque reduction) NT tests were as given by Earley *et al.*, (1967) but modified to include use of "accessory factor" of normal serum. Microimmunofluorescence (IF) tests were based on a modification of Wang's (1971) method for immunotyping trachoma-inclusion conjunctivitis chlamydiae (Cory *et al.*, unpublished). In the latter tests, the indirect technique was employed, in which viral antigens produced in tissue culture were exposed to serial dilutions of specific hyperimmune mouse fluids. Resulting antigen-antibody complexes were further conjoined with commercially available fluorescein-labeled, antimouse goat serum.

All serological tests were performed with the same lots of antisera or immune ascites fluids. All of these hyperimmune fluids were produced by Casals' method of multiple intraperitoneal injections of killed followed by live viruses over a 4-week period (Hammon and Sather, 1969). Mice were injected with Sarcoma 180 cells at or near the time of final vaccination. Ascites fluid was extracted 1 to 2 weeks later and the mice were then exsanguinated for serum.

C. Calculations of Antigenic Relationships

Titer ratios were derived by dividing the titer of an immune fluid tested against a heterologous virus or antigen by the titer of the same fluid tested against its homolog. The geometric mean of two titer ratios,*

$$r = \sqrt{r_1 \times r_2}\ ,$$

expresses the extent of the cross relationship between the two

*r_1 = *the heterologous to homologous titer ratio of virus 1;* r_2 = *that of virus 2.*

strains. These calculations were detailed by Archetti and Horsfall (1950) in a study of antigenic variation among influenza A viruses and, more recently, were used by Calisher *et al.* (1969) to distinguish among Simbu group arboviruses.

In all calculations of the present study, no effort was made to exclude values of borderline significance or below for various reasons. For example, the above formula is unworkable when zero appears in any part and seemingly insignificant values, if replicable, may be highly significant. No replications of CF and IF tests were made; in these tests, zeros were replaced by the arbitrarily chosen number 0.1. The NT tests were performed twice and results (dex reduction of titer) were arithmetically averaged.

A technique of numerical taxonomy, cluster analysis by a simple single linkage method, attributed to Sneath by Lessel and Holt (1970), was employed to derive indices of similarity among viruses. In this, pairs and clusters of viruses having the highest mutual similarity based on degree of cross reactivity were identified and listed serially until all viruses had been linked at some percentage level. Pairs and clusters of viruses and their linkages were depicted as dendrograms or tree diagrams, which in this study are not intended to convey phylogenetic relatedness, but only phenetic similarity as determined by serological cross tests.

III. TESTS

A. Neutralization Tests

Results of these cross tests (Tables I and II; Fig. 1) reveal that a number of antigens are shared among members of the Hughes group. Sapphire II virus from *Argas cooleyi* was at least similar to the other viruses of the group, all of which are from ticks of the *Ornithodoros capensis* complex. Most closely related to Sapphire II virus were Raza and Hughes viruses, which shared less than 40% of NT antigens with it. Although Hughes virus is also distinct in cross-NT test, the interrelationships of the remaining viruses are not as clear. For example, Punta, Salinas and Zirqa viruses are nearly indistinguishable (74% of antigens shared) and Farallon is quite similar to Raza and Soldado viruses (over 60% of antigens shared). These relationships, shown as a simple bar diagram in Fig. 1, become more apparent when results are subjected to cluster analysis and depiction as a dendogram (Fig. 2). This diagram has the advantage of showing linkages among pairs or clusters of viruses.

TABLE I

Cross Relationships of Some Hughes Group Viruses in Neutralization Tests

Virus	Hyperimmune serum titer[a]						
	Hughes	*Sapphire II*	*Raza*	*P. Salinas*	*Farallon*	*Soldado*	*Zirqa*
Hughes	*2.9*	1.2	1.0	0.9	0.6	0.9	1.4
Sapphire II	1.1	*≥3.2*	1.1	0.5	0.2[b]	0.6	1.5
Raza	1.3	1.3	*≥2.9*	≥2.8	1.4	1.9	3.7
Punta Salinas	0.5	0.03[b]	1.2	*≥5.0*	0.9	0.6	≥4.2
Farallon	1.0	0.8	1.5	1.7	*≥1.9*	1.9	≥2.1
Soldado	1.7	0.4	1.1	2.1	1.1	*≥2.9*	≥3.0
Zirqa	0.9	0.1[b]	0.5	≥2.7	0.7	1.1	*≥4.1*

[a] *Dex reduction of titer (PFU) in Vero cells; average of two titrations.*

[b] *Values insignificant; i.e., no reduction of titer in duplicate tests (see text).*

TABLE II
Cross Relationships of Some Hughes Group Viruses in Neutralization Tests

	Value of r^a						
	1. *Hughes*	2. *Sapphire II*	3. *Raza*	4. *P. Salinas*	5. *Farallon*	6. *Soldado*	7. *Zirqa*
1. Hughes	*1*	0.38	0.39	0.18	0.33	0.43	0.33
2. Sapphire II		*1*	0.39	0.03	0.16	0.16	0.10
3. Raza			*1*	0.48	0.62	0.50	0.39
4. Punta Salinas				*1*	0.40	0.30	0.74
5. Farallon					*1*	0.64	0.43
6. Soldado						*1*	0.52
7. Zirqa							*1*

[a] $r=\sqrt{r^1 \times r^2}$; r^1=*heterologous to homologous titer reduction ratio of virus 1 with virus 2;* r^2=*heterologous to homologous titer reduction ratio of virus 2 with virus 1.*

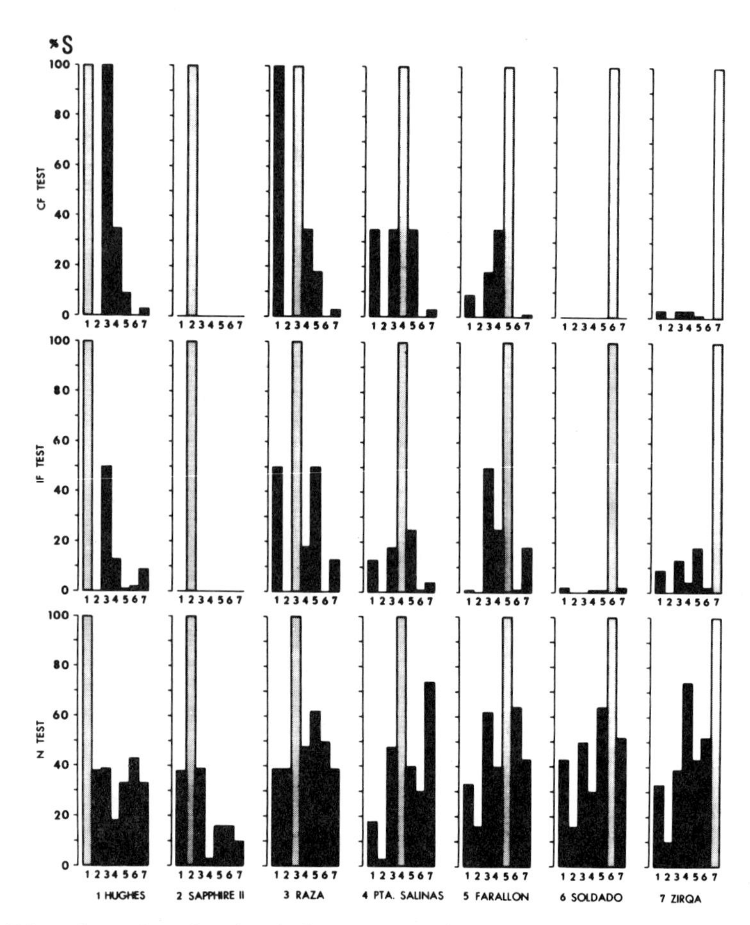

Fig. 1. Serological interrelationsips of seven Hughes group viruses as determined by CF, IF, and NT. Ordinate, percentage similarity (%) based on proportion of antigens shared. Abscissa, identity of virus as indicated by number. Shaded bars, homologous antigen-antiserum reactivity. Clear bars, heterologous antigen-antiserum reactivity relative to homologous reactivity.

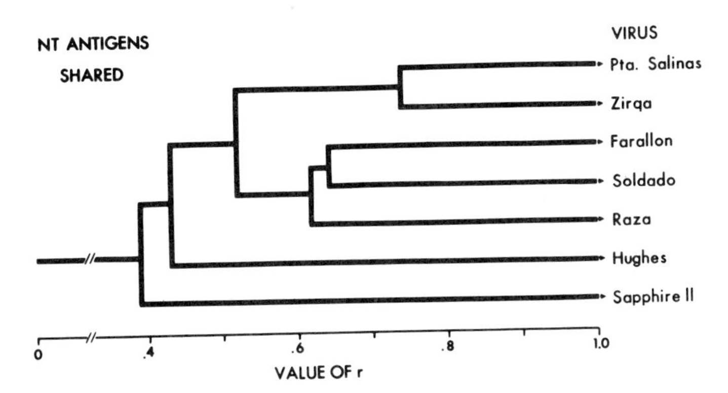

Fig. 2. Proportion of NT antigens shared between Hughes group viruses; phenogram depicting similarity clusters and linkages.

B. Complement-Fixation Tests

Cross-CF tests failed to show any significant relationship between Sapphire II virus and the others of the group (Tables III and IV; Fig. 1). Similarly, Soldado virus antigen and antiserum of high homologous titer fixed only an insignificant amount of complement in the presence of heterologous reagents of the group. However, among the remaining five viruses, various degrees of relationship were seen, with Zirqa virus sharing the least antigens (less than 3%), and Hughes and Raza viruses indistinguishable from each other. (However, it should be noted that had the homologous systems been titrated to end-points, some degree of differences between Hughes and Raza viruses might have been seen). These interrelationships are also particularly obvious when a cluster analysis is performed and the results are plotted as a dendogram (Fig. 3).

TABLE III

Cross Relationships of Some Hughes Group Viruses in Complement-Fixation Tests

Antigen	*Hyperimmune serum titer*[a]						
	Hughes	*Sapphire II*	*Raza*	*P. Salinas*	*Farallon*	*Soldado*	*Zirqa*
Hughes	*≥256*	2	≥256	64	2	0	4
Sapphire II	2	*≥256*	0	0	0	0	0
Raza	≥256	2	*≥256*	128	8	0	4
Punta Salinas	128	2	64	*≥256*	8	0	8
Farallon	64	2	128	≥256	*64*	2	16
Soldado	2	2	0	0	0	*≥256*	0
Zirqa	8	0	8	4	0	0	*≥128*

[a]*Highest dilution of hyperimmune serum fixing complement at optimal antigen dilution. Significant titers, ≥4.*

TABLE IV

Cross Relationships of Some Hughes Group Viruses in Complement-Fixation Tests

Virus	Value of r^a						
	1. Hughes	*2. Sapphire II*	*3. Raza*	*4. P. Salinas*	*5. Farallon*	*6. Soldado*	*7. Zirqa*
1. Hughes	*1*	<0.01	1	0.35	0.09	<0.01	0.03
2. Sapphire II		*1*	<0.01	<0.01	<0.01	<0.01	<0.01
3. Raza			*1*	0.35	0.18	<0.01	0.03
4. Punta Salinas				*1*	0.35	<0.01	0.03
5. Farallon					*1*	<0.01	0.01
6. Soldado						*1*	<0.01
7. Zirqa							*1*

[a] $r = \sqrt{r^1 \times r^2}$; r^1 = *heterologous to homologous titer-ratio of virus 1 with virus 2;* r^2 = *heterologous to homologous titer-ratio of virus 2 with virus 1.*

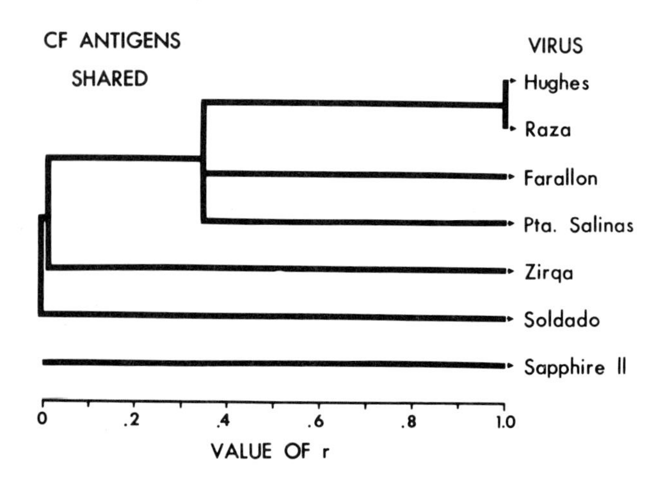

Fig. 3. Proportion of CF antigens shared between Hughes group viruses; phenogram depicting similarity clusters and linkages.

C. Microimmunofluorescence Tests

Results of these tests (Tables V and VI; Fig. 1) are comparable with those of the CF tests, except that Hughes and Raza viruses, though similar, are mutually distinguishable, sharing 50% of IF antigens. Especially comparable is the unrelatedness of both Sapphire II and Soldado viruses to any of the others in the group. The former failed to relate to any other Hughes group agents except that its antiserum reacted at a low level (1:8) with Soldado antigen of high reactivity (Table V). This reaction was not a reciprocal one and the resulting *r* value was insignificant (less than 1%) (Table VI). The reactions of Soldado virus with four Hughes group agents were also minimal. Low-grade reciprocal reactions of Soldado virus were seen with Hughes and Zirqa viruses (r = 2%), and with Punta Salinas virus (r = 1%) and an equally low grade nonreciprocal reaction for Soldado virus was noted with Farallon virus (r = 1%). Cluster analysis resulted in a dendrogram (Fig. 4) that is quite similar to that of the CF test (Fig. 3). In the IF test, as in the CF test (Fig. 3), Farallon virus was seen to be most similar to the Hughes and Raza complex. However, here, in contrast to the CF results, Farallon virus is distinct from Punta Salinas virus.

TABLE V

Cross Relationships of Some Hughes Group Viruses in Immunofluorescence Tests

Antigen	*Hyperimmune serum titer*[a]						
	Hughes	*Sapphire II*	*Raza*	*P. Salinas*	*Farallon*	*Soldado*	*Zirqa*
Hughes	*128*	0	256	64	0	4[a]	32
Sapphire II	0	*128*	0	0	0	0	0
Raza	8	0	*64*	32	32	0	8
Punta Salinas	16	0	128	*512*	32	4	32
Farallon	16	0	16	32	*32*	4	32
Soldado	4	8	0	4	0	*256*	4
Zirqa	4	0	16	4	4	4	*128*

[a]*Reciprocal of highest serum dilution giving a positive reaction. Serum titers of 4 indicate weakly positive reaction at 1:8 dilution of serum.*

TABLE VI
Cross Relationships of Some Hughes Group Viruses in Immunofluorescence Tests

Virus	Value of r[a]						
	1. Hughes	*2. Sapphire II*	*3. Raza*	*4. P. Salinas*	*5. Farallon*	*6. Soldado*	*7. Zirqa*
1. Hughes	*1*	<.01	.50	.13	.01	.02	.09
2. Sapphire II		*1*	<.01	<.01	<.01	<.01	<.01
3. Raza			*1*	.18	.50	<.01	.13
4. Punta Salinas				*1*	.25	.01	.04
5. Farallon					*1*	.01	.18
6. Soldado						*1*	.02
7. Zirqa							*1*

[a] *See footnote, Table IV.*

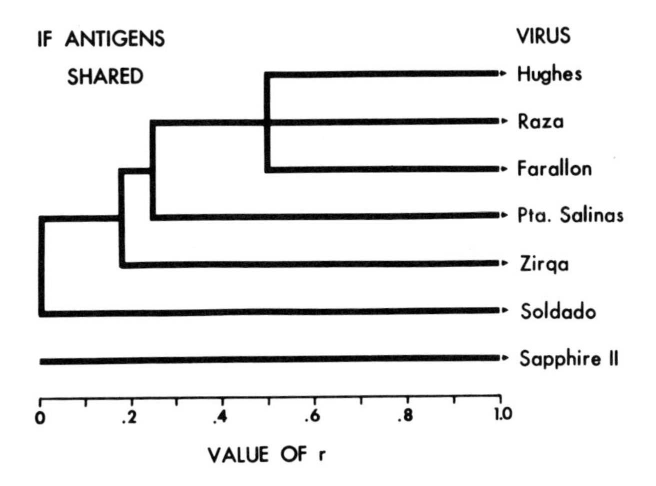

Fig. 4. Proportion of IF antigens shared between Hughes group viruses; phenogram depicting similarity clusters and linkages.

IV. DISCUSSION AND CONCLUSIONS

The present study is not intended as a definitive one but only as a preliminary and provisional effort toward resolving the status and interrelationships of the Hughes group agents, and particularly their associations with vector species. Because of the preliminary nature of the data additional studies are indicated before broad generalizations can be made. For example, different lots and preparations of antigens and immune fluids should be tested for purposes of comparison. Also, efforts should be made to encompass endpoint titration values, which values will aid in detection of dissimilarities among closely related viruses. The need to correct for zero values in the calculation of r may falsely indicate low degrees of cross reactivity, particularly in the NT test. Thus, an arithmetic, rather than geometric, mean may be appropriate in determining relationships of viruses that react in one direction only. These qualifications notwithstanding, some inferences of the present study are apparent. Here, hyperimmune fluids prepared by multiple vaccinations were employed. Because such fluids can generally be expected to react broadly with other viruses of a given serogroup, the present results should reveal the maximum cross reactivity of the Hughes group reagents. If so, it may be concluded that cross-CF and IF tests offer means for distinguishing among the various Hughes group viruses.

Accordingly, it might be preferable to employ the NT test (in which all members are seen to share at least 38% of antigens) for preliminary grouping of isolates. This apparent broad reactivity of the NT test and the relative specificity of CF reactions is uncommon among the arboviruses (Casals, 1966) and may provide further evidence of dissimilarity between Hughes group agents and other taxons. (It should be noted that no attempt was made here to modify the reactivity of the NT reaction through use of an antiserum prepared by minimal vaccination. Such a procedure might result in more specificity of the NT test for Hughes group viruses.)

Some tentative conclusions regarding the infragroup classification of the various viruses studied are also offered. All three tests confirm that Sapphire II virus, from inland bird ticks (*A. cooleyi*), is at most only remotely related to the other members of the group. Recognition of this division at the subgroup level is herewith proposed (e.g., Hughes subgroup and Sapphire II subgroup). The individuality of Soldado virus is also evident in cross-CF and IF tests indicating that it may eventually be found desirable to recognize an additional subdivision (e.g., serological complex) delineating this virus of *Ornithodoros* species principally including *O. maritimus* and *O. capensis*, from the complex of Hughes-like viruses in other *Ornithodoros* species. Present data do not permit accurate evaluation of the latter serological complex. However, some trends are obvious, inviting speculation that the apparent subdivisions based on antigenic relationships may correlate with distribution and vector associations of the viruses. For example, (1) a North American subcomplex or "superspecies" of viruses consisting of Hughes, Raza, and Farallon may exist in *O. denmarki* and species near *denmarki;* (2) a South American subcomplex, presently monotypic for Punta Salinas virus, may exist in *O. amblus* and sibling forms; and (3) an Asian subcomplex, also monotypic and presently represented by Zirqa virus, may be found in *O. muesebecki* and other tick species yet to be recognized. Resolution of these conjectures depends not only on an adequate understanding of Hughes group virus interrelationships, but also on a secure knowledge of species and distribution of *Ornithodoros (Alectorobius)* ticks. Advances in the latter area are now being made through the use of scanning electron microscopy of both immature and adult stages of these ticks by Hoogstraal (U.S. Naval Medical Unit No. 3) and Clifford and Keirans (RML) (Hoogstraal *et al.*, 1974, 1976). It should be noted that definitive identification of a tick or ticks yielding virus is seldom done. Instead, typical specimens are usually selected from the series to be tested. Thus, with ticks of a confusing morphological complex, such as that of *O. capensis,* the actual virus donor may go unrecognized. One approach to a better understanding of virus interrelationships (not only of the Hughes group) may lie in the direction barely touched upon here: numerical taxonomy. If enough viral parameters--qualitative and quantitative, serological, biological, and otherwise--are treated statistically so as to identify correlations among the parameters, a dendrogram could result that is phyletic as well as phenetic. When this is done it should be possible to recognize the clusters and "operational taxonomic units" as distinct if not natural units and to identify stem viruses. The implications of this for zoonosology, epidemiology, and control of arboviral disease are obvious.

Acknowledgments

We are indebted to Dr. Carl Eklund, deceased, for helpful comments. Mr. Ed Patzer and Mr. Harry Meibos provided expert technical assistance.

References

1. Aitken, T. H. G., Jonkers, A. H., Tikasingh, E. S., and Worth, C. B. (1968). *J. Med. Ent. 5*,501-503.
2. Archetti, I., and Horsfall, F. L. (1950). *J. Exp. Med. 92*,441-462.
3. Buckley, S. M. (1964). *Proc. Soc. Exp. Biol. Med. 116*, 354-358.
4. Buckley, S. M. (1971). *Proc. Int. Symp. Tick-borne Arboviruses, Smolenice 1969*,43-52. Pub. House of Slovak Acad. Sci., Bratislava.
5. Buckley, S. M., and Clarke, D. H. (1970). *Proc. Soc. Exp. Biol. Med. 135*,533-539.
6. Calisher, C. H., Kokernot, R. H., de Moore, J. F., Boyd, K. R., Hayes, J., and Chappell, W. A. (1969). *Am. J. Trop. Med. Hyg. 18*, 779-787.
7. Casals, J. (1966). *9th Int. Congr. Microbiol., Moscow 1966*, 441-452.
8. Casals, J. (1970). *Misc. Publ. Entomol. Soc. Am. 6*, 327-329.
9. Clifford, C. M., Thomas, L. A., Hughes, L. E., Kohls, G. M., and Philip, C. B. (1968). *Am. J. Trop. Med. Hyg. 17*, 881-885.
10. Clifford, C. M., Yunker, C. E., Thomas, L. A., Easton, E. R., and Corwin, D. (1971). *Am. J. Trop. Med. Hyg. 20*, 461-468.
11. Converse, J. D., Hoogstraal, H., Moussa, M. I., Feare, C. J., and Kaiser, M. N. (1975). *Am. J. Trop. Med. Hyg. 24*, 1010-1018.

12. Converse, J. D., Hoogstraal, H., Moussa, M. I., and Evans, D. E. (1976). *Acta Virol. 20*,243-246.
13. Early, E., Peralta, P. H., and Johnson, K. M. (1967). *Proc. Soc. Exp. Biol. Med. 125*,741-747.
14. Feare, C. J. (1976). *Ibis 118*,112-115.
15. Hahon, N., and Hankins, W. A. (1970). *Appl. Microbiol. 19*, 224-231.
16. Hammon, W. McD., and Sather, G. E. (1969). *In* "Diagnostic Procedures for Viral and Rickettsial Infections" (E. H. Lennette and N. J. Schmidt, eds.), 4th ed., pp. 227-280. Amer. Publ. Health Assoc., New York.
17. Hoogstraal, H. (1973). *In* "Viruses and Invertebrates" (A. J. Gibbs, ed.), pp. 349-390. North Holland Publ. Co., Amsterdam and London.
18. Hoogstraal, H., Oliver, R. M., and Guirgis, S. S. (1970). *Ann. Entomol. Soc. Am. 63*,1762-1768.
19. Hoogstraal, H., Kadarsan, S., Kaiser, M. N., and Van Peenen, P. F. D. (1974). *Ann. Entomol. Soc. Am. 67*, 224-230.
20. Hoogstraal, H., Clifford, C. M., Keirans, J. E., Kaiser, M. N., and Evans, D. E. (1976). *J. Parasitol. 62*,799-810.
21. Hughes, L. E., Clifford, C. M., Thomas, L. A., Denmark, H. A., and Philip, C. B. (1964). *Am. J. Trop. Med. Hyg. 13*,118-122.
22. Johnson, H. N., and Casals, J. (1972). *Proc. 5th Symp. Study of the Role of Migrating Birds in the Distribution of Arboviruses*, Publ. House "Nauka," Novosibirsk.
23. Jonkers, A. H., Casals, J., Aitken, T. H. G., and Spence, L. (1973). *J. Med. Ent. 5*,517-519.
24. Keirans, J. E., Yunker, C. E., Clifford, C. M., Thomas, L. A., Walton, G. A., and Kelly, T. C. (1976). *Experientia 32*,453.
25. Lessel, E. F., and Holt, J. G. (1970). *In* "Methods for Numerical Taxonomy" (W. R. Lockhart and J. Liston, eds.), pp. 50-62. Am. Soc. Microbiol., Washington.
26. Radovsky, F. J., Stiller, D., Johnson, H. N., and Clifford, C. M. (1967). *J. Parasitol. 53*,890-892.
27. Stim, T. B. (1969). *J. Gen. Virol. 5*,329-338.
28. Theiler, M., and Downs, W. G. (1973). *In* "The Arthropod-Borne Viruses of Vertebrates." Yale Univ. Press, New Haven and London.

29. Varma, M. G. R., Bowen, E. T. W., Simpson, D. I. H., and Casals, J. (1973). *Nature 244*,452.
30. Wang, S. P. (1971). *Excerpta Med.,* 273-288 *(Proc. Symp. Boston 1970)*.
31. Williams, R. E., and Moussa, M. I. (1972). *J. Egyptian Publ. Health Assoc. 47*,25-35.
32. Yunker, C. E. (1975). *Med. Biol. 53*,302-311.
33. Yunker, C. E., Clifford, C. M., Thomas, L. A., Cory, J., and George, J. E. (1972). *Acta Virol. 16*,415-521.

Chapter 15

THE STRUCTURE OF THE GENOME OF UUKUNIEMI VIRUS

Ralf F. Pettersson

I. INTRODUCTION

Recently a large number of both tick- and mosquito-borne arboviruses have been grouped in the Bunyaviridae family (Porterfield *et al.*, 1975/76). Many of these are antigenically related and fall into the Bunyamwera supergroup. In addition, a large number are still unclassified and have been included in the Bunyaviridae family as possible members, mainly because of similar morphology and morphogenesis. The structure of one of these, Uukuniemi virus, the prototype of the Uukuniemi group and a tick-borne virus, has been extensively studied during the last few years as a model for the bunyaviruses.

The virus particles consist of a lipoprotein envelope and a ribonucleoprotein core. Two glycoproteins, G_1 and G_2 (MW ≃ 65,000 and 75,000) form the surface projections. They are clustered to form subunits, which are arranged in an icosahedral surface lattice consistent with a $T = 12$ symmetry (Pettersson *et al.*, 1971; von Bonsdorff and Pettersson, 1975). The core consists of three sizes of apparently circular ribonucleoprotein strands, each containing one species of single-stranded RNA of

ISBN 0-12-429765-X

negative polarity and multiple copies of the N protein (MW ≃ 25,000) (Pettersson and Kääriäninen, 1973; Pettersson and von Bonsdorff, 1975; Ranki and Pettersson, 1975). Also associated with the ribonucleoproteins seems to be a large minor polypeptide, L (MW ≃ 180,000-200,000) (N. Guttman and R. Pettersson, unpublished data). Thus, there appear to be four structural polypeptides and three sizes of RNA in purified Uukuniemi virions.

The finding of three sizes of circular ribonucleoproteins raised two important questions: (1) Does each RNA species contain unique genetic information or do they share sequences? (2) Are the RNA species after removal of the protein still circular, and if so, how is the circularization accomplished? In the following, experiments to answer these questions are described. The results indicate that the RNA species contain different genetic information with no or very little overlapping, and that the RNA segments have a circular configuration under nondenaturing conditions due to base-pairing of inverted complementary sequences at the 3' and 5' ends of the molecules.

II. EVIDENCE FOR THREE UNIQUE RNA SEGMENTS IN VIRIONS

A. Fingerprinting Analyses

The nucleic acid extracted from plaque-cloned Uukuniemi virus grown at low multiplicity of infection has consistently revealed three stable species of single-stranded RNA sedimenting at 29 S (large = L), 22 S (medium = M), and 17 S (small = S) in a sucrose gradient (Pettersson and Kääriäinen, 1973). There are in principle three different ways in which the genetic information of the three RNA species could be arranged (Fig. 1). First, each RNA could contain completely different sequences (alternative 1). Second, S and M could be subsets of L (alternatives 2a and b) similar to the defective nucleic acids found in a variety of viruses. Only two examples of such a situation are depicted in Fig. 1. Third, there could be a partial overlapping of sequences between the ends of the RNAs (alternative 3). The extent and location of such overlapping sequences could of course vary in a number of ways. Figure 1 only shows one situation. To distinguish between these possibilities the pattern of radioactively labeled ribonuclease T1 oligonucleotides ("fingerprints") derived from each RNA was compared (Pettersson *et al.*, 1977). The virus was grown in the presence of ^{32}P-orthophosphate, purified, and the labeled RNA species fractionated on a sucrose gradient. The RNAs were digested with RNase T1 to generate oligonucleotides of various lengths

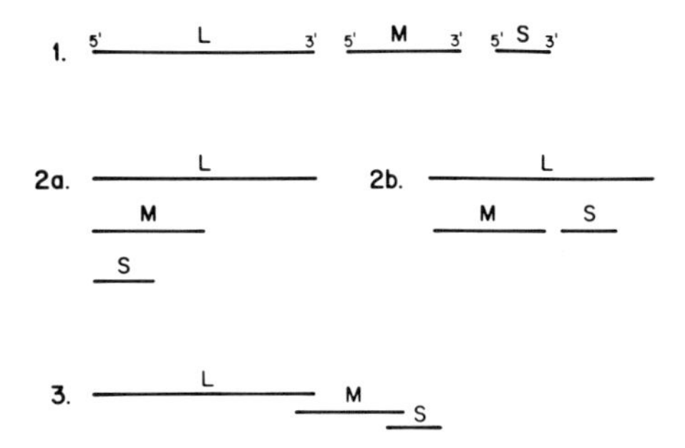

Fig. 1 Possible arrangement of Uukuniemi virus genome. L, large; M, medium; S, small RNA species. (1) A segmented genome with no overlapping of sequences. (2a) The S sequences are present in M and L, and the M in L and partly in S. (2b) The M and S sequences are present in L, but unrelated to each other. (3) A segmented genome with partial terminal overlapping.

all terminating in a guanosine residue. The oligonucleotides were then fractionated by two-dimensional gel electrophoresis according to De Wachter and Fiers (1972). In this system, electrophoresis in the first dimension fractionates the oligonucleotides mainly according to their base composition and in the second dimension on the basis of chain length. The oligonucleotides are observed as distinct spots on the autorodiograms of the gels. The procedure in general separates the long (more than about 12 nucleotides) unique oligonucleotides from the bulk of short nonunique oligonucleotides. Since the probability that a long oligonucleotide occurs twice in the same molecule or any other nonrelated RNA molecule is very small, one can with this technique compare the relationship between nucleotide sequences of different RNA species.

Figure 2 shows the fingerprints obtained from Uukuniemi virus L, M, and S RNAs by this method. The unique oligonucleotides are seen in the lower two-thirds of the autoradiograms. In each case a completely different pattern was observed. Careful analysis of the spots indicated that at least 15 of them present in M RNA did not overlap or coincide with any of the oligonucleotides derived from L, and at least 30 spots in L did not overlap any of the M spots. Similarly, at least five spots were unique for the S RNA. The spots in the L and M fingerprints had intensities roughly corresponding to their size. The S fingerprint, however, displayed spots of varying intensities. The weaker spots represent oligonucleotides derived from contaminating L and M sequences, whereas the stronger ones represent S-specific oligonucleotides. The uniqueness of the oligonucleotides derived from each RNA was confirmed by

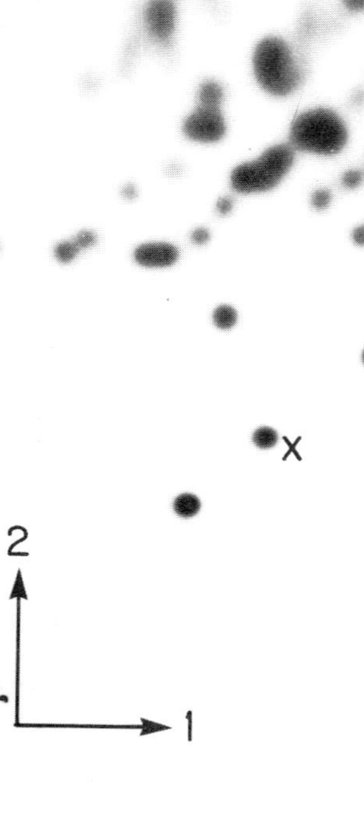
2
1
L

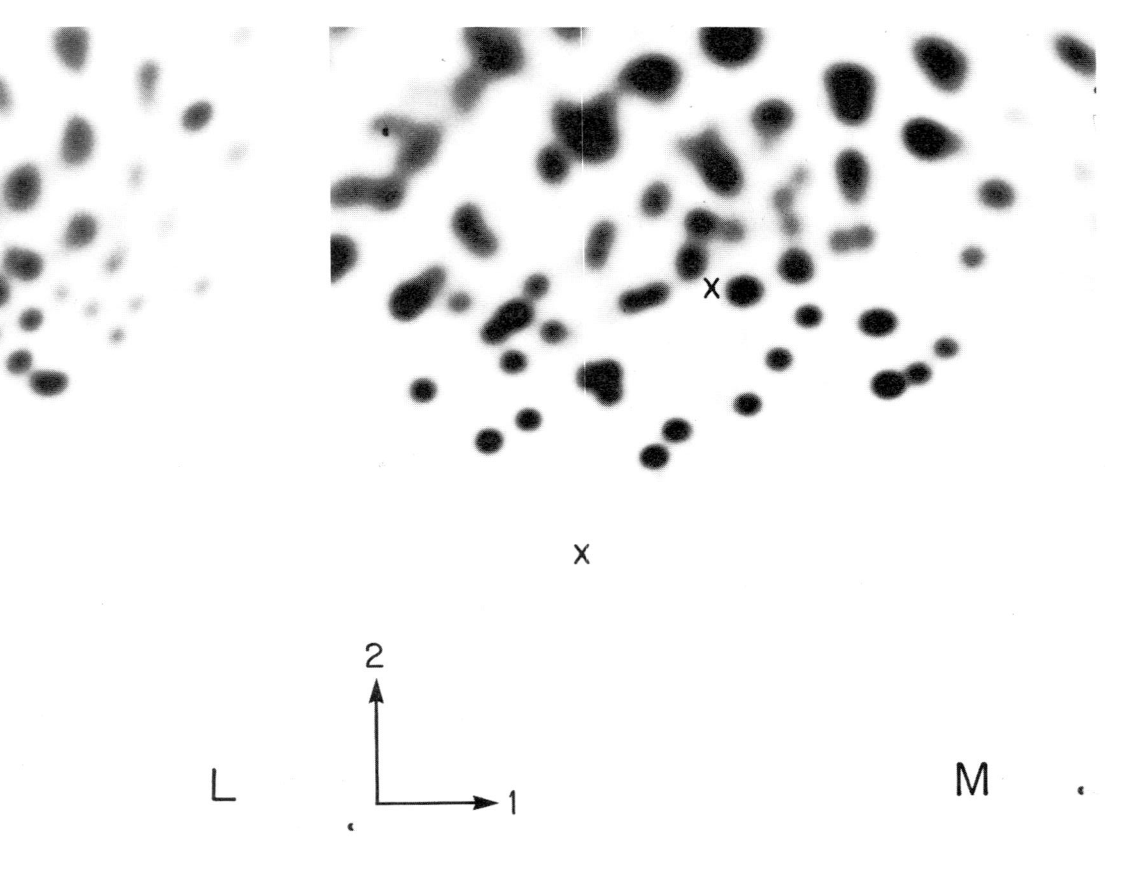
2
1
M

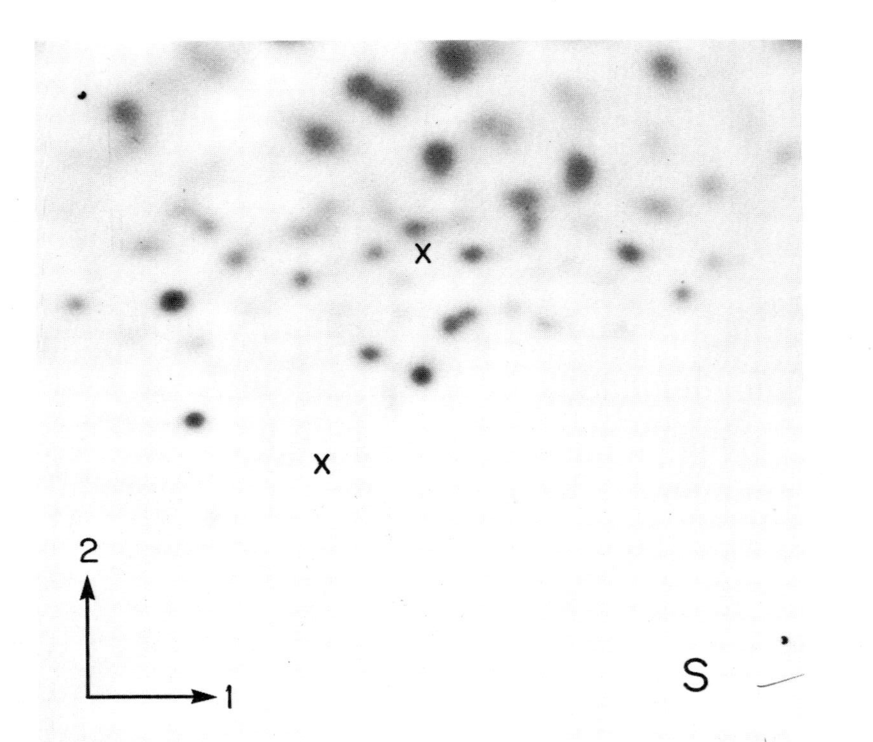

Fig. 2 Two-dimensional gel electrophoresis of ^{32}P-labeled T1-oligonucleotides of Uukuniemi virus L, M, and S RNAs. The electrophoresis was run as described by De Wachter and Fiers (1972) and previously (Pettersson et al., 1977). The first dimension was at 10% acrylamide, 6 M urea, and pH 3.5, and the second dimension at 22% acrylamide, pH 8.0. The upper × represents the position of bromophenol blue dye and the lower × that of the xylene cyanol dye. The spots were visualized by autoradiography. (Adapted from Pettersson et al., 1977, by permission of MIT Press.)

eluting the oligonucleotides and analyzing their base composition. The results excluded the possibility that M and S RNAs are subsets of L RNA (Fig. 1, alternatives 2a and b) since such a situation would mean that all unique oligonucleotides present in the M and S fingerprints should also be present in the L fingerprint. This method would not, however, detect a small overlap at the ends of the molecules (alternative 3). Based on the results one can say that if any overlapping of sequences occurs it is likely to be less than about 250 nucleotides. Irrespective of whether partial overlapping occurs, it is clear that each RNA segment contains unique nucleotide sequences.

B. Complexity Analysis

The fingerprinting technique described above can also be used to calculate the complexity (chain length) of a given RNA molecule. The unique oligonucleotide, which is known to occur only once per molecule, is cut out, the radioactivity quantitated, and the chain length determined. Knowing the input radioactivity in the RNA, the recovery of the oligonucleotides in the gel spots, and the degree of contamination with other RNAs one can then estimate the chain length of the RNA molecule (Pettersson *et al.*, 1977). Using this method, values of 8053 ± 1,324 (L), 3474 ± 541 (M), 1938 ± 406 (S) nucleotides were obtained. They correspond to molecular weights of 2.7×10^6, 1.2×10^6, and 0.5×10^6, respectively. Since these values are in agreement with the molecular weights obtained by other methods (see below) each size class of RNA consists of only one species of molecule.

III. CIRCULAR AND LINEAR FORMS OF THE RNA SEGMENTS

A. Electron Microscopy of RNA

Since the ribonucleoproteins had previously been found by electron microscopy to have a circular configuration it was of interest to know whether the RNA species after deproteinization also were circular, and if so how this configuration was maintained. To answer these questions an electron microscopic study was undertaken (Pettersson and Hewlett, 1976; Hewlett *et al.*, 1977). Purified virions were deproteinized by sodium dodecyl sulfate-phenol extraction and the mixture of L, M, and S RNAs was spread for electron microscopy by a modified Kleinschmidt technique (Davis *et al.*, 1971). Initial experiments where the RNA was spread from 50% formamide (FA) indicated that

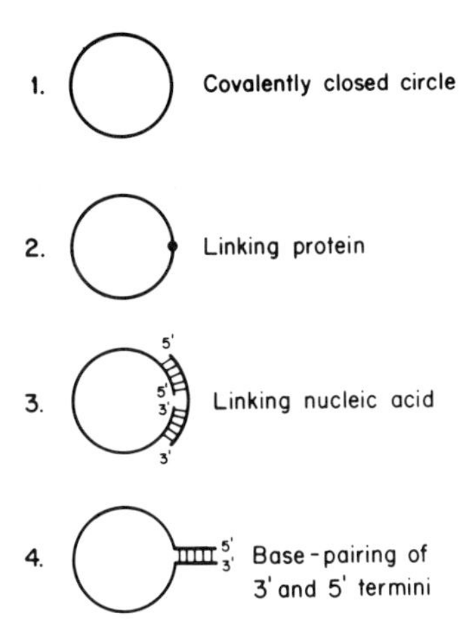

Fig. 3 Possible models for circularization of single-stranded RNA molecules.

the RNA molecules indeed were circular (Fig. 4B). There are several ways by which circularization of single-stranded RNA could occur (Fig. 3, alternatives 1-4). First, the circles could be closed by covalent interaction of the 3' and 5' ends of a linear molecule. Second, a protein could link the ends together. Third, a small piece of RNA or DNA could connect the ends by hydrogen-bonding of complementary sequences. Fourth, the circular forms could be maintained by intramolecular base pairing between inverted complementary sequences at the 3' and 5' ends of the molecules.

To distinguish between these possibilities the RNA was spread under increasingly denaturing conditions to denature any hydrogen-bonding between complementary sequences. Three different methods were used: spreading from formamide (FA) concentrations ranging from 40 to 99% or from 4 *M* urea/80% FA, and after treatment of the RNA with glyoxal. In Fig. 4A-5 the RNA was spread from FA concentrations ranging from 40 to 99%. As the FA concentration was increased, the molecules became more and more extended as a result of denaturation of the secondary structure. When spread from 50 or 60% FA, 50 to 70% of the molecules appeared to be circular (Fig. 4B,C). At 70% FA about 40 to 50% were still circular (Fig. 4D). At 85% FA the fraction of circles dropped to about 30% (Fig. 4E). Since none of the above conditions were fully denaturing, the RNA was finally heated at 60°C for 15 minutes in 99% FA and then immediately spread from 60% FA. Under these conditions no circles were seen (Fig. 4F). The RNA was also spread from 4 *M* urea/80% FA, another moderately denaturing condition. Five

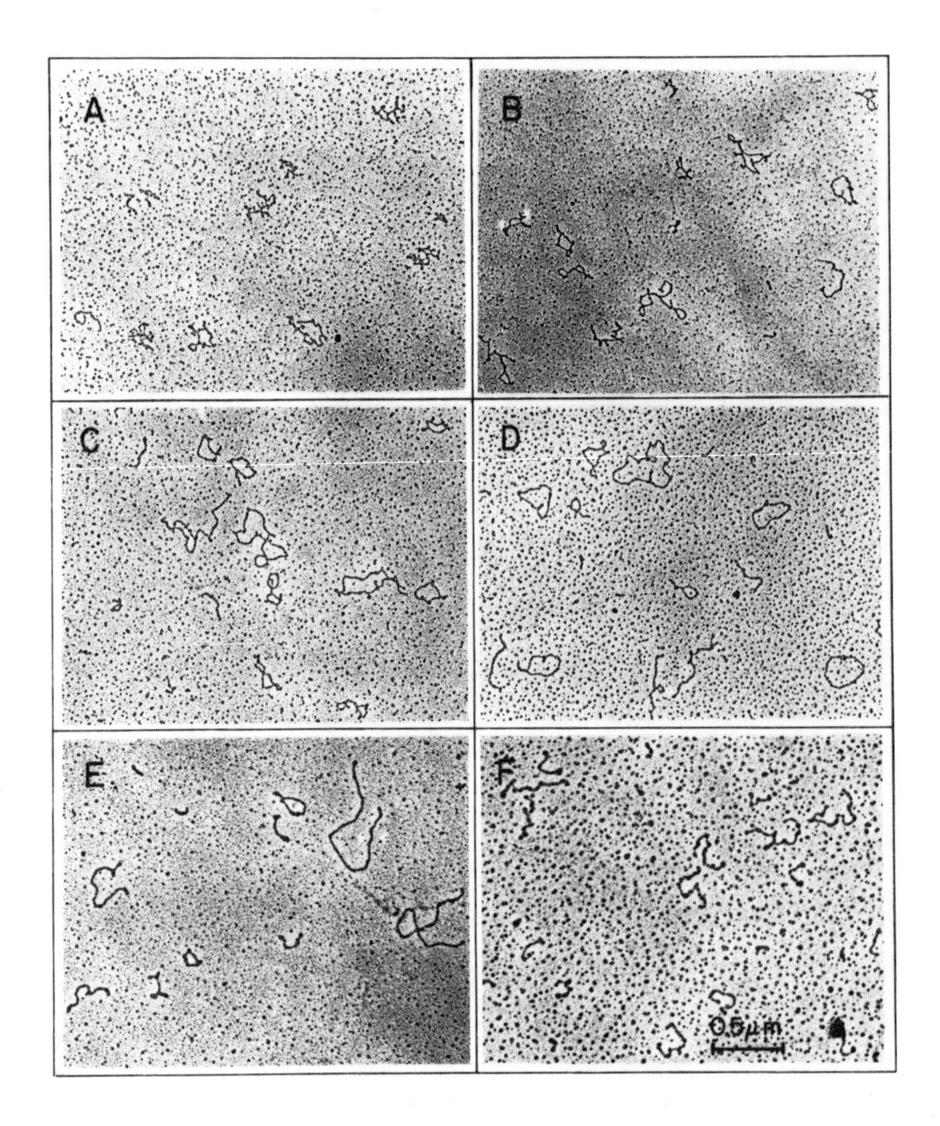

Fig. 4 Electron micrographs of Uukuniemi virus RNA spread from increasing concentrations of FA. The RNA was spread from 40% (A), 50% (B), 60% (C), 70% (D), and 85% (E) onto a hypophase containing 30% less FA and 1/10 of the electrolyte concentration in the spreading solution. RNA was also heated at 60°C for 15 minutes in 99% FA prior to spreading from 60% FA (F). [Reprinted from Hewlett et al., 1977, by permission of the American Society of Microbiology.]

to 30% of the molecules were scored as circles. The third method involved treatment of the RNA with 0.5 *M* glyoxal for 15 minutes at 37°C prior to spreading from 50% FA. Glyoxal is a compound that adds to guanosine residues and especially blocks G-C hydrogen bonding (Hsu *et al.*, 1973). Less than 5% of the molecules were scored as circles under these conditions. Histograms of the RNA molecules spread under the various conditions revealed three size classes of molecules and showed no significant degradation of the molecules. Single-stranded poliovirus 35 S RNA spread under the above conditions displayed only linear molecules.

These results suggested that the RNA species were not circularized by means of covalent interactions. As an independent piece of evidence for a noncovalent closure of the circles we have shown that the 5'-terminal structure of each RNA species is pppAp (Pettersson *et al.*, 1977). Treatment of the RNA with proteinase K, a non-specific protease, or deoxyribonuclease, did not significantly reduce the fraction of circular molecules, suggesting that neither a linking protein nor a DNA fragment were involved in the circle formation. On the other hand, ribonuclease A degraded all molecules to undetectable fragments. The circles could be reformed when allowed to self-anneal following complete denaturation. Based on these results we conclude that the Uukuniemi virus RNA species are maintained in a circular form by hydrogen-bonding between inverted complementary sequences at 3' and 5' ends of the molecules (Fig. 3, alternative 4).

The types of circular molecules observed are shown in Fig. 5. Many of the circles have one or more panhandles, whereas some appear smooth. Whether any of the visible panhandles in fact represent the base-paired ends is uncertain. They could also represent internal secondary structures unrelated to the closure of circles.

B. Length Measurement of RNA

Histograms of molecules spread by the urea/formamide or glyoxal mothods revealed three length classes of molecules (Table I). When poliovirus RNA was used as an external length standard (2.5×10^6 daltons) mean values of about 1.9×10^6 (L), 0.9×10^6 (M), and 0.4×10^6 (S) daltons were obtained for the three RNA species. These values are all somewhat lower than those obtained by other methods (see above and Pettersson *et al.*, 1977). The length of the circular molecules spread from 70% FA also fell into three classes. In each size class the mean length of the circular molecules was the same as that of the linear ones.

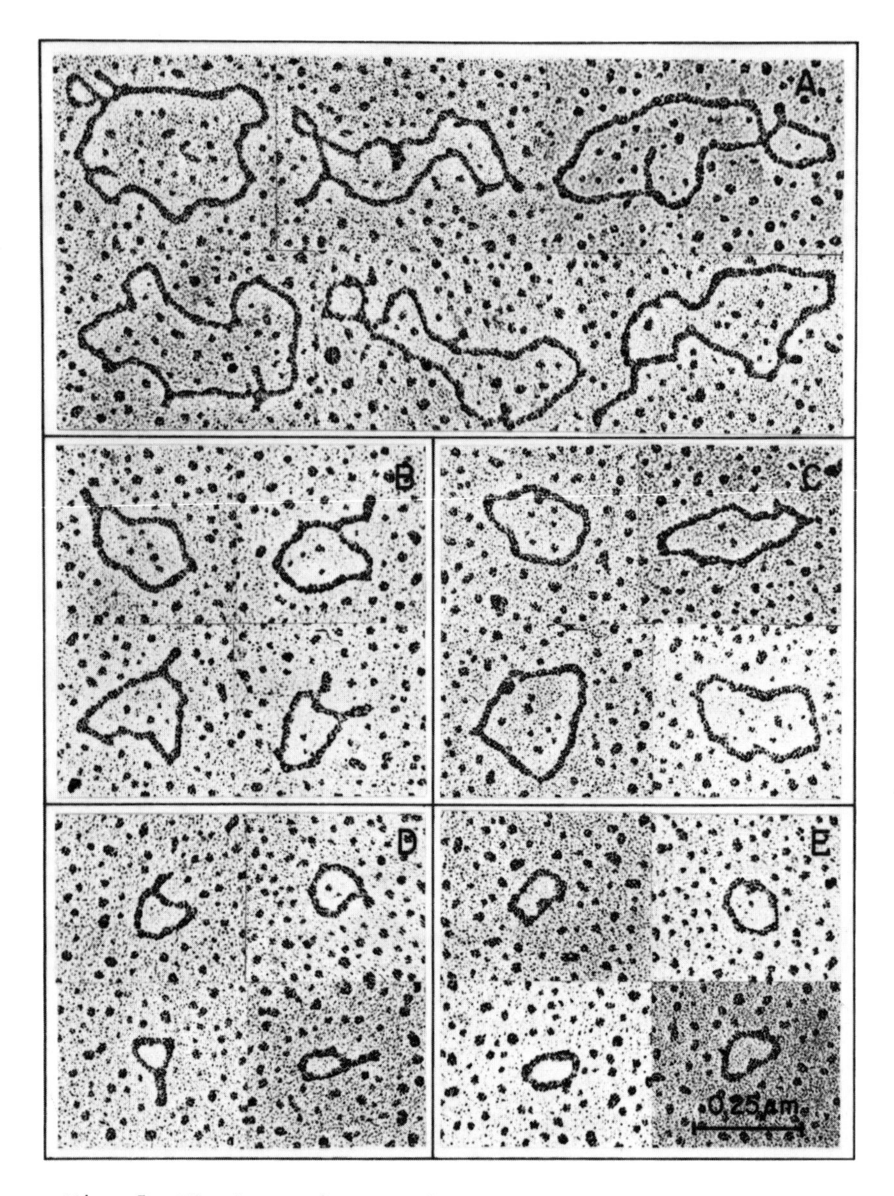

Fig. 5 Electron micrographs of selected circular Uukuniemi virus RNA molecules. The RNA was spread from 70% FA onto a 40% hypophase. Shown are L (A), M (B), and S (D) RNA molecules with distinct panhandles. Many of the M and S molecules were also seen without any apparent panhandles (C and E). [Reprinted from Hewlett et al., 1977, by permission of the American Society of Microbiology.]

TABLE I
Lengths and Molecular Weights of Uukuniemi Virus RNA Species[a]

Method and RNA	*Number of molecules measured*	*Number of molecules in peak*	*Length (m) mean ± SD*[b]	*Molecular weight ×10⁶ ± SD*[b,c]
Urea/Formamide				
L		40	1.59 ± 0.06	1.76 ± 0.07
M	} 458	179	0.74 ± 0.06	0.82 ± 0.07
S		188	0.34 ± 0.07	0.34 ± 0.07
Polio	153	84	2.27 ± 0.10	
Glyoxal				
L		41	1.65 ± 0.07	1.97 ± 0.08
M	} 441	160	0.80 ± 0.05	0.95 ± 0.06
S		192	0.40 ± 0.06	0.47 ± 0.07
Polio	206	74	2.09 ± 0.07	

[a] *Reprinted from Hewlett et al., 1977, by permission of the American Society of Microbiology.*

[b] *SD, standard deviation.*

[c] *Calculated from the length of the similarly treated and spread poliovirus RNA standard and assuming it has a molecular weight of* 2.5×10^6 *daltons.*

IV. CONCLUSIONS

The present results show that the genome of one bunyavirus, Uukuniemi virus, consists of three unique segments. This means that the three RNA segments probably are transcribed into complementary RNA segments each encoding different polypeptides. This situation would be analogous to that in the orthomyxovirus group (Palese, 1977). Multiple species of RNA have been found also in several other bunyaviruses, for instance, Lumbo virus (Bouloy *et al.*, 1973/74), La Crosse virus (Obijeski *et al.*, 1976), snowshoe hare virus (Gentsch and Bishop, 1976; Clewley *et al.*, 1977), and Inkoo (R. Pettersson and P. Saikku, unpublished data). Recent results by Clewley *et al.* (1977) have shown that the three RNA species of snowshoe hare virus and La Crosse virus also are unique. Thus it appears that the bunyaviruses in general possess a segmented genome. The Bunyaviridae family consists of a large number of serologically related and unrelated arboviruses. The segmentation of the genome offers one explanation for the evolution of such a diverse group of viruses. The ecology of these viruses, the occurrence of chronic infections of ticks and mosquitoes, the possibility of trans-ovarial transmission and simultaneous infection of the vector with two different bunyaviruses would all greatly facilitate situations in which reassortment of RNA segments could occur. In fact, such a reassortment of RNA segments has recently been demonstrated under laboratory conditions between temperature sensitive (ts) mutants of snowshoe hare virus (Gentsch and Bishop, 1976) and more recently between ts mutants of snowshoe hare and La Crosse viruses, two serologically related viruses (J. Gentsch, presented at the 77th Annual Meeting of the American Society for Microbiology, 1977).

The existence of stable circular forms of Uukuniemi virus ribonucleoprotein and ribonucleic acid species is an interesting though not unique finding among animal RNA viruses. Sindbis virion RNA can also form circles that are easily denatured by moderate formamide concentration (Hsu *et al.*, 1973). Recently very stable circular forms of RNA have been demonstrated for Lumbo virus, another bunyavirus (Samso *et al.*, 1975) and of Sendai RNA from defective interfering particles (Kolakovsky, 1976). In addition circular ribonucleoproteins have been demonstrated for La Crosse virus (bunyavirus) (Obijeski *et al.*, 1976), measles virus (paramyxovirus) (Thorne and Demott, 1976), and Tacaribe virus (arenavirus) (Palmer *et al*, personal communication). Furthermore, three nucleotides at the 3' and 5' ends of vesicular stomatitis RNA are complementary, suggesting that it can circularize (Banerjee *et al.*, 1977). Thus it appears that circular forms of single-stranded RNA are widespread among animal viruses. Uukuniemi virus RNAs and nucleoproteins form fairly stable circles, suggesting that the circular forms actually may serve as template for transcription and/or replication. An inverted complementary sequence at the ends of an RNA implies that the RNA of opposite polarity will also have the same structure at its respective ends. This means that the same recognition sequences for the replicase/transcriptase would be present at the 3' end of the negative and positive strands. What specific role the circular form and the complementary ends play in the life cycle of these viruses remains to be seen.

Acknowledgements

This work was partially performed at the Department of Biology and the Center of Cancer Research of the Massachusetts Institute of Technology, Cambridge, Massachusetts.

References

Banerjee, A. K., Abraham, G., and Colonno, R. J. (1977). *J. Gen. Virol. 34,* 1-8.

Bouloy, M., Krams-Ozden, S., Horodniceanu, F., and Hannoun, C. (1973/74). *Intervirology 2,* 173-180.

Clewley, J., Gentsch, J., and Bishop, D. H. L. (1977) *J. Virol. 22,* 459-468.

Davis, R. W., Simon, M., and Davidson, N. (1971). *In* "Methods in Enzymology" (L. Grossman and K. Moldave, eds.), Vol. 21, pp. 413-428. Academic Press, New York.

De Wachter, R., and Fiers, W. (1972). *Ann. Biochem. 49,* 184-197.

Gentsch, J., and Bishop, D. H. L. (1976). *J. Virol. 20,* 351-354.

Hewlett, M. J., Pettersson, R. F., and Baltimore, D. (1977). *J. Virol. 21,* 1085-1093.

Hsu, M. -T., Kung, H. -J., and Davidson, N. (1973). *Cold Spring Harbor Symp. Quant. Biol. 39,* 943-950.

Kolakovsky, D. (1976). *Cell 8,* 547-555.

Obijeski, J. F., Bishop, D. H. L., Palmer, E. L., and Murphy, F. A. (1976). *J. Virol. 20,* 664-675.

Palese, P. (1977). *Cell 10,* 1-10.

Pettersson, R. F., and Hewlett, M. J. (1976). *In* "Animal Virology" (D. Baltimore, A. S. Huang, and C. F. Fox, eds.), pp. 515-527. Academic Press, New York.

Pettersson, R. F., and Kääriäinen, L. (1973). *Virology 56,* 608-619.

Pettersson, R. F., and von Bonsdorff, C. -H. (1975). *J. Virol. 15,* 386-392.

Pettersson, R. F., Kääriäinen, L., von Bonsdorff, C. -H., and Oker-Blom, N. (1971). *Virology 46,* 721-729.

Pettersson, R. F., Hewlett, M. J., Baltimore, D., and Coffin, J. M. (1977). *Cell 11*, 51-64.
Porterfield, J. S., Casals, J., Chumakov, M. P., Gaidamovich, S. Ya., Hannoun, C., Holmes, I. H., Horzinek, M. C., Mussgay, M., Oker-Blom, N., and Russell, P. K. (1975/76). *Intervirology 6*, 13-24.
Ranki, M., and Pettersson, R. F. (1975). *J. Virol. 16*, 1420-1425.
Samso, A. M., Bouloy, M., and Hannoun, C. (1975). *C. R. Acad. Sci. Ser. D. 280*, 779-782.
Thorne, H. V., and Dermott, E. (1976). *Nature 264*, 473-474.
von Bonsdorff, C. -H, and Pettersson, R. F. (1975). *J. Virol. 16*, 1296-1307.

Chapter 16

REPLICATION OF ARBOVIRUSES IN ARTHROPOD *IN VITRO* SYSTEMS

Mary Pudney, C. J. Leake, and M. G. R. Varma

I. INTRODUCTION

The usefulness of arthropod *in vitro* systems, particularly cell lines, is already becoming obvious from the literature (Singh, 1971, 1972; Yunker, 1971; Rehacek, 1972; Dalgarno and Davey, 1973; Buckley, 1976; Rehacek, 1976). This chapter will describe some of the work being carried out in our laboratory using these systems.

Mosquito and Tick Cell Lines

In our laboratory we have established and maintain cell lines from culicine mosquitoes, *Aedes aegypti* (LSTM-AA-20A), *Aedes malayensis* (LSTM-AM-60), and *Aedes pseudoscutellaris*

ISBN 0-12-429765-X

(LSTM-AP-61) (Varma and Pudney, 1969; Varma *et al.*, 1974); anopheline mosquitoes, *Anopheles stephensi* (LSTM-AS-43) and *Anopheles gambiae* (LSTM-AG-ff) (Pudney and Varma, 1971; Marhoul and Pudney, 1972); ixodid ticks, *Rhipicephalus appendiculatus* (LSTM-RA-243) (Varma *et al.*, 1975); and a triatomid bug, *Triatoma infestans* (LSTM-TI-32) (Pudney and Lanar, 1977), which provide considerable scope for comparative studies of arbovirus replication *in vitro*.

II. COMPARATIVE ASPECTS OF ARBOVIRUS GROWTH

A. Sensitivity

Logically the examination of arbovirus replication in cell lines must start with preliminary studies of growth or no growth. The results we have obtained are summarized in Table I. The tick-borne flaviviruses LGT and LI and the unclassified QRF and ZIR, as well as the phlebotomine-borne bunyavirus SFS, failed to grow in all the mosquito lines tested. The most broadly susceptible cell lines were the *A. malayensis* (AM-60) and *A. pseudoscutellaris* (AP-61), supporting the growth of all 18 of the mosquito-borne viruses tested. The alphavirus ONN and the flavivirus DEN-2 failed to grow in the *A. aegypti* (AA-20A) cell line. The anopheline lines AG-55 (*A. gambiae*) supported the growth of 13 of the viruses tested, and AS-43 (*A. stephensi*) only 11, including the mosquito-borne bunyaviruses, which grew quite well in this cell line. Buckley *et al.* (1976) have observed that the sodium deoxycholate-sensitive (i.e., enveloped) tick-borne and phlebotomus-borne viruses failed to grow in Singh's *Aedes* cell lines, and we have confirmed this for our *Aedes* and *Anopheles* lines.

All the tick-borne viruses tested to date have multiplied in the *R. appendiculatus* cell lines, although to varying degrees. The viruses that multiplied are the flaviviruses LGT and LI, the bunyaviruses DUG and GAN, and the unclassified QRF, ZIR, PS, SOL, and KET. Several mosquito-borne viruses also multiplied in the tick cells, including the alphaviruses GET, SF, ONN, and WHA, and the flavivirus WN. But the flavivirus JE and the three mosquito-borne bunyaviruses GER, TAH, and CE as well as the rhabdoviruses CHP and PIRY failed to replicate in these cells.

Minimum infectious dose experiments, using a number of viruses, have shown a widely varying pattern. The results obtained with the alphavirus SF, the flavivirus WN, and the bunyavirus CE are shown in Table II. In order of sensitivity the cell lines were AP-61, AM-60, AG-55, AA-20A, and then AS-43. It is

TABLE I
Comparative Growth of Arboviruses of Five Taxons in Six Arthropod Cell Lines[a]

Virus taxon.	*Proven or suspected vector*	*AA-20A*	*AS-43*	*AG-55*	*AM-60*	*AP-61*	*RA-243*[b]
Alphavirus	Mosquito	6/7	2/7	5/7	7/7	7/7	5/5
Flavivirus	Mosquito	4/5	3/5	3/5	5/5	5/5	1/2
	Tick	0/2	0/2	0/2	0/2	0/2	2/2
Bunyavirus	Mosquito	5/5	5/5	4/5	5/5	5/5	0/3
	Phlebotomus	0/1	0/1	0/1	0/1	0/1	NT
	Tick/mosquito	NT	NT	NT	NT	NT	2/2
Rhabdovirus		1/1	1/1	1/1	1/1	1/1	0/2
Unclassified	Tick	0/2	0/2	0/2	0/2	0/2	5/5

[a]*NT, Not tested. Number positive/number tested.*
[b]*Results obtained using three tick cell lines, RA-243, RA-219, and RA-257 from R. appendiculatus.*

TABLE II
Minimum Infectious Doses for SF, WN, and CE in Five Mosquito Cell Lines[a]

Virus	*Dilution*	*AA-20A*	*AS-43*	*AG-55*	*AM-60*	*AP-61*
SF	neat	+	-	+	+	+
Seed titer	-5	+	-	+	+	+
7.0 dex pfu/ml	-6	-	-	+	+	+
Vero cells	-7	-	-	-	+	+
WN	-6	+	+	+	+	+
Seed titer	-7	-	-	-	+	+
6.7 dex pfu/ml	-8	-	-	-	+	+
PS cells	-9	-	-	-	+	+
CE	-3	+	+	+	+	+
Seed titer	-4	+	+	+	+	+
6.9 dex pfu/ml	-5	+	+	-	+	+
Vero cells	-6	+	+	-	+	+
	-7	-	-	-	+	-

[a]*+, Virus replication; -, no replication.*

obvious that in comparative studies, one of the major determinants of susceptibility is the innate difference between individual cell lines.

In comparisons of susceptibility of various cell lines, variations in strain and passage history of the virus used must also be taken into consideration. However, comparison of these studies with those in the most widely used cell lines, Singh's *A. albopictus* and *A. aegypti* (Singh, 1967) and Peleg's *A. aegypti* (Peleg, 1968) reveals a high degree of correlation, suggesting considerable specificity (Table III).

B. Growth Kinetics

The growth kinetics of 20 different viruses has been compared in the cell lines, and consistent results were frequently obtained in the 60 and 61 cells, similar to those recorded in Singh's *A. albopictus* cells. The results in the other cell lines were more variable.

TABLE III
Comparison of Growth/No Growth Studies in Six Aedes Cell Lines[a]

Virus	*AA-20A*	*S.AA*	*P.AA*	*SAAL*	*AM-60*	*AP-61*
SIN	+	+	+	+	+	+
SF	+	+	+	+	+	+
CHIK	+	+	NT	+	+	+
ONN	0	0	NT	+	+	+
JE	+	0	+	+	+	+
WN	+	+	+	+	+	+
DEN-2	0	0	+	+	+	+
YF	+	+	NT	+	+	+
LGT	0	0	NT	0	0	0
CE	+	+	+	+	+	+
CHP	+	+	NT	+	+	+
QRF	0	0	NT	0	0	0

[a]*S.AA, Singh's A. aegypti; P.AA, Peleg's A. aegypti; SAAL, Singh's A. albopictus. +-virus replication 0-no replication NT-not tested*

Examples will be given of representative viruses, and the growth compared to that obtained in the cold-blooded vertebrate cell line from *Xenopus laevis* (Pudney *et al.*, 1973a), which is maintained at the same temperature as the mosquito cells and has proved useful for arbovirus studies (Leake *et al.*, 1977).

1. Alphaviruses

Using SF virus in the AM-60 and AP-61 cells there was a rapid rise in extracellular titer to 7.5 dex pfu/ml by 24 hours, followed by a rapid fall. In contrast the rise in titer in the AA-20A cells was more gradual, reaching a peak of 5.5 dex pfu/ml 14 days after infection. There was also good growth in the anopheline line AG-55, with the peak 5.5 dex pfu/ml in 3-5 days, followed by a gradual decline. The other anopheline cell line did not support the growth of SF (Fig. 1).

In the tick cell lines the alphaviruses showed variable patterns of growth. Thus ONN and WHA reached fairly low titers, whereas SIN and SF grew well. In all three tick cell lines SF reached 6.7 logs 1-2 days after infection, giving a similar pic-

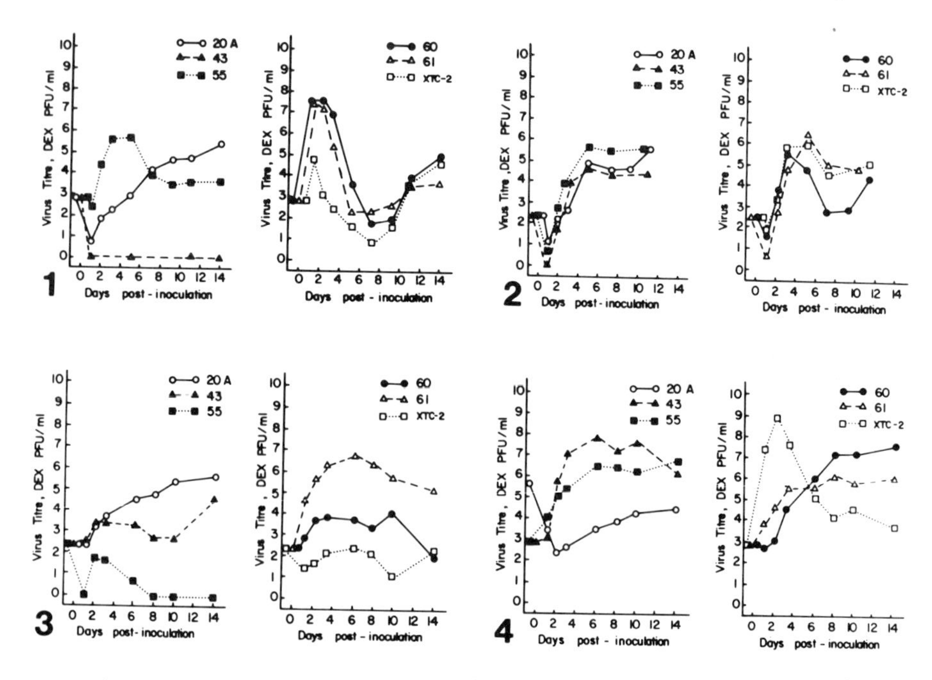

Fig. 1 Comparative growth of viruses from four taxons in five mosquito cell lines. (1) SF alphavirus, (2) WN flavivirus, (3) CE bunyavirus, (4) CHP rhabdovirus.

ture to that obtained in the AM-60 and AP-61 cells, except that the drop was more marked in the tick cells (see Fig. 2). No CPE was seen.

2. *Flaviviruses*

Generally the rate of flavivirus growth was slower than the alphaviruses, particularly in the AM-60 and AP-61 cells. For example, WN multiplied in all the cell lines giving very similar results. Generally there was a rise in titer to a peak of 5.6 dex pfu/ml, from 3-5 days after infection. This was maintained in the AA-20A, AS-43, and AG-55 cells, but fell in the AM-60 and AP-61 cells--these two lines also gave a CPE with this virus (see Fig. 1). WN reached a peak of 3.6 dex pfu/ml in the tick cells after 10 days, whereas the tick-borne flaviviruses LGT and LI reached 5.5 dex pfu/ml by the fourth day and then fell by about 2 logs during the remainder of the experiment (Fig. 2).

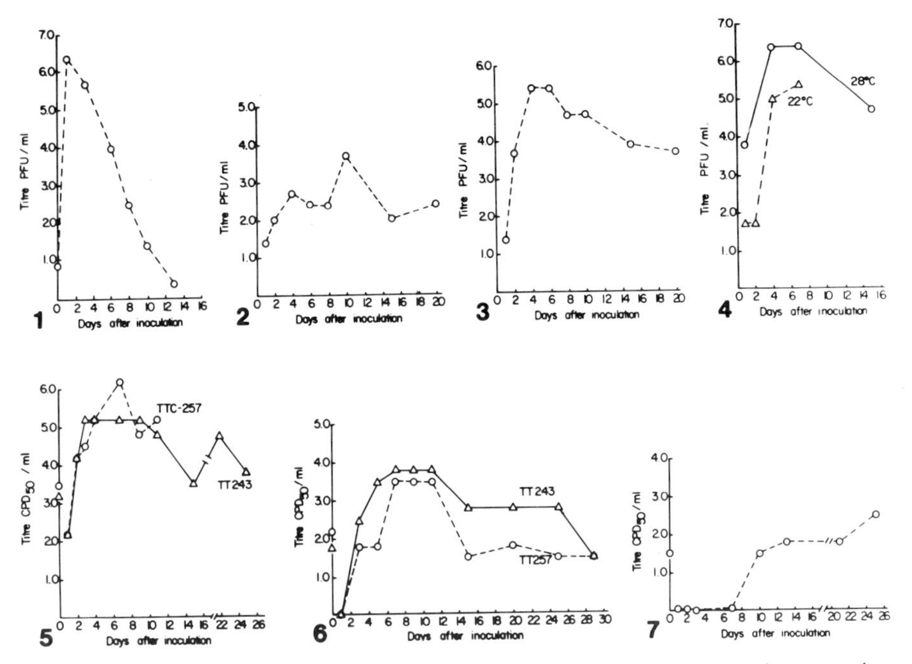

Fig. 2 Growth of mosquito-borne and tick-borne viruses in the RA-243 cell line. (1) SF, (2) WN, (3) LGT, (4) LI, (5) QRF, (6) KET, (7) ZIR.

3. *Bunyaviruses*

These showed variable growth patterns in the different cell lines. CE is the example described, which grew well in the AP-61 cell line, showing a progressive rise in titer to a peak of 6.5 dex pfu/ml, 6 days after infection. In contrast, in the related AM-60 cell line, the virus titer remained fairly constant at about 4.0 dex pfu/ml until 10 days after infection, and then dropped. In the AA-20A cell line, the titer rose gradually and was still rising at 14 days. In the AS-43 cells the virus remained constant at about 3.5 dex pfu/ml and then began rising, whereas the growth was poor in the AG-55 cells (Fig. 1). The three mosquito-borne bunyaviruses tested, GER, CE, and TAH, did not replicate in the tick cells. West (1974) successfully propagated the bunyavirus DUG, which has been isolated from both mosquitoes and ticks in our *Boophilus microplus* cells (Pudney *et al.*, 1973b), and we have now shown that both GAN and DUG will replicate in the RA-243 cells. DUG reached 4 dex pfu/ml by 6 days, which was then maintained for 22 days, whereas GAN reached a slightly higher peak of 5.4 dex pfu/ml by the third day, which fell slightly and was maintained at about 3.5 dex pfu/ml until 22 days. No CPE was produced.

4. *Rhabdoviruses*

To date we have only tested CHP in the mosquito cell lines. Unlike the results with the other viruses, the AS-43 cell line supported the best virus growth, producing 7 dex pfu/ml after 3 days. The titer was then maintained at this level. In the other anopheline cell line (AG-55) a slower rise was seen, and the lower peak of 6.5 dex pfu/ml was reached 6 days after infection and was also maintained. The rise was even more gradual in the AM-60 and AP-61 cells, reaching 7 dex pfu/ml by 8 days in the first, and 5.6 dex pfu/ml from 3 days in the latter. The AA-20A cell line was markedly less sensitive and an inoculum of 5.5 dex/bottle was needed to successfully establish an infection, and a slow increase in titer was produced (Fig. 1).

Neither CHP or PIRY grew in the tick cell lines. CHP has been isolated from phlebotomines and shown to be transmitted by mosquitoes in the laboratory, whereas PIRY does not as yet have a proven vector.

5. *Unclassified*

All the viruses of this group so far tested have grown in the tick cells, although the growth kinetics were quite variable. QRF, although isolated from argasid ticks grew well, with a peak titer of 5.6 dex CPD_{50}/ml on day 3-7, which gradually fell, although 4 dex of virus was still being produced by day 26 (Fig. 2). The RA-243 cell line was sensitive to infection with QRF virus, and could be infected with a 10^{-7} dilution of the seed virus. However, all the Hughes group viruses tested appeared to have a lag phase that varied from 7 days for ZIR, 6 days for PS and 3 days for SOL, before extracellular virus could be detected. This was then followed by a gradual slow rise, with low peak titers (2.4 dex pfu/ml ZIR, day 26 see Fig. 2; 4.0 dex pfu/ml SOL day 22; 5.5 dex pfu/ml PS on day 14.

It is obvious from these and more extensive studies (Leake, in preparation) that there is considerable variation in the growth pattern of a particular virus in various cell lines. The interesting fact is the close agreement in the growth pattern with previously published studies with SIN, CHIK, SF, JE, WN, DEN-2, and CE virus in Singh's *A. albopictus* cells. It would appear that individual growth patterns are characteristic of various viruses in specific or closely related cell lines.

III. PERSISTENT INFECTIONS

This topic has been covered in detail by other authors, particularly in relation to SIN in Singh's *A. albopictus* cells (Stollar and Thomas, 1975). Buckley (1976) concluded that inapparent, persistent infection of mosquito cell cultures may be induced with any arbovirus capable of replication in a given cell system. We have established persistent infections in the AM-60 and AP-61 cells with DEN-2, JE, and WE as well as the alphaviruses CHIK, SIN, and SF. Attenuation during persistence has been reported for CHIK, SF, and DEN-2 (Buckley, 1976).

IV. USE AS A DIAGNOSTIC TOOL

The observations of Singh and Paul (1968) that a CPE could be produced in Singh's *A. albopictus* cells by the mosquito-borne flaviviruses JE, WN, and DEN-1, -2, -3, and -4, suggested the possibility of direct virus isolations in these cells. The CPE was characterized by syncytia formation (Paul *et al.*, 1969) as compared to the moderate to marked cell destruction normally produced by arboviruses in vertebrate cell cultures (Scherer and Syverton, 1954; Buckley, 1964; Karabatsos and Buckley, 1967). Buckley (1969) confirmed the production of a CPE with WN virus in these cells, but failed to obtain a CPE with JE virus or with YF and SLE virus. In contrast, Sweet and Unthank (1971) obtained CPE with SLE as well as with Den-1, -2, -3, and -4. The nature of the surface of the culture vessel appeared to play an important part in the production of a CPE. Paul *et al.* (1969) and Suitor and Paul (1969) have observed that the CPE produced by DEN-2 in the *A. albopictus* cells is enhanced when the cells are grown on plastic tissue culture containers.

A. Cytopathic Effect

With the establishment of two cell lines from *A. malayensis* and *A. pseudoscutellaris* (Varma *et al.*, 1974) that are related to *A. albopictus*, the number of cell lines in which a cytopathic effect could be produced has been increased. Preliminary studies (Varma *et al.*, 1974) demonstrated that syncytia formation could be induced by WN, JE, and DEN-2 in the AP-61 cells and by WN and JE in the AM-60 cells. In later subcultures of the AM-60 cells, the CPE produced showed a more focal death consisting of rounding up of the cells followed by detachment from the surface of the culture vessel. Also in later subculture levels a CPE was produced with DEN-2 virus, which had not been observed previously. When cells were stressed by growing them on plastic in a mammalian medium, CPE could be obtained with JE, WN, DEN-1, DEN-2, TEM, ZIKA, and NTA, of the viruses so far tested. It was obvious from these experiments that a considerable number of factors are involved in the production of a CPE in mosquito cells.

1. Innate characteristic of a cell line. CPE so far has only been produced in *A. albopictus*, AM-60, and AP-61 cell lines, although plaques have been produced by JE in *Culex tritaeniorhyncus* (Hsu *et al.*, 1975).

2. Surface on which cells are growing. Considerable enhancement of CPE has been shown in cells growing on plastic, and in some cases no CPE is seen at all on glass.

3. Passage levels of cells. It was found that ZIKA virus would produce a CPE in low passage level AP-61 cells, passage levels 30-50, but not in high passage levels (i.e., passage 197).

4. Passage history of virus. This appeared to be of particular importance to the alphaviruses as we obtained clear syncytia formation with SIN and GET isolated from pools of field collected mosquitoes in the AP-61 cells. Previous tests with mouse brain material had proved negative although GET virus had only undergone five mouse brain passages. Syncytia formation from mosquito material was also produced only in low passage levels of the mosquito cells. This widens the scope of these cells for the isolation of arboviruses, as well as emphasizing the importance of virus passage history on the production of a CPE. Mouse brain passages of TEM (p15), ZIKA (p5), and Ntaya (p5) all produced CPE in the AP-61 cells although this has not been reported in the *A. albopictus* cells. Varma *et al.* (1976) described studies on the AP-61 cell line with three YF virus strains. While the cells supported replication of all three strains, the production of CPE varied, being clear with the wild-type strain, late and inconsistent with the French neurotropic strain, and absent with the 17D vaccine strain.

There appear to be two types of CPE produced in the AP-61 cells by the viruses so far tested. JE, WN, DEN-1, DEN-2 produced progressive syncytia formation while Tembusu, Zika, Ntaya, and YF (wild type) produced cell death leading to the rounding up and floating off of cells (Fig. 3).

At present we are studying a syncytial type of CPE produced by a large range of alphaviruses in low passage level AP-61 cells grown on plastic at 32°C. The AP-61 cells cannot be continuously maintained at this temperature, but will survive for up to 10 days in a healthy condition, although they can be grown continuously at 30°C. Nine alphaviruses, three bunyaviruses, and a CAL group virus have all produced syncytia within 2-5 days

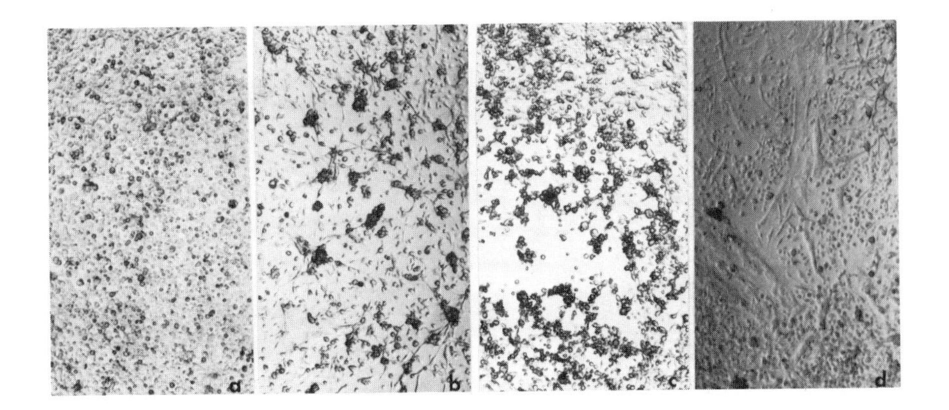

Fig. 3. Cytopathic effect produced in AP-61 cells. (a) Uninfected control A. pseudoscutellaris cells, (b) NTA 7 days after infection, (c) YF (TR 4205) 6 days after infection, (d) GET (Primary isolation) 6 days after infection.

at 32°C (Fig. 4). The syncytia formation has been specifically blocked by immune sera. The picornavirus Nodamura produced an atypical CPE in the AP-61 cells at 32°C. Although the growth of the virus was not examined, it does grow in the *A. albopictus* cells without production of a CPE (Bailey *et al.*, 1975). However, under certain conditions (low FCS concentration, use of old cells, in certain plastic containers) a spontaneous syncytia formation can occur in the AP-61 cells at 32°C after about 10 days (Table II). Despite this, it appears that this system could be used for virus isolation procedures, and could cut down the time required for identification purposes. Further studies are in progress.

B. Primary Isolations

Although CPE is generally absent in mosquito cell lines infected with arboviruses, the sensitivity of these cells to infection with arboviruses indicates their usefulness for isolation purposes. The only recorded direct isolations are by Singh and Paul (1969) and Singh (1972), who isolated the four serotypes of DEN virus from human sera and mosquitoes, in the *A. albopictus* cells, and concluded that the cell line proved a very convenient tool for the rapid isolation and identification of the DEN virus serotypes during recent epidemics in India. Chappell *et al.* (1971) also showed that *A. albopictus* cells were superior to both infant mice and LLC-MK2 cells for the

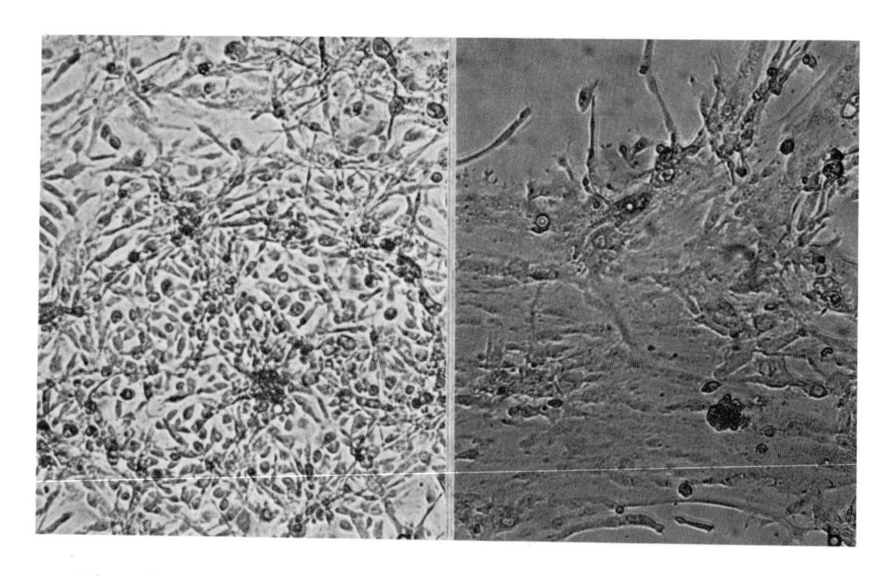

Fig. 4 Cytopathic effect produced at 32°C. (a) Control AP-61 cells sc42 3 days at 32°C, (b) ONN 3 days after infection at 32°C.

isolation of DEN-2 virus from human blood and naturally infected mosquitoes during the 1969 epidemic in Puerto Rico. We have isolated DEN-2 from human serum collected during the 1969 Puerto Rico epidemic in the AP-61 cells (Varma *et al.*, 1976). Sweet and Unthank (1971) observed that SLE virus in a known positive infectious field collected mosquito pool produced a cytopathic effect in the *A. albopictus* cells. All previous work using mouse-brain material had shown growth but not CPE.

Using the AP-61 cells grown at 28°C we have isolated three strains of SIN, five strains of GET, as well as 14 strains of JE and 12 of Tembusu. Three pools that were proved to be positive for GET by mouse inoculation gave doubtful results, as did three shown to be JE. In comparison, the original isolations in mice, several years previously, had been 5 SIN, 8 GET, 18 JE and 12 TEM.

Although using these old mosquito pools, the mosquito cells did not prove any more sensitive than suckling mice, we did show that alphaviruses could also be isolated directly in mosquito cells. The majority of work with arboviruses is carried out with mouse-adapted or mouse-passaged strains, and we now have 34 strains of arboviruses that have not been passaged in vertebrate cells. The fact that the viral envelope is derived from host cell membrane underlines the importance of these isolations, as mosquito-propagated virus may have altered infectivity for heterologous host cells and vice versa. The DEN-2 strain (PR-159) we isolated in mosquito cells could not be detected in pig kidney (PS) cells, and had a titer 3 dex lower in suckling mice than by AP-61 syncytia formation. The syncytia formation could be blocked by the use of DEN-2 antiserum prepared against mouse brain virus. The other isolates have not yet been tested in this way.

Although the highly attenuated 17D strain of YF did not produce a CPE in the AP-61 cells, we were able to isolate several strains of YF in these cells (Varma *et al.*, 1976). Pools of naturally infected mosquitoes, human liver samples, and sentinel monkey serum were sent to us under code from MARU by Dr. Karl Johnson. All five samples had previously been tested in the *A. albopictus* cells (in which a CPE was not produced), Vero cells, and intracerebrally in suckling mice. All samples produced a distinct CPE in the AP-61 cells consisting of foci of dark granular cells, which could be observed as early as 4 days after infection and progressed rapidly to complete cell destruction involving the whole cell sheet. The endpoint was usually reached in 6-10 days, and in one case in 13 days. The titers obtained in the AP-61 cells were of the same order as those obtained in Vero cells and higher than those obtained in the ATC-15 cells, and suckling mice.

The CPE could be specifically blocked using antiserum produced against 17D YF. Buckley (1969) had obtained multiplication of the Asibi and Couma strains of YF virus in the *A. albopictus* cells with a sporadically occurring CPE, which she rated as negative. Yunker and Cory (1975) obtained clearly defined plaques with the 17D strain in the ATC-15 cells incubated at $37^{o}C$ under an agarose overlay, but no previous CPE had been seen.

During isolation of viruses from the mosquito pools collected in Kenya, a CPE was obtained with a pool of *Mansonia uniformis* mosquitoes. This was a cell degeneration type of CPE, not syncytia, which appeared on the eleventh day after infection. Further passage in mosquito cells consistently produced a CPE, but no effect could be produced in suckling mice or the mamalian cell systems used (PS, Vero, and BHK). Under the electron microscope a viruslike particle was seen with a diameter of 30 nm, in the range of the picornaviruses. On purification it was shown to contain a single-stranded RNA, and two polypeptides (Pudney *et al.*, to be published). The virus is resistent to ether and is stable to acid and alkali. It is very thermostable with no loss in titer after 21 days in medium at $28^{o}C$, and was still detectable at 92 days. The RNA is infectious, in that it causes CPE in the AP-61 cells. This small RNA virus has been called Kawino. The virus does not grow in BHK, LLC-MK_2 cells, or the amphibian line XTC-2, although there does appear to be some maintenance in PS cells. It grows in the AP-61 cells, the AM-60 cells, and the *A. albopictus* cells, but not in the

AA-20A, AS-43, and RA-219 cells. A CPE was only produced in the AP-61 cells where the highest titers of released virus were reached on day 3 (5.3 dex pfu/ml). The virus appears to be closely cell-associated, as after cell disruption a virus titer of 10^8 dex pfu/ml was obtained 24 hours after infection. Plaque assay could be carried out in the AP-61 cells under CMC or agarose overlay.

Pools of *M. uniformis* mosquitoes from a laboratory colony did not produce a CPE in the AP-61 cells, and preliminary experiments have failed to show replication of Kawino virus in the colony mosquitoes after membrane feeding.

Kawino is distinct from another picornavirus, Nodamura, isolated from *Culex tritaeniorhyncus* mosquitoes (Scherer and Hurlbut, 1967) as Nodamura has been shown to contain two species of RNA (Newman and Brown, 1973). Nodamura virus has been shown to grow in the *A. albopictus* cell line and Porterfield's *A. aegypti* cells (Bailey *et al.*, 1975) and also BHK-21 cells without producing a CPE. The growth of Nodamura virus has not been examined in the AP-61 cells.

C. Plaquing

The practical use of mosquito cells in arbovirus studies was shown in 1969 by Suitor, who demonstrated that JE virus could produce plaques in the *A. albopictus* cells. This was later confirmed by Cory and Yunker (1972) using a different strain of JE and they also obtained plaques with the flaviviruses WN, YF, and DEN-1, -2, -4 and the rhabdovirus VSI. Following an extensive study (Yunker and Cory, 1975) these workers showed that 30 of the 124 virus strains they tested would produce plaques in *A. albopictus* cells incubated at 35-37 C under agarose overlay. All these positive viruses were proven or suspected to be mosquito-borne. The majority of the viruses used had previously failed to produce CPE in the *A. albopictus* cells. Representatives from six serological groups produced plaques in the cells although this varied from group to group. Thus only one out of nine alphaviruses tested, the mosquito-borne CHIK, produced plaques, whereas 14 of the 16 flaviviruses did so. Most of the Bunyamwera group viruses produced plaques with the exception of the California group viruses. Negative results were obtained with all the tick-borne and vector-unassociated viruses, two strains of insect pathogen, and six nonarboviruses.

Hsu *et al.* (1975) have described plaque formation by JE virus in *C. tritaeniorhyncus* cells, although CPE had not previously been seen.

Yunker and Cory (1975) have suggested that the lack of reports on arbovirus plaquing in mosquito cells is due partly to the lack of an obvious CPE and partly to the technical difficulties of preparing satisfactory monolayers of cells.

These technical difficulties were similarly encountered in studies with our AM-60 and AP-61 cell lines, and initial results were very variable. However, using Yunker and Cory's basic technique, employing L-15 medium and a low-gel point agarose (Seaplaque agarose: Marine Colloids Ltd.) we have now obtained consistent results. So far plaques have been produced in the AP-61 cells at $28^{\circ}C$ with DEN-1, DEN-2 (strains New Guinea and PR159), TEM, YF (primary isolates), ZIKA, JE, and NTA (Fig. 5). Plaques could be obtained with DEN-3, CHIK, SF, and SIN by increasing the temperature to $32^{\circ}C$.

Fig. 5 Plaque production in AP-61 cells. (a) DEN-2 (PRI59), (b) NTA, (c) ZIKA, (d) JE, (e) DEN-1, (f) TEM, (g-j) Wild YF: (g) 303165 Haemogogus lucifer, (h) 303547 H. lucifer, (i) 311376 human liver suspension, (j) 902677 Sentinel rhesus monkey serum. Agarose removed and cells stained with napthalene black.

V. CONTAMINATION OF CELL LINES

Contamination can be considered in two main areas.

A. Cross Contamination of Cell Lines

Evidence of extrinsic cell contamination of an established insect cell subline (Grace's *A. aegypti* cell line) has been demonstrated by Greene and Charney (1971) using isoenzyme analysis. Using these techniques we have shown that all our cell lines are distinct (Pudney and Swindlehurst, unpublished).

B. Virus Contamination

Hirumi (1976) has presented a comprehensive review of the possible sources of contamination in arthropod cell lines. In his most extensive study he demonstrated the presence of no fewer than five types of viruslike particles. These resembled toga-, parvo-, orbi-, picorna-, and bacterial viruses, and he feels that the toga-viruslike particle may be responsible for syncytia formation in the *A. albopictus* cells.

Other evidence for viruslike contamination of arthropod cell lines has been reported by Filshie *et al.* (1967), Suitor and Paul (1969), Pudney *et al.* (1971), Cunningham *et al.* (1975), Stollar *et al.* (1975), and Hirumi (1976).

Because of the suggested link between syncytia formation in mosquito cells and the presence of contaminating viruses (Hirumi, 1976), we are thoroughly examining the AP-61 cells for the prsence of viral contamination. The results so far have proved negative. Electron-microscopic examination of both normal and DEN-2 stressed 61 cells has not shown the presence of contaminating viruslike particles. Inoculation of disrupted AP-61 cells into the PS, Vero, BHK-21, LLC-MK_2, AM-60, XTC-2 cell lines or suckling mice proved negative.

VI. CONCLUSIONS

Previous studies of arbovirus replication have now been extended by the use of our mosquito and tick cell lines. Innate differences between cell lines were obvious, and of particular interest was the similarity of results obtained with the AM-60, AP-61 and the related *A. albopictus* cell line. The range of viruses found to produce a cytopathic effect in mosquito cells has been extended using the AP-61 cells under specified conditions. Isolations from field material in mosquito cells have been carried out and provided strains of arboviruses that have not been passed in vertebrate cells. The increasing use of mosquito cell lines has already proved the usefulness of these systems as laboratory tools for the study of arbovirus replication. With the availability of tick cell lines, similar *in vitro* studies can now be carried out with the numberous tick-borne arboviruses.

References

Bailey, L., Newman, J. F. E., and Porterfield, J. S. (1975). *J. Gen. Virol. 26,* 15-20.

Buckley, S. M. (1964). *Proc. Soc. Exp. Biol. Med. 116,* 354-358

Buckley, S. M. (1969). *Proc. Soc. Exp. Biol. Med. 131,* 625-630.

Buckley, S. M. (1976). *In* "Invertebrate Tissue Culture Research Applications" (K. Maramorosch, ed.), pp. 201-229. Academic Press, New York.

Chappell, W. A., Calisher, C. H., Toole, Roberta, F., Maness, Kathryn C., Sasso, Donna R., and Henderson, B. E. (1971). *Appl. Microbiol. 22,* 1100-1103.

Cory, J., and Yunker, C. E. (1972). *Acta Virol. Prague 16,* 90.

Cunningham, A., Buckley, S. M., Casals, J., and Webbs, R. (1975). *J. Gen. Virol. 27,* 97-101.

Dalgarno, L., and Davey, M. W. (1973). *In* "Viruses and Invertebrates" (A. J. Gibbs, ed.), pp. 245-270. North Holland Publ. Co. Amsterdam and London.

David-West, T. S. (1974). *Arch. Ges. Virusforsch. 44,* 330-336.

Filshie, B. K., Grace, T. D. C., Poulson, D. F., and Rehacek, J. (1967). *J. Invert. Pathol. 9,* 271-273.

Greene, A. E., and Charney, J. (1971). *Curr. Top. Microbiol. Immunol. 55,* 51-61.

Hirumi, H. (1976). *In* "Invertebrate Tissue Culture Research Applications" (K. Maramarosch, ed.), pp. 233-268. Academic Press, New York.

Hsu, S. H., Wong, W. J., and Cross, J. H. (1975). *Southeast Asian J. Trop. Med. Publ. Health 6,* 1-2.

Karabatsos, N., and Buckley, S. M. (1967). *Am. J. Trop. Med. Hyg. 16,* 99-105.

Leake, C. J., Varma, M. G. R., and Pudney M. (1977). *J. Gen. Virol. 35,* 335-339.

Marhoul, Z., and Pudney, M. (1972). *Trans. Foy. Soc. Trop. Med. Hyg. 66,* 183-184.

Newman, J. F. E., and Brown, F. (1973). *J. Gen. Virol. 21,* 371-384.

Paul, S. D., Singh, K. R. P., and Bhat, U. K. M. (1969). *Ind. J. Med. Res. 57,* 339-348.

Peleg, J. (1968). *Virology 35*, 617-619.
Pudney, M., and Lanar, D. (1977). *Ann. Trop. Med. Parasital 71*, 109-118.
Pudney, M., and Varma, M. G. R. (1971). *Exp. Parasitol. 29*, 7-12.
Pudney, M., McCarthy, D., and Shortridge, K. F. (1971). *Proc. 3rd Int. Coll. Invert. Tissue Culture, Smolenice*, 337-345.
Pudney, M., Varma, M. G. R., and Leake, C. J. (1973a). *Experientia 29*, 466-467.
Pudney, M., Varma, M. G. R., and Leake, C. J. (1973b). *J. Med. Ent. 10*, 493-496.
Rehacek, J. (1972). *In* "Invertebrate Tissue Culture" (C. Vago, ed.), pp. 279-320. Academic Press, New York.
Rehacek, J. (1976). *In* "Invertebrate Tissue Culture. Applications in Medicine, Biology and Agriculture" (E. Kurstak and K. Maramorosch, eds.), pp. 21-33. Academic Press, New York.
Scherer, W. F., and Hurlbut, H. S. (1967). *Am. J. Epidem. 86*, 271-285.
Scherer, W. F., and Syverton, J. T. (1954). *Am. J. Pathol. 30*, 1075.
Sinarachatanant, P., and Olson, L. C. (1973). *J. Virol. 12*, 275-283.
Singh, K. R. P. (1967). *Curr. Sci. 36*, 506-508.
Singh, K. R. P. (1971). *Curr. Top. Microbiol. Immunol. 55*, 127-133.
Singh, K. R. P. (1972). *Advan. Virus Res. 17*, 187-206.
Singh, K. R. P., and Paul, S. D. (1968). *Curr. Sci. 37*, 65-67.
Singh, K. R. P., and Paul, S. D. (1969). *Bull. Wld. Hlth. Org. 40*, 982-983.
Stollar, V., and Thomas, V. L. (1975). *Virology 64*, 367-378.
Suitor, E. C. Jr. (1969). *J. Gen. Virol. 5*, 545-546.
Suitor, E. C. Jr., and Paul, F. J. (1969). *Virology 38*, 382-485.
Sweet, B. H., and Unthank, H. D. (1971). *Curr. Top. Microbiol. Immunol. 55*, 150-154.
Varma, M. G. R., and Pudney, M. (1969). *J. Med. Ent. 6*, 432-439.
Varma, M. G. R., Pudney, J., and Leake, C. J. (1974). *Trans. R. Soc. Trop. Med. Hyg. 68*, 374-382.
Varma, M. G. R., Pudney, M., and Leake, C. J. (1975). *J. Med. Ent. 6*, 698-706.
Varma, M. G. R., Pudney, M., Leake, C. J., and Peralta, P. H. (1975/76). *Intervirology 6*, 50-56.
Yunker, C. E. (1971). *Curr. Top. Microbiol. Immunol. 55*, 113-126.
Yunker, C. E., and Cory, J. (1975). *Appl. Microbiol. 29*, 81-89.

Chapter 17

SUSCEPTIBILITY OF A TICK CELL LINE *(DERMACENTOR PARUMAPERTUS* NEUMANN) TO INFECTION WITH ARBOVIRUSES

U. K. M. Bhat and C. E. Yunker

I. INTRODUCTION

Despite continuing success in the establishment of cell lines from mosquitoes (Hink, 1976), a serially propagating cell line from ticks was not forthcoming until 1975, when Varma *et al.* (1975) reported the establishment of three cell lines from *Rhipicephalus appendiculatus* Neumann (Ixodidae). Mosquito cell lines have been used advantageously in several basic and diagnostic virological and rickettsial studies [e.g., as useful systems for isolation of certain arboviruses and rickettsiae (Chappel *et al.*, 1971; Cory *et al.*, 1975; Singh and Paul, 1969; and Varma *et al.*, 1975-1976), to characterize unidentified viruses and determine their probably arthoropod vectors (Buckley, 1971; Buckley *et al.*, 1976; Singh, 1971), in the production of superior serological reagents (Ghosh *et al.*, 1973; Pavri and Ghosh, 1969; Yunker *et al.*, 1970),and in the attenuation of arbo-

ISBN 0-12-429765-X

viruses (Banerjee and Singh, 1969; Buckley, 1973, Peleg, 1971)]. Similar applications of tick cell lines await evaluation of their pathogen-supporting abilities. Recently we established a continuous cell line from the ixodid tick *Dermacentor parumapertus* Neumann (Bhat and Yunker, 1977). Its susceptibility to infection with a number of arboviruses of various groups is reported here.

II. MATERIALS AND METHODS

A. Cell Culture

Cells from the continuous line of *D. parumapertus* (RML-14) from passages 10 to 17 were used. Details of maintenance of this cell line and of its culture medium were given earlier (Bhat and Yunker, 1978). Cells from 8- to 10-day-old stock culture flasks were mechanically detached and suspended in the medium. One-ml aliquots of the suspension, containing approximately 1 to 2 × 10^6 cells, were distributed among batches of 16 × 125 mm plastic screw-cap culture tubes (LUX Scientific Corp., Newbury Park, California) that had been pretreated by adding 1 ml culture medium and allowing this to remain for at least 24 hours. Cultures were incubated at 29 ± 1°C. On days 3 and 6 the existing medium in each culture tube was replaced with 1 ml fresh medium. A monolayer was formed in about 6 days.

B. Virus Inoculation and Assay

Twenty-four virus strains used in the present study were from the Rocky Mountain Laboratory collection (Table I). Stock viruses, all as mouse brain suspensions, were diluted 1:100 or 1:1000 in the culture medium before inoculation (Yunker and Cory, 1975). To study viral growth curves, existing medium was removed from 20 8-day-old tube cultures of *D. parumapertus* cells and each tube was inoculated with 1 ml of the virus suspension. A sample of the inoculum was stored at -65°C for titration at a later date. Two cultures inoculated with each virus were harvested on postinoculation(p. i.) days 0(2h after inoculation), 1, 2, 4, 6, 8, 10, and 15 as follows: cultures were frozen and thawed thrice, centrifuged at 200 xg for 5 min, and the supernatants were pooled and stored at -65 C. On p. i. day 10, after sampling was done, the medium in each of the remaining culture tubes was replaced with 1 ml of fresh medium.

Control tubes (without cells) received 1 ml virus suspension; these were harvested in a similar manner on p.i. days 4 and 10. Generally, virus-infected tissue culture fluids were titrated in African green monkey kidney (Vero strain CCL-81) cells by plaque enumeration. Powassan and Silverwater viruses were similarly assayed in a stable line of pig kidney (PS) cells (obtained from Dr. J. S. Porterfield, Medical Research Council, London, U.K.). Identities of selected viruses that grew in the tick cells were confirmed by plaque-reduction (neutralization) tests in Vero cells (Early *et al.*, 1967).

III. RESULTS

Viruses tested in *D. parumapertus* cell cultures included members of the genera Alphavirus, Flavivirus, Orbivirus, some "possible members" of the genus Bunyavirus (Porterfield *et al.*, 1975-1976), and some yet unclassified viruses. Mosquitoes or ticks are proven or suspected vectors of most of these viruses, although a few are apparently not vector-borne (Table I). Several of these viruses multiplied to relatively high levels, some multiplied less well, and a few failed to grow altogether. Cytopathic effect was not apparent in cultures infected with any of the viruses. Typically, growth curves showed an initial lag phase that was followed by a logarithmic increase in virus titer. Thereafter, amounts of virus recovered stabilized, although in a few instances virus concentrations continued to increase until the tests were terminated.

A. Alphavirus

Chikungunya, a mosquito-borne virus, was the only member of this genus tested in *D. parumapertus* cells, and it showed significant growth (Fig. 1). A maximum titer of 7.5 dex was attained on the fourth p.i. day.

B. Flavivirus

Eight strains of viruses of this genus were tested (Fig. 1). Of these, two strains of West Nile virus [one of mosquito origin (strain 7259-60; size of plaques produced in Vero cells 2-3 mm), the other isolated from ticks (strains Ar108-60; size of plaques produced in Vero cells 8-12 mm)] and one strain each of Central European tick-borne encephalitis, Omsk hemorrhagic fever, Powassan, and Langat viruses (all tick-borne) multiplied

TABLE I
Viruses Tested in the D. parumapertus *Cell Line*

Virus[a]	*Genus*[b]	*Sero-group*[c]	*Vector*[d]	*Strain designation*	*Mouse passage*
Chikungunya[e]	Alphavirus	A	Mosquito	23161	174
West Nile[e]	Flavivirus	B	Mosquito	7259-60	27
West Nile[e]	Flavivirus	B	Mosquito	Ar108-60	7
TBE[e]	Flavivirus	B	Ixodid tick	Hypr	2
OHF	Flavivirus	B	Ixodid tick	Guriev	1
Powassan	Flavivirus	B	Ixodid tick	M794	7
Langat[e]	Flavivirus	B	Ixodid tick	TP21	9
Modoc	Flavivirus	B	Not vector-borne	M544	4
Rio Bravo	Flavivirus	B	Not vector-borne	M64	15
Dugbe[e]	Bunyavirus[f]	NSD	Ixodid tick	Ar1792	16
Hazara	Bunyavirus[f]	CCHF	Ixodid tick	JC280	9
Silverwater	Bunyavirus[f]	KSO	Ixodid tick	131	8
Bhanja	Bunyavirus[f]	UNG	Ixodid tick	G690	20
Sunday Canyon	Bunyavirus[f]	UNG	Argasid tick	RML52301-11	6
Kemerovo	Orbivirus	KEM	Ixodid tick	R10	8

Tribec[e]	Orbivirus	KEM	Ixodid tick	VR468	18
CTF[e]	Orbivirus	UNG	Ixodid tick	SS-18	8
CTF[e]	Orbivirus	UNG	Ixodid tick	Florio	30
Hughes[e]	UNCL	HUG	Argasid tick	Original	11
Soldado	UNCL	HUG	Argasid tick	TRVL52214	18
SapphireII	UNCL	HUG	Argasid tick	RML52301-14	3
Midway	UNCL	NYM	Argasid tick	47153	8
Quaranfil	UNCL	QRF	Argasid tick	Ar1113	4
Dhori[e]	UNCL	UNG	Ixodid tick	611313	5

[a] *CTF, Colorado tick fever; OHF, Omsk hemorrhagic fever; TBE, Central European Tick-borne encephalitis.*

[b] *UNCL, Unclassified.*

[c] *CCHF, Crimean-Congo hemorrhagic fever; HUG, Hughes; KEM, Kemerovo; KSO, Kaisodi; NSD, Nairobi sheep disease; NYM, Nyamanini; QRF, Quaranfil; UNG, ungrouped.*

[d] *Known or suspected.*

[e] *Identity confirmed by plaque-reduction test after growth in* D. parumapertus *cells.*

[f] *Possible members of the genus Bunyavirus (Porterfield et al., 1975-1976).*

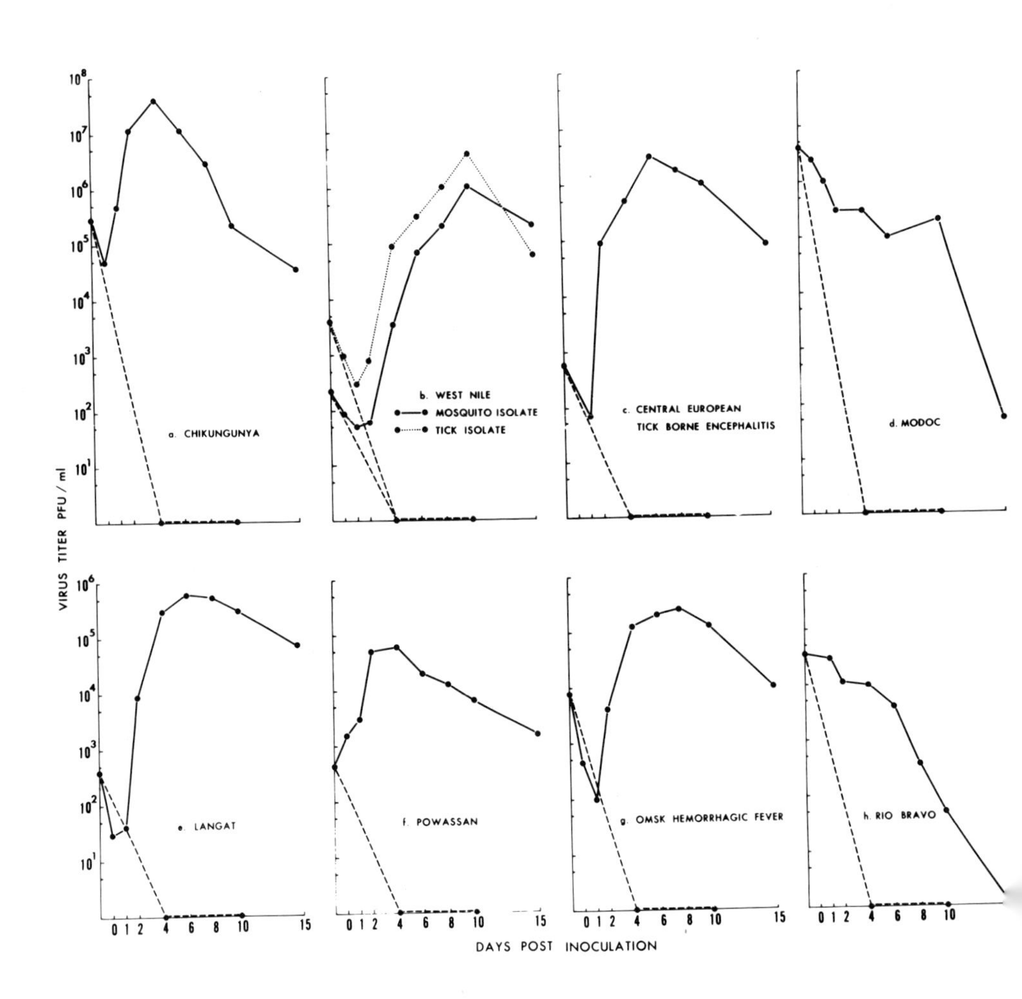

Fig. 1. Multiplication of certain Alpha- and Flaviviruses in the D. parumapertus cell line. Two infected cultures harvested, pooled, and titrated in Vero cells on days 0 (2 hours after inoculation), 1, 2, 4, 6, 8, 10, and 15, except Powassan virus titrated in pig kidney cells. Titer of inoculum indicated on ordinate. ————, Experimental; ----, control.

to relatively high levels. A maximum titer of 4.8-6.6 dex was attained between p.i. days 4 and 10 with those viruses. Their growth curves were typical except in the case of Powassan, in which a lag phase was not detected. Despite differences in origin, in passage history, and in plaque sizes produced by the two strains of West Nile virus, both grew equally well in these cultures. Langat virus (strain TP21) grew well in *D. parumapertus* cells, as it did in the *Rhipicephalus appendiculatus* cell line (Varma *et al.*, 1975). Two other Flaviviruses, Modoc and Rio Bravo, failed to multiply. Their titers dropped following inoculation.

C. Bunyavirus

Five possible members of this genus were tested in *D. parumapertus* cells (Fig. 2). All were tick-borne viruses and represented various serogroups or were ungrouped. Hazara (Crimean-Congo hemorrhagic fever serogroup) and Dugbe (Nairobi sheep disease serogroup) viruses multiplied well, and maximum titers of 5.5 and 5.4 dex were attained on p.i. days 15 and 6 respectively. Hazara virus, which was inoculated at concentration too low to detect in Vero cells, remained undetected in *D. parumapertus* cells up to the fourth p.i. day. It was first detected on the sixth p.i. day, and thereafter its titer increased rapidly up to p.i. day 15 when the experiment was terminated. Dugbe virus has previously been shown to propagate in a serially passed strain of cells from the tick *Boophilus microplus* (David-West, 1974). Silverwater (Kaisodi serogroup), Bhanja, and Sunday Canyon (serologically ungrouped) viruses persisted in *D. parumapertus* cells, but their growth, if any, remained undetected due to high doses of inocula that were used.

D. Orbivirus

Four strains of tick-borne viruses of the genus Orbivirus were tested (Fig. 2). Kemerovo and Tribec viruses (both Kemerovo serogroup) and two strains of Colorado tick fever virus (CTF serogroup) multiplied well. Maximum titers of 5.1 dex for Kemerovo and 5.3 dex for Tribec viruses were obtained on p.i. days 10 and 4, respectively. Whereas these viruses exhibited the typical growth curve terminating in a plateau phase, both strains of Colorado tick fever virus continued to increase in titer up to p.i. day 15 when the experiment was terminated. At this time maximum yields were 6.5 dex (Florio strain) and 5.7 dex (strain SS-18).

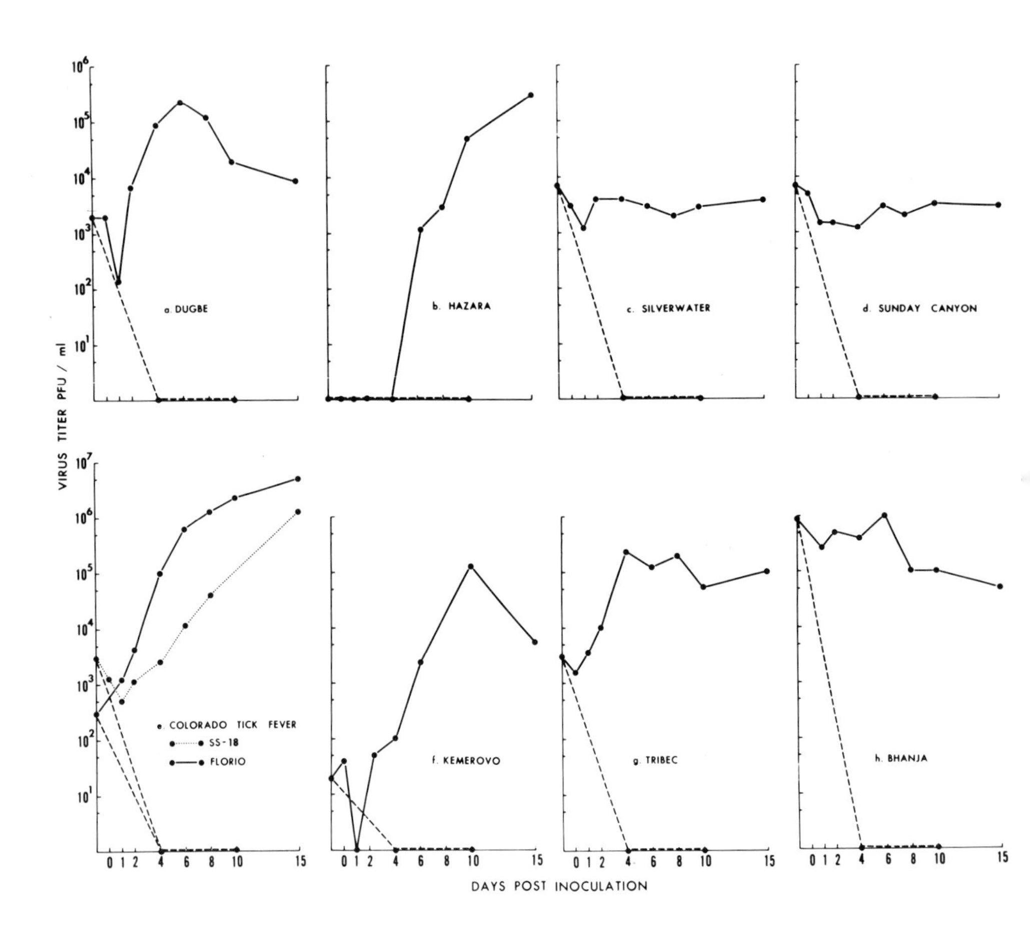

Fig. 2. Multiplication of certain Bunya- and Orbiviruses in the D. parumapertus *cell line. Two infected cultures harvested, pooled, and titrated in Vero cells on days 0 (2 hours after inoculation), 1, 2, 4, 6, 8, 10, and 15, except Silverwater virus titrated in pig kidney cells. Titer of inoculum indicated on ordinate. ——, Experimental; ---, control.*

E. Unclassified Viruses

Six tick-borne viruses of undetermined taxonomic position were tested in *D. parumapertus* cells (Fig. 3). Of these, Hughes (Hughes serogroup), Midway (Nyamanini serogroup), and Dhori (serologically ungrouped) viruses multiplied well, whereas Soldado (Hughes serogroup) virus showed only a low level of multiplication. Maximum yields were 5.5 dex for Hughes, 2.5 dex for Midway, 6.1 dex for Dhori, and 3.7 dex for Soldado viruses on p.i. days 15, 15, 4, and 8 respectively. In contrast, Sapphire II (Hughes serogroup) and Quaranfil (Quaranfil serogroup) viruses showed only a gradual decrease in titer following inoculation. A strain of the latter has previously been found to multiply well in cell cultures of the tick *R. appendiculatus* (Varma *et al.*, 1975).

IV. DISCUSSION

We have evaluated the ability of a newly established line of tick cells to support growth of a number of arboviruses. Included were some mosquito-borne and non-vector-borne agents, as well as 18 distinct entities representing approximately one-fourth of the viruses known or suspected to be tick-borne. Our results may be compared with similar evaluations of arbovirus growth in mosquito and tick cells grown *in vitro*. The most extensively tested mosquito cell line is that of *Aedes albopictus* (ATC-15) developed by Singh (1967). Its response to viral infection depends primarily upon the vector of the virus and is also correlated with the possession or lack by the virion, of essential lipids (Buckley *et al.*, 1976). Thus, ATC-15 cells readily support growth of mosquito-borne viruses regardless of their sensitivity to lipid solvents, but are refractory to tick-borne viruses except Orbiviruses, which are relatively resistant to lipid solvents (Buckley *et al.*, 1976; Singh, 1971; Yunker 1971). In contrast, tick cells, whether as primary cultures (Rehacek, 1971a,b; Yunker, 1971) or as serially passaged strains or lines (David-West, 1974; Varma *et al.*, 1975) permit growth of both mosquito- and tick-borne viruses, irrespective of essential lipids. Our results, which reveal that the *D. parumapertus* cell line is broadly receptive to infection with both mosquito- and tick-borne viruses, are consistent with these studies. Thus, all findings to date would indicate that type of vector and presence or absence of essential lipids are not eminent factors in determining infectivity of viruses for tick cells.

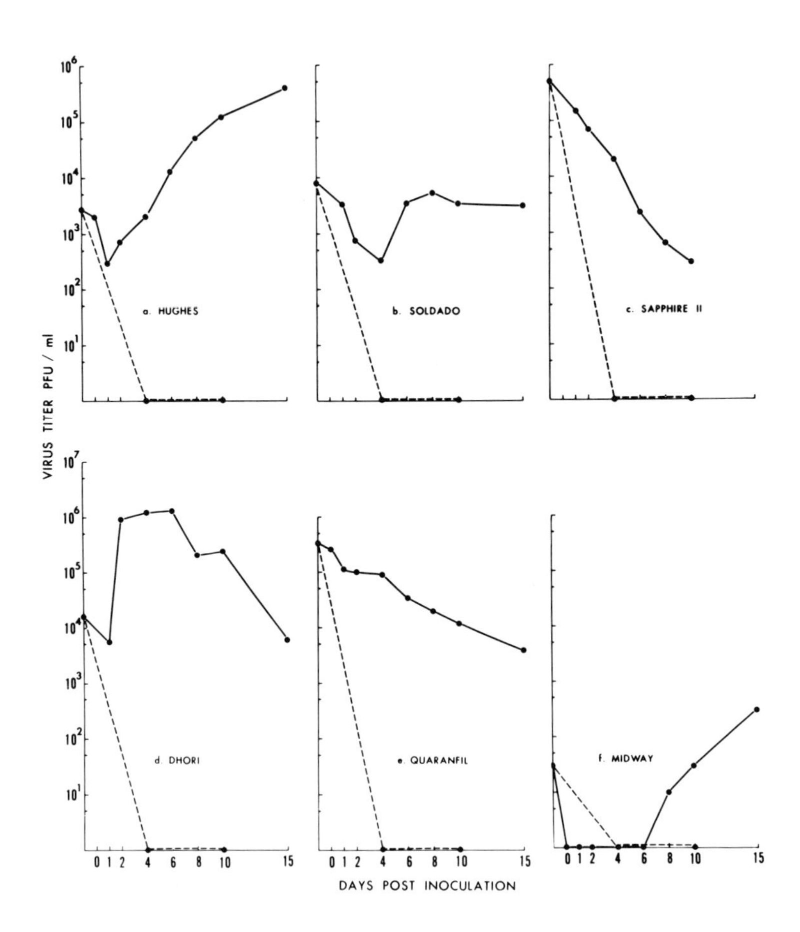

Fig. 3. Multiplication of certain unclassified arboviruses in the D. parumapertus *cell line. Two infected cultures harvested, pooled, and titrated in Vero cells on days 0 (2 hours after inoculation), 1, 2, 4, 6, 8, 10, and 15. Titer of inoculum indicated on ordinate. ____, Experimental; ---, control.*

Although broad, the susceptibility of tick cells to vector-borne animal viruses does not extend to those nominally classified as arboviruses but that apparently do not utilize an intermediate vector. Modoc and Rio Bravo viruses, both antigenically allied with the Flaviviruses and capable of direct transmission among vertebrate hosts, failed to multiply in *D. parumapertus* cells, as they have in insect tissue and cell culture systems (Buckley, 1969; Yunker and Cory 1968). Thus, tick cell lines, like those from mosquitoes, may be useful in the identification, classification, and characterization of viruses.

Some differences in growth were seen among the various tick-borne viruses tested in *D. parumapertus* cells. Whereas all viruses associated with ixodid ticks multiplied well or were maintained at high levels in these cells, only one argasid virus, Hughes, grew well; others grew only poorly or not at all. The reason for these differences is not evident. However, argasid viruses, whose exposure to vectors and hosts is limited (Hoogstraal, 1973), may be more highly adapted than ixodid viruses, which are generally associated with a wide range of vectors and hosts. This may account for variations observed in the susceptibility of *D. parumapertus* cells to various tick-borne arboviruses. However, it is necessary to test more of the approximately 100 viruses associated with ticks, particularly those from argasids, in *D. parumapertus* cells before such a hypothesis can be adequately judged.

The high degree of sensitivity of tick tissue and cell cultures for detection of certain arboviruses has previously been noted (Rehacek and Kozuch, 1964; Yunker and Cory, 1967). Primary cultures from *Hyalomma asiaticum* and *Dermacentor andersoni* were able to take up and amplify Central European tick-borne encephalitis and Colorado tick fever viruses, respectively, that had been diluted beyond the expected level of detectability in conventional systems. Our results with Hazara virus provide another example of the sensitivity of the tick cell culture system and suggest that it may be used advantageously in the primary isolation of certain viruses.

Acknowledgments

We thank Jack Cory, Rocky Mountain Laboratory, for helpful advice and assistance. Thanks are also due to Mrs. Alma Dinehart Smith for preparing the graphs.

References

Banerjee, K., and Singh, K. R. P. (1969). *Ind. J. Med. Res.* 57, 1003-1005.

Bhat, U. K. M., and Yunker, C. E. (1977). *J. Parasitol. 63,* 1092-1098.

Buckley, S. M. (1969). *Proc. Soc. Exp. Biol. Med. 131,* 625-630.

Buckely, S. M. (1971). *In* "Current Topics in Microbiology and Immunology" (E. Weiss, ed.), Vol. 55, pp. 133-137, Springer-Verlag, New York.

Buckley, S. M. (1973). *Proc. 3rd Int. Conf. Invertebrate Tissue Culture, Smolenice 1971,* pp. 307-326, Bratislava, Slovak Academy of Sciences.

Buckely, S. M., Hayes, C. G., Maloney, J. M., Lipman, M., Aitken, T. H. G., and Casals, J. (1976). *In* "Invertebrate Tissue Culture, Applications in Medicine, Biology, and Agriculture" (E. Kurstak and K. Maramorosch, eds.), pp. 3-19, Academic Press, New York.

Chappell, W. A., Calisher, C. H., Toole, R. F., Maness, K. C., Sasso, D. R., and Henderson, B. D. (1971). *Appl. Microbiol. 22,* 1100-1103.

Cory, J., Yunker, C. E., Howarth, J. A., Hokama, Y., Hughes, L. E., Thomas, L. A., and Clifford, C. M. (1975). *Acta Virol. 19,* 443-445.

David-West, T. S. (1974). *Arch. Ges. Virusforsch. 44,* 330-336.

Early, E., Peralta, P. H., and Johnson, K. M. (1967). *Proc. Soc. Exp. Biol. Med. 125,* 741-747.

Ghosh, S. N., Tongaonkar, S. S., and Dandawate, C. N. (1973). *Curr. Sci. 42,* 286-288.

Hink, W. F. (1976). *In* "Invertebrate Tissue Culture. Research Applications" (K. Maramorosch, ed.) pp. 319-369, Academic Press, New York.

Hoogstraal, H. (1973). *In* "Virus and Invertebrates" (A.J. Gibbs, ed.), pp. 349-390. North-Holland Publ. Co., Amsterdam.

Pavri, K. M., and Ghosh, S. N. (1969). *Bull. Wld. Hlth. Org. 40,* 984-986.

Peleg, J. (1971). *In* "Current Topics in Microbiology and Immunology" (E. Weiss, ed.), Vol. 55, pp. 155-161, Springer-Verlag, New York.

Porterfield, J. S., Casals, J., Chumakov, M. P., Gaidamovich, S. Y., Hannoun, C., Holmes, I. H., Horzinek, M. C., Mussgay, J., Oker-Blom, N., and Russell, P. K. (1975-1976). *Intervirology 6,* 13-24.

Rehacek, J. (1971a). *Ann. parasitol. (Paris) 46,* 197-231.

Rehacek, J. (1971b). *In* "Invertebrate Tissue Culture" (C. Vago, ed.), Vol. II, pp. 280-321. Academic Press, New York.

Rehacek, J., and Kozuch, O. (1964). *Acta Virol. 8,* 470-471.

Singh, K. R. P. (1967). *Curr. Sci. 36,* 506-508.

Singh, K. R. P. (1971). *In* "Current Topics in Microbiology and Immunology" (E. Weiss, ed.), Vol. 55, pp. 127-133. Academic Press, New York.

Singh, K. R. P., and Paul, S. D. (1969). *Bull. Wld. Hlth. Org. 40,* 982-983.

Varma, M. G. R., Pudney, M., and Leake, C. J. (1975). *J. Med. Entomol. 11,* 698-706.
Varma, M. G. R., Pudney, M., Leake, C. J., and Peralta, P. H. (1975-1976). *Intervirology 6,* 50-56.
Yunker, C. E. (1971). *In* "Current Topics in Microbiology and Immunology" (E. Weiss, ed.), Vol. 55, pp. 113-126. Academic Press, New York.
Yunker, C. E., and Cory, J. (1967). *Exp. Parasit. 20,* 267-277.
Yunker, C. E., and Cory, J. (1968). *Am. J. Trop. Med. Hyg. 17,* 889-893.
Yunker, C. E., and Cory, J. (1975). *Appl. Microbiol. 29,* 81-89.
Yunker, C. E., Ormsbee, R. A., Cory, J., and Peacock, M. G. (1970). *Acta Virol. 14,* 383-392.

Chapter 18

OBSERVATIONS RELATED TO CYTOPATHIC EFFECT IN *AEDES ALBOPICTUS* CELLS INFECTED WITH SINDBIS VIRUS

Victor Stollar, Keith Harrap, Virginia Thomas, and Nava Sarver

I. INTRODUCTION

The interaction between arboviruses and their hosts is a complex one. Not only is there a mandatory and alternative transmission between the vertebrate host and the arthropod host, but the outcome of infection is different in the two cases. Thus many arthropod-borne togaviruses that can cause serious illness in man or other mammals replicate in the mosquito vector without any apparent ill effects. Similarly in tissue culture systems, alphaviruses [Sindbus virus (SV), Semliki forest virus (SFV), or Ross River virus (RRV)] a subgroup of the family Togaviridae, produce a rapid cytopathic effect (CPE) in cultured vertebrate cells but have not been reported to cause cell damage in mosquito cells even though they replicate to high titer (Peleg, 1969; Stevens, 1970; Raghow *et al.*, 1973a).

The availability of established lines of mosquito cells (Singh, 1967; Peleg, 1968) has made it possible to set up model tissue culture systems (Stollar *et al.*, 1975, 1976) with which to examine such questions as (1) those selective pressures exerted on the replication of arboviruses in the vertebrate and insect host and (2) the basis for the production of CPE in cells of vertebrate origin but not in cells derived

ISBN 0-12-429765-X

from mosquitoes.

Knowing why these two different kinds of cells respond differently to virus infection would be very instructive for our understanding of viral virulence and might also explain the means by which some cells can maintain their integrity and at the same time support extensive viral replication.

We shall describe here two facets of the interaction between Sindbis virus and *Aedes albopictus* cells, which are related to these questions: first, electron-microscopic studies of the morphogenesis of Sindbis virus, and second, experiments with clones of *A. albopictus* cells, which are an exception to the general rule and show marked cytopathic effect after infection with Sindbis virus.

There have been several electron-microscopic studies that describe the development of alphaviruses in cultured *A. albopictus* cells. Raghow *et al.* (1973a,b) studied the morphogenesis of two alphaviruses, Ross River virus (RRV) and Semliki Forest Virus (SFV) in cultured *A. albopictus* cells. In RRV infections, membrane-bounded cytoplasmic inclusions were observed, some of which contained nucleocapsids early in infection but enveloped virus particles at later times. At later times nucleocapsids appeared to bud through the bounding inclusion membrane. Occasionally nucleocapsids were found free in the cytoplasm but budding was seen only infrequently from the plasma membrane. Late in the infection, a second type of membrane-bounded structure was observed, which contained enveloped virus particles but differed from the cytoplasmic inclusions described above. In SFV infections free nucleocapsids were never observed, and only a few cells contained small cytoplasmic inclusions with enveloped virus. Other cells were observed, which contained large vacuoles similar to the "type-1 cytopathic vacuoles" of SFV-infected mammalian cells (Grimly *et al.*, 1968). Thus, in contrast to RRV, no really significant ultrastructural changes occurred in SFV-infected cells during the time of rapidly increasing virus titer. It was therefore presumed that much of the SFV matured early at the cell membrane and that the process was not readily detected by electron microscopy. Similarly, with RRV infections these workers favored nucleocapsid budding through the plasma membrane as the prime method of infective extracellular virus production. No release of virus from the inclusions was observed and no inclusions were seen near the cell membrane as would be expected if release by "reverse phagocytosis" was occurring. Indeed it was felt that virus maturing in these inclusions was subsequently destroyed by lysosomal-like activity. An attractive feature of this scheme is that it suggests a means (fusion of virus-containing vacuoles with lysozomes) whereby mosquito cells can restrict and control the process of viral multiplication.

A different interpretation of viral morphogenesis was made by Gliedman *et al.* (1975) and Brown *et al.* (1976) from their observations on *A. albopictus* cells infected with Sindbis virus. They found very little if any virus "budding" at the plasma membrane; nor did they see free nucleocapsids in the cytoplasm. They did observe, however, many large and complex vesicles, which contained not only virus particles but also nucleocapsids, membranes, and ribosomes. It was inferred from these observations that the whole process of virus biosynthesis and assembly took place in vesicles, that envelopes were acquired in association with intravesicular membranes, and that virus was released from vesicles by "reverse phagocytosis." Furthermore, the segregation of the viral synthetic processes within vesicles and apart from the rest of the cell could in their view explain why alphaviruses can replicate in mosquito cells without inducing any cytopathic effect. Their results, however, differ from our findings (see below), in which nucleocapsids were clearly visible in the cytoplasm and were observed budding through the plasma membrane. We suggest, therefore, that alternative reasons must be sought to explain the absence of CPE in mosquito cells infected with alphaviruses.

II. MORPHOGENESIS OF SINDBIS VIRUS IN *A. ALBOPICTUS* CELLS

A. Electron Microscopy

Mosquito cells harvested and fixed 18 and 21 hours after infection showed considerable numbers of virus particles distributed on the cell surface. This extracellular virus was always rather indistinct in outline, because of its association with amorphous electron-dense material (Figs. 1 and 2). This effect might be due to certain components of the culture medium adhering to extracellular virus causing it to aggregate and stick to the cell surface. Many cells were observed in which virus nucleocapsids around 28 nm in diameter were clearly visible in the cytoplasm (Fig. 1). Often in such cases cells were surrounded by mature virus of indistinct outline as described above. "Budding" of virus particles was observed at the plasma membrane in many of the infected cells examined (Figs. 2,3,4,5,) with nucleocapsids in various stages of association with the plasma membrane.

Many mature enveloped virus particles were also seen in vacuoles within the cells (Fig. 2,6). Such vacuoles were found both close to the cell surface and with their membranes obviously contiguous with the plasma membrane suggesting movement of the vacuole to the cell surface leading to the release of the

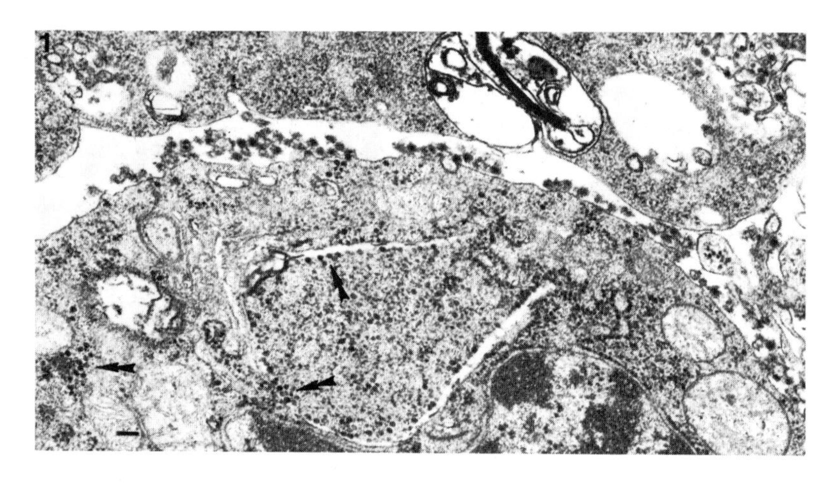

Fig. 1. Sindbis virus infection in A. albopictus *cells. (at 21 hr) Enveloped virus particles can be seen outside the cell and nucleocapsids (arrowed) are clearly visible in the cytoplasm. Bar = 100 nm.*

The cells shown in Figures 1 through 11 are A. albopictus *AI cells (see Table I, text, Igarashi and Stollar, 1976, and Sarver and Stollar, 1977). Cells were infected at multiplicities greater than 10 pfu/cell.*

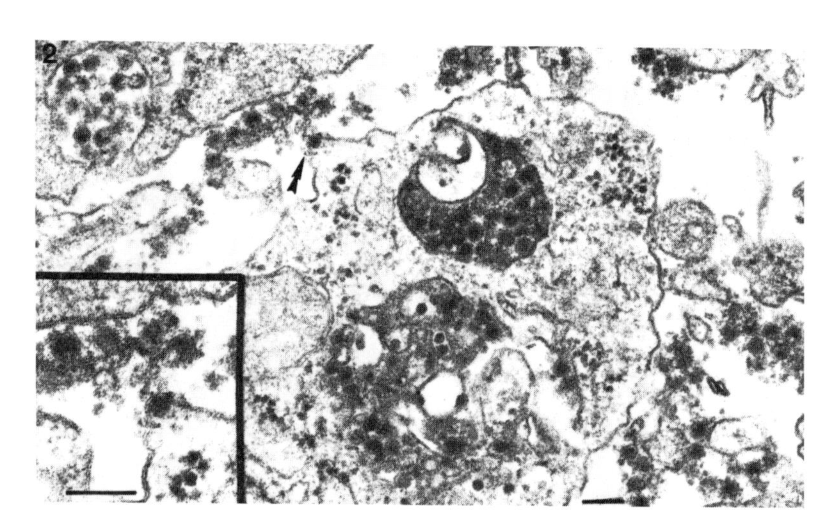

Fig. 2. Two methods of Sindbis virus release from A. albopictus *cells. In the same cell virus can be seen (1) budding from the plasma membrane (arrowed and inset) and (2) maturing into vacuoles. A virus-containing vacuole in an adjacent cell has a membrane contiguous with the plasma membrane (top left). Bar, 100 nm.*

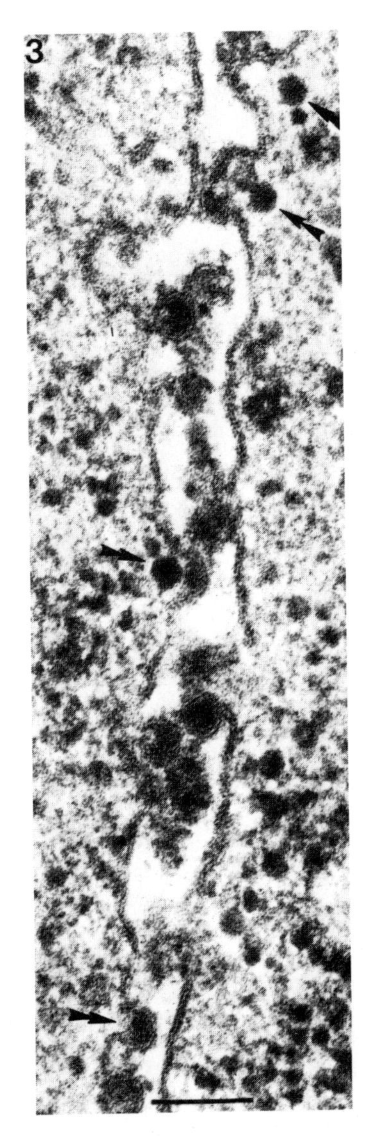

Fig. 3. Stages of "budding" of Sindbis virus nucleocapsids (arrowed) through the plasma membranes of two adjacent A. albopictus cells. Bar, 100 nm.

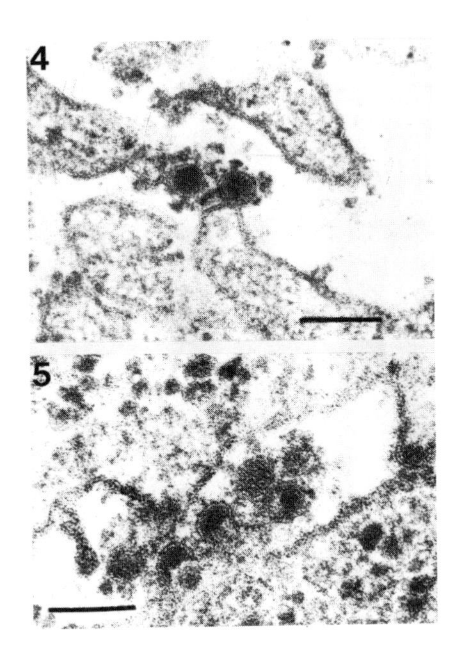

Figs. 4 and 5. Two examples of Sindbis virus budding from the plasma membrane of an A. albopictus cell. Bar, 100 nm.

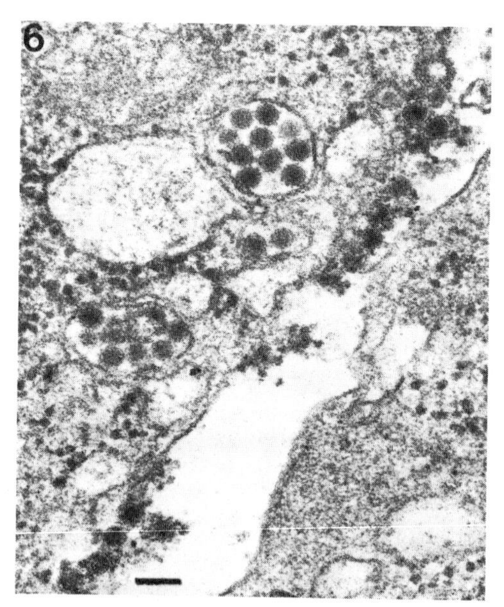

Fig. 6. Sindbis virus in vacuoles in an A. albopictus *cell. Other virus particles can be seen at the cell surface. Bar, 100 nm.*

enclosed virus particles by "reverse phagocytosis." In general, the virus particles found within vacuoles were free of any associated debris and were clearly outlined. Our observations suggest that mature, virus particles can be released from infected mosquito cells in two ways: (1) by "budding" through the plasma membrane as is commonly observed in vertebrate cells and (2) by budding into vacuoles followed by reverse phagocytosis and expulsion of mature virions from vacuoles. It is interesting to note that both methods of virus release were seen in the same cell (Fig. 2).

What appear to be morphogenetic aberrations were observed in some cells so that two, three, or four nucleocapsids were seen apparently surrounded by a single membrane (Fig. 7). These "multicore" aberrant forms were observed only in intracytoplasmic vacuoles. An alternative explanation of these observations is that virus nucleocapsids are contained in membrane protusions within the vacuole. However, similar membrane forms lacking nucleocapsids were not found. The proximity of other individual nucleocapsids to the vacuole membrane suggest that budding into the vacuole is a likely maturation step, but convincing stages of such budding were not observed, and it is possible therefore that nucleocapsids are enclosed by membrane at the inner surface of the vacuole in a more complex way, than that observed at the surface of the cell, perhaps by proliferation or convolution of the membrane. Mock-infected cells were also examined but no

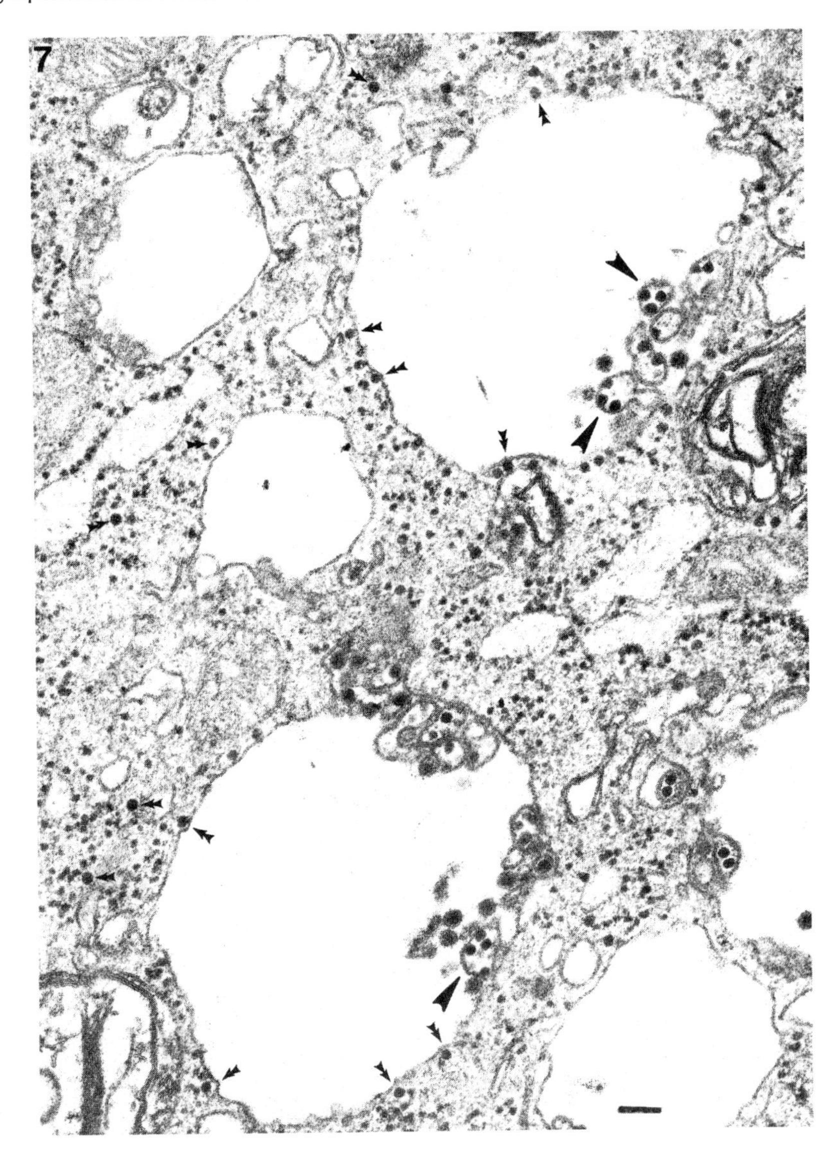

Fig. 7. Possible aberrant morphogenesis of Sindbis virus at the membrane of intracytoplasmic vacuoles (large arrows). Several free nucleocapsids are visible (small arrows). Bar, 100 nm.

virus-like particles were seen and there were no observations suggesting viral morphogenesis.

Although extensive vacuolation is typical of SV-infected cells, we do not think that the presence of cytoplasmic vacuoles results only from viral infection. It may well be a rather non-specific response to a broad variety of adverse environmental changes. This is also suggested by our observations from time to time of moderate vacuolization in our normal or stock *A. albopictus* cells. We have no evidence at present that these cells are contaminated with any virus(es).

B. Immunoperoxidase Staining of SV-Infected Cells

These experiments were undertaken with the hope of comparing, at the ultrastructural level, the localization of Sindbis virus antigens in BHK-21 and in mosquito cells. The procedure was first evaluated by light microscopy. Figure 8 shows that uninfected BHK-21 cells that were reacted with anti-SV globulin, or infected cells that were reacted with normal gamma globulin gave no reaction. Infected cells, on the other hand, reacted with anti-SV gamma globulin gave a marked and clear localization of viral antigens predominantly in the perinuclear region. Similar results were obtained in the case of infected *A. albopictus* cells (Fig. 9). Precise localization of the antigen was more difficult because the cells are smaller. However, in a number of cells discrete foci of stained material were seen. These could represent cytoplasmic inclusions or large vesicles containing viral antigens.

Cells examined in this way by light microscopy were also examined by electron microscopy. The methods required for fixation and penetration of antibody are not optimal for preservation of the cellular structure. Nevertheless, Fig. 10 shows that the reaction occurred only in infected *A. albopictus* cells and was confined to the cytoplasm. Within the cytoplasm there was prominent staining not only of the vesicles but also of the plasma membrane (Figs. 10 and 11). Localization of viral antigen in the plasma membrane would be consistent with viral maturation or budding at the cell surface. Similar results were obtained with SV-infected BHK cells.

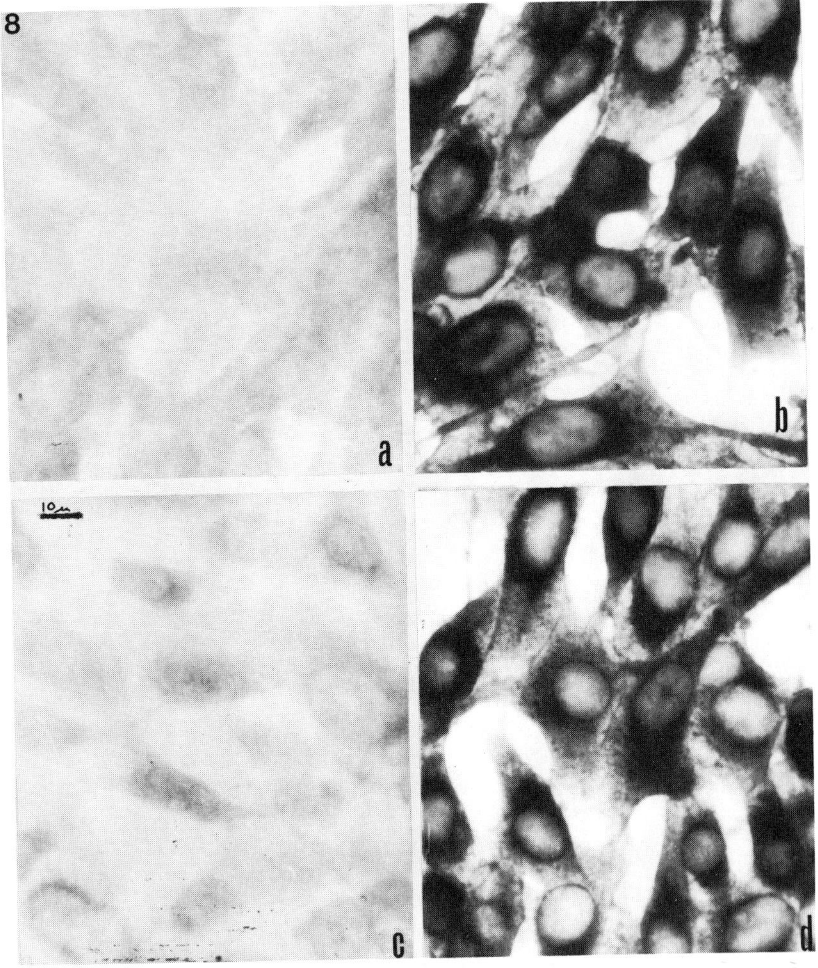

Fig. 8. Immunoperoxidase reaction on mock-infected and SV-infected BHK cells, as seen by light microscopy. (a) Mock-infected cells reacted with anti SV γ globulin; (b) SV-infected cells reacted with anti SV γ globulin; (c) SV-infected cells reacted with normal rabbit γ globulin; (d) SV-infected cells reacted with anti-SV γ globulin. Reactions (a) and (b) were carried out in one experiment and reactions (c) and (d) in a separate experiment. Bar, 10μm.

Immunoperoxidase staining was carried out by the three-step method of Dougherty et al. *(1972) with slight modifications (Thomas and Stollar, 1974). The infection of cells (input MOI > 500 pfu/cell) and the immunoperoxidase staining were both carried out in eight chamber Lab-Tek tissue culture slides. One further modification was the exposure of fixed cells to dilute fetal calf serum before the reaction with the anti-Sindbis virus globulin.*

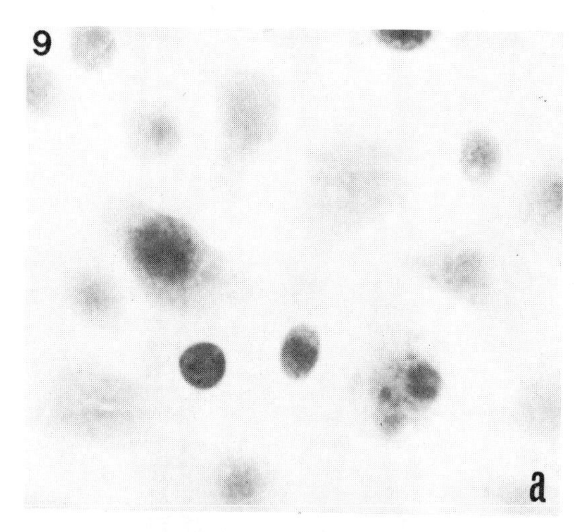

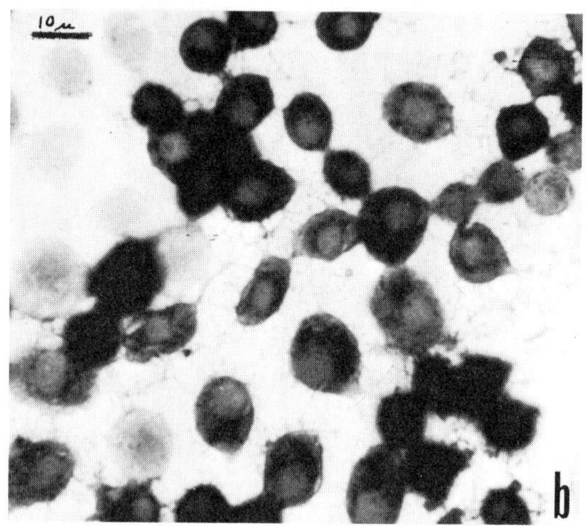

Fig. 9. Immunoperoxidase reaction on mock-infected and SV-infected A. albopictus *cells as seen by light microscopy. (a) mock-infected cells reacted with anti-SV γ globulin; (b) SV-infected cells reacted with anti-SV γ globulin. Bar, 10μm.*

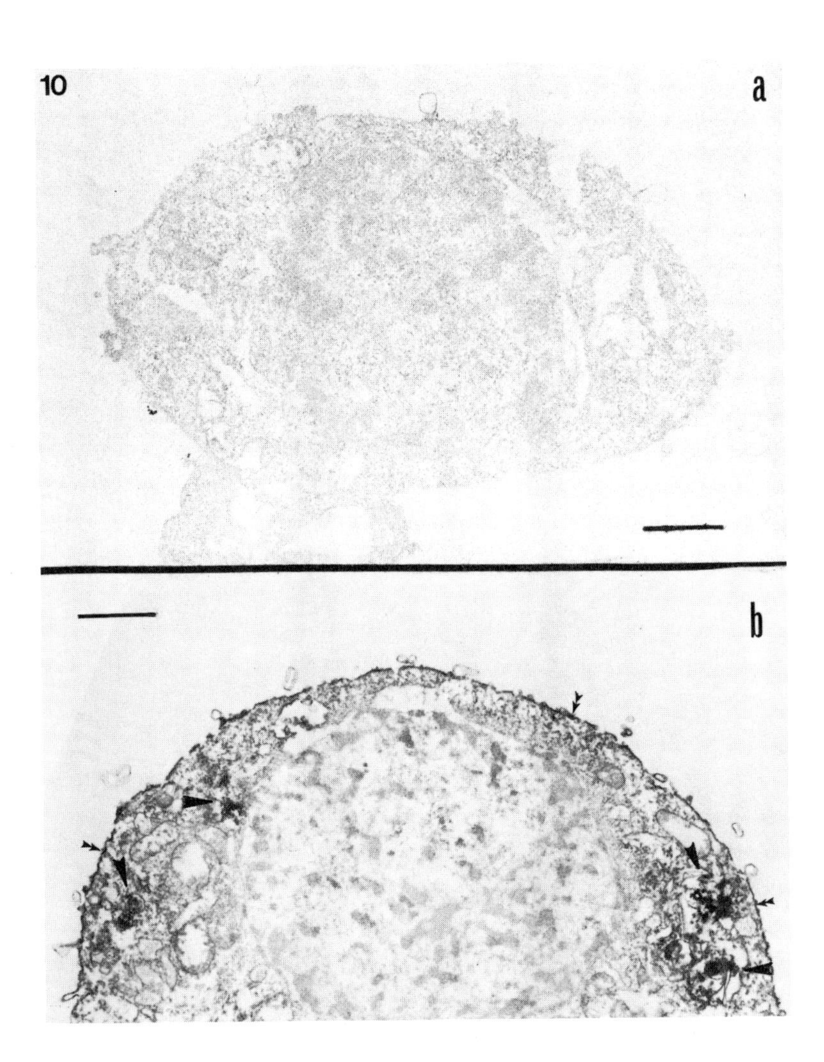

Fig. 10. Immunoperoxidase reaction as seen by electron microscopy of (a) mock-infected and (b) Sindbis virus-infected A. albopictus *cells. Electron-dense reaction is visible in the cytoplasm and on the plasma membrane of the infected cell. Mock-infected and infected cells were reacted with anti-SV γ globulin. Photographic processing was identical in (a) and (b). Bar, 1 μm.*

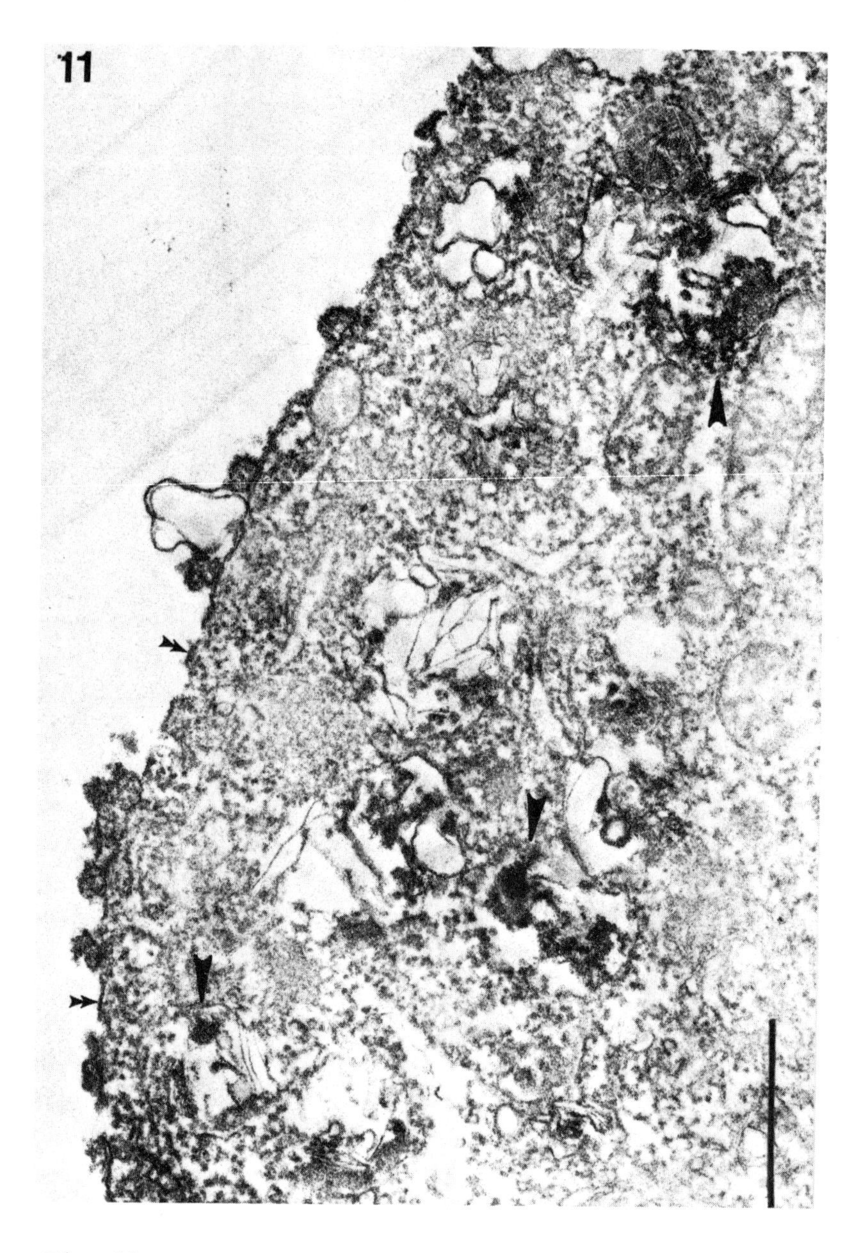

Fig. 11. Part of an infected A. albopictus *cell showing immunoperoxidase-labeled antigen in cytoplasmic foci (large arrows) and on the plasma membrane (small arrows). Reaction was with anti-SV γ globulin. Bar 1 μm.*

III. CLONES OF *A. albopictus* CELLS THAT SHOW CYTOPATHIC EFFECT AFTER INFECTION WITH SINDBIS VIRUS

During the course of cloning experiments a number of *A. albopictus* cell clones were obtained that showed a distinct and marked CPE after infection with SV. As discussed earlier, these findings differ markedly from previous observations made in our laboratory and by other investigators.

Table I summarizes the responses of various cloned and uncloned cell populations to infection with Sindbis virus.

TABLE I
Cytopathic Response of Various A. albopictus *Cell Strains to Sindbis Virus and Eastern Equine Encephalitis Virus*[a]

Time after infection (hours):		24		48	
Temperature (deg):		28	34	28	34
Cell strain	Infected with				
F-33	SV	0	0	0	0
F-33	uninfected	0	0	0	0
AI	SV	0	0	0	0
AI	uninfected	0	0	0	0
LT	SV	0	0	0	0
LT	uninfected	0	0	0	0
LT (C-4)	SV	0	0	0-1+	1+
LT (C-4)	uninfected	0	0	0	0
LT (C-7)	SV	1+	4+	3+	4+
LT (C-7)	uninfected	0	0	0	0
LT (C-19)	SV	2+	4+	3+	4+
LT (C-19)	uninfected	0	0	0	0
LT (C-22)	SV	0	0	0	0
LT (C-22)	uninfected	0	0	0	0
AIS C-3	SV	0	0	0	0
AIS C-3	uninfected	0	0	0	0
LT (C-7)	EEEV	1+	4+	2+	4+
AIS C-3	EEEV	0	0	0	0

[a]*Cells were infected with SV or EEEV at an approximate input multiplicity of 100 pfu/cell. For further details about the cells see text and Sarver and Stollar (1978).*

The F-33, AI, and LT cultures are all uncloned populations, grown in MM, MM·E, and E medium, respectively. MM medium is the medium described by Mitsuhashi and Maramorosch (1964). E medium is Eagle's medium (Eagle, 1959) containing nonessential amino acids, and MM·E medium contains 1 volume of MM medium and 9 volumes of E medium. All media were supplemented with fetal calf serum (see also Sarver and Stollar, 1978). The various cloned lines were all maintained in E medium.

None of the uncloned cells (F-33, AI, or LT) manifested CPE after infection with SV. The cloned cells showed varying responses ranging from no CPE (C-22 and AIS C-3) to mild and transient CPE (C-4) to marked and extensive CPE (C-7 and C-19). Two other points emerge from this table: first, the CPE was more marked when cells were maintained after infection at 34° rather than at 28°, and second, the differential sensitivity of C-7 and AIS C-3 cells was not restricted to SV but was also seen after infection with eastern equine encephalitis virus.

Figure 12 illustrates the virus-induced CPE in LT C-7 cells. At 28° two changes were noted. The cells were rearranged into

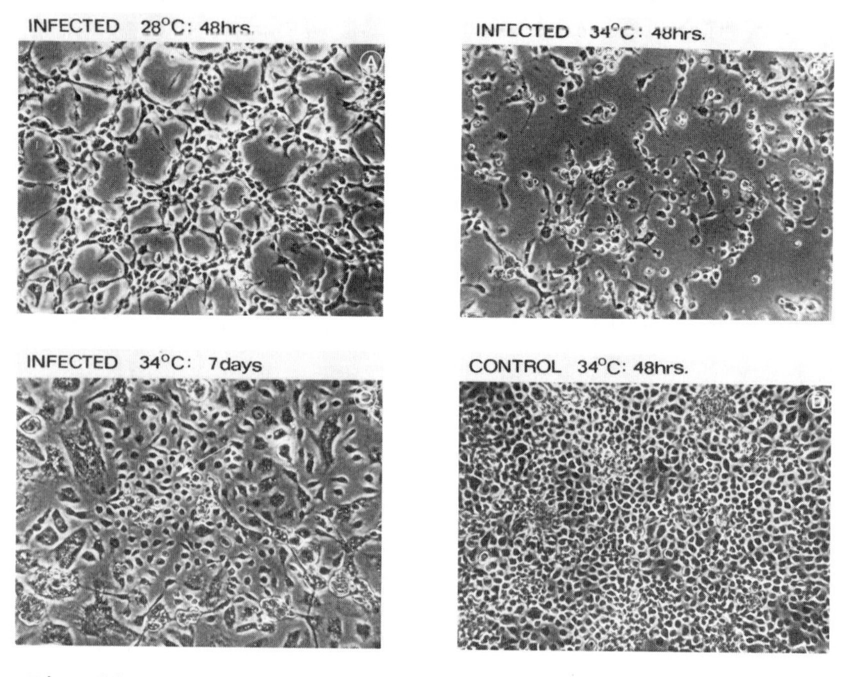

Fig. 12. Light microscopy of mock-infected and SV-infected A. albopictus *LT (C-7) cells. (A) 48 hours after infection at 28°C; (B) 48 hours after infection at 34°C; (C) 7 days after infection at 34°C; note the region of normal-looking cells (arrow); (D) mock-infected cells after 48 hours at 34°C. Phase contrast X160.*

a networklike pattern and the individual cells tended to change in morphology from epithelioid to fibroblastic. At 34° the CPE was much more marked and was characterized by extensive cell destruction. Although Fig. 12B shows cells at 34° 48 hours after infection, such CPE was generally evident by 24 hours or earlier. Control uninfected C-7 cells are shown for comparison. SV-infected AIS C-3 cells had a similar appearance. Typically, cultures were able to recover from the acute cytopathic effects and as might be expected, more quickly at 28° than at 34°. Figure 12C, for example, shows a small focus of normal-appearing cells 7 days after the initial infection.

In order to further characterize the virus-induced CPE, comparative studies were carried out with the LT C-7 and the AIS C-3 cell cultures, selected as prototypes of CPE-susceptible and CPE-resistant clones, respectively.

We considered the possibility that C-7 cells were killed because they had multiple metabolic and physiological deficiencies and were therefore in a generally debilitated state. However, since uninfected C-7 cells grew at a rate very similar to that of AIS C-3 cells, this seemed unlikely.

Virus yields from the two cell types were compared both at 28 and at 34° (Sarver and Stollar, 1978). Although in some cases yields of SV tended to be slightly higher from the C-7 cells than from the AIS C-3 cells, the differences were usually not greater than two or threefold and usually less. After infection with EEEV the yields were very similar from the two cell types. We therefore concluded that there was not any significant correlation between the occurrence of CPE and the amount of infectious extracellular virus released.

We then turned to an examination of virus-induced RNA synthesis in the two cell types (Fig. 13). At 28° in the C-7 cells the rate of RNA synthesis, as measured by incorporation of ^{3}H-uridine into TCA-precipitable material, was about double that in the AIS C-3 cells. At 34° viral RNA synthesis was the same in AIS C-3 cells as at 28°, whereas in the LT C-7 cells it increased another twofold. At 37° viral RNA synthesis was still unchanged in the AIS C-3 cells, but in the LT C-7 cells there was a further slight increase. In mock-infected cells actinomycin-resistant RNA synthesis remained at a low basal level. Summarizing these results, we observed that in the CPE-resistant, AIS C-3 cells viral RNA synthesis was unaffected by temperature; in contrast, in the CPE-sensitive LT C-7 cells viral RNA synthesis increased with increasing temperature just as did the virus-induced cytopathic effect.

Finally, we turned to the possible effects of virus infection on host biosynthetic processes. Figure 14 illustrates the effects on host RNA synthesis. Host RNA synthesis was determined by measuring only incorporation into the nucleus and by using short pulses (15'). During a 15-minute pulse nearly

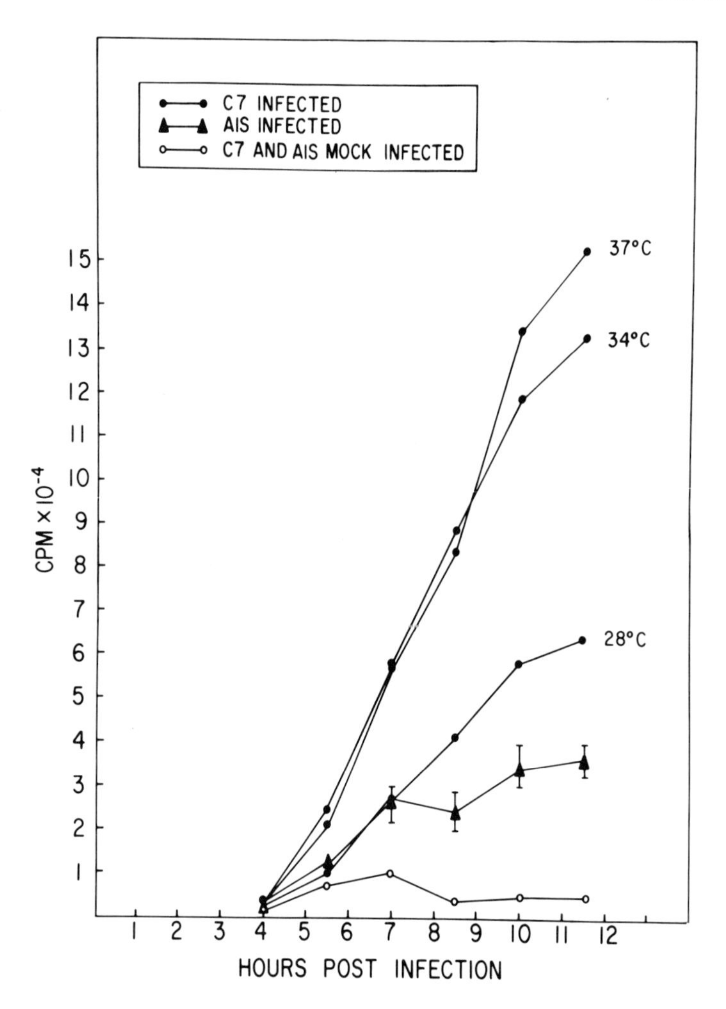

Fig. 13. Virus-directed RNA synthesis in SV-infected A. albopictus *LT (C-7) and in* A. albopictus *(AIS C-3) cells at 28, 34, and 37°C.*

Monolayers (∿3-10^6 cells) in 60-mm plates were infected and incubated under 1.9 ml of E medium. Two hours after infection, all plates were treated with Actinomycin D (5 μg/ml final concentration) for 60 minutes, after which time [5-^{3}H]-uridine (>20 Ci/mM, 10 μCi/ml final concentration) was added. At the indicated times, duplicate plates were harvested and acid-precipitable (5% TCA) counts (see Sarver and Stollar, 1977) were determined on duplicate 200 μl samples. Infected C-7 cells, temperature as indicated in figure ●. Infected AIS C-3 cells, at 28, 34, and 37°; Δ, all values were within the range shown; O, mock-infected C-7 and AIS C-3 cells at 28, 34, and 37°.

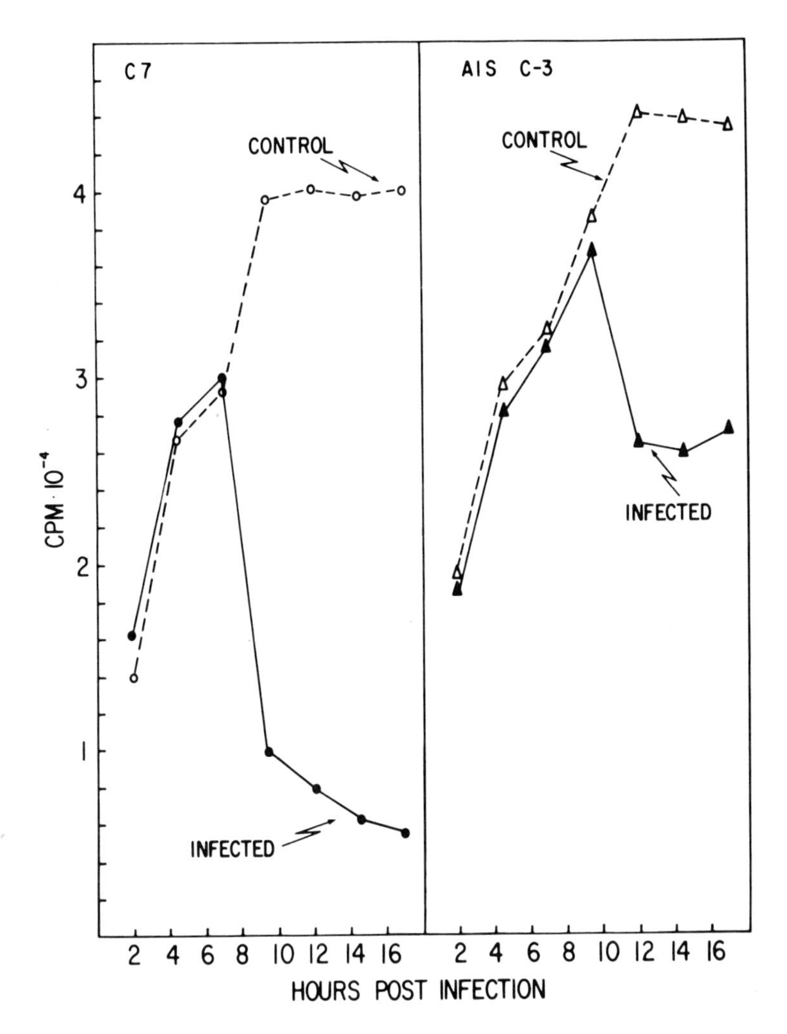

Fig. 14. Host RNA synthesis in SV-infected and mock-infected C-7 and AIS C-3 cells at 34°. 60-mm plates were seeded with C-7 or AIS C-3 cells and incubated at 28°. Two days later monolayers were infected as described and incubated at 34° under 2 ml of E medium. At the indicated times, individual cultures were pulsed with [5-^{3}H]-uridine (>20 Ci/mM, 10 µCi/ml final concentration) for 15 minutes at 34°. Pulses were terminated by chilling the monolayers and washing them three times with cold PBS. Nuclei were then isolated and acid precipitable (15% TCA) counts were determined on duplicate 400-µl samples. Values plotted represent TCA-precipitable cpm × 10^{-4} per 400-µl sample.

all of the newly synthesized host-specific RNA should remain in the nucleus, whereas all viral RNA synthesis should be confined to the cytoplasm. In the LT C-7 cells beginning at 9 hours after infection there was an abrupt and marked reduction in host RNA synthesis. By 14-16 hours the level in infected cells was only 10-15% of that in mock-infected or control cells.

In the AIS C-3 cells on the other hand, although there was a depression of host RNA synthesis, it occurred later, was less severe, and as shown in other experiments was only transient in duration. Subsequent experiments have shown that CPE-resistant clones can be obtained that show no inhibition of host RNA synthesis following infection with Sindbis virus.

IV. DISCUSSION AND CONCLUSION

The comparative study of togavirus replication in vertebrate and mosquito cells provides an unusual opportunity to examine critically those factors in the virus-cell interaction that determine whether or not infection leads to cell death. The availability of cell populations that differ so markedly not in their yields of virus but in their response to infection compels us to ask, "What are the specific properties of the host cell that render them sensitive or resistant to virus-induced cytopathic effect?"

As described above, one of the more fascinating observations concerns the covariation in the *A. albopictus* LT C-7 cells of viral RNA synthesis and CPE, both of which increase as the temperature is increased. One possible explanation of these findings is as follows: (1) the host cell contributes a specific protein subunit to the viral RNA polymerase (Clewley and Kennedy, 1976); (2) in CPE-susceptible mosquito cells there is an alteration of this host-cell specific protein; (3) this alteration results in disturbed regulation and overproduction of viral RNA; (4) this specific host cell protein may also be an important component of a cellular RNA polymerase. If so, the same alteration that results in increased viral RNA synthesis might render host cell RNA synthesis increasingly susceptible to shut-off following virus infection. Purification and characterization of the viral RNA polymerase would make it possible to test these ideas.

Turning to our electron-microscopic observations, it is clear that they differ from those of Gliedman *et al.* (1975), who observed subviral and viral particles only within vesicles and were unable to observe virus budding at the cell surface. It is strange but noteworthy that a given virus-cell system can apparently behave in such distinctive ways in different laboratories.

These differences clearly might be related to the history and properties of the cell strains or the virus strains that were used. One obvious difference is that Gliedman *et al.* grew *A. albopictus* cells in MM medium, whereas the cells that we examined by electron microscopy had been adapted to a 1:9 mixture of MM medium and E medium (see above). Such differences in the medium might markedly alter the rates of certain critical metabolic processes and thus influence the rate of synthesis, assembly, and processing of nucleocapsids. These effects could result both from differences in the content of small organic compounds and from differences in the salt concentrations and therefore osmotic pressure. For example, MM medium has an osmotic pressure of about 410 milliosmoles/kg (V. Stollar, unpublished observations) whereas the value for Eagle's medium is only 290 milliosmoles/kg.

With the evidence available at present, we favor the interpretation that the budding of free cytoplasmic nucleocapsids at the plasma membrane is a primary morphogenetic system, perhaps favored by certain culture methods, whereas the intravesicular morphogenesis is a later secondary system that ultimately predominates as persistently infected cultures are established. It seems, however, that certain factors not known precisely, could tip the balance so that intravesicular morphogenesis predominates even early after infection.

Since we were unable to find any differences in viral morphogenesis in CPE-susceptible and CPE-resistant cells and since in our experiments free nucleocapsids were seen both in cytoplasm and budding through the plasma membrane, sequestration of viral synthesis and assembly in cytoplasmic vesicles does not seem to be an adequate explanation for the lack of cytopathic effect in infected mosquito cells. However, the fact that, like the whole mosquito infected with virus, mosquito cells in culture are usually unaffected by alpha virus infection suggests that the survival of the infected mosquito vector in nature can be explained by events at the cellular level.

Acknowledgments

These investigations were supported by Grant No. AI 11290 from the National Institute of Allergy and Infectious Diseases, and by the United States-Japan Cooperative Medical Science Program through Public Health Services Grant No. AI 05920.

References

Brown, D. T., Smith, J. F., Gliedman, J. B., Riedel, B., Filtzer, D., and Renz, D. (1976). *In* "Invertebrate Tissue Culture: Applications in Medicine, Biology, and Agriculture" (E. Kurstak and K. Maramorosch, eds.). Academic Press, New York.

Clewley, J. P., and Kennedy, S. I. T. (1976). *J. Gen. Virol. 32,* 395-411.

Dougherty, R. M., Marucci, A. A., and Distefano, H. S. (1972). *J. Gen. Virol. 15,* 149-162.

Eagle, 1959. *Science 130,* 432-437.

Gliedman, J. B., Smith, J. F., and Brown, D. T. (1975). *J. Virol. 16,* 913-926.

Grimley, P. M., Berezeski, I. M., and Friedman, R. M. (1968). *J. Virol. 2,* 1326-1338.

Igarashi, A., and Stollar, V. (1976). *J. Virol. 19,* 398-408.

Mitsuhashi and Maramorosch. (1964). *Contr. Boyce Thompson Inst. 22,* 435-460.

Peleg, J. (1968). *Virology 35,* 617-619.

Peleg, J. (1969). *J. Gen. Virol. 5,* 463-471.

Raghow, R. S., Grace, T. D. C., Filshie, B. K., Bartley, W., and Dalgarno, L. (1973a). *J. Gen. Virol. 21,* 109-122.

Raghow, R. S., Davey, M. W., and Dalgarno, L. (1973b). *Arch. gEsamte Virusforschung 43,* 165-168.

Sarver, N., and Stollar, V. (1977). *Virology 80,* 390-400.

Singh, K. R. P. (1967). *Curr. Sci. 36,* 506-508.

Stevens, T. M. (1970). *Proc. Soc. Exp. Biol. Med. 134,* 356-361.

Stollar, V., Shenk, T. E., Koo, R., Igarashi, A., and Schlesinger, R. W. (1975). *In* "Pathobiology of Invertebrate Vectors of Disease," *Ann. N. Y. Acad. Sci. 266,* 214-231.

Stollar, V., Shenk, T. E., Koo, R., Igarashi, A., and Schlesinger, R. W. (1976). *In* "Invertebrate Tissue Culture: Applications in Medicine, Biology, and Agriculture" (E. Kurstak and K. Maramorosch, eds.). Academic Press, New York.

Thomas, V. L., and Stollar, V. (1974). *Proc. Electron Microscopy Soc. America.*

Chapter 19

ECOLOGICAL MARKERS OF ARBOVIRUSES

S. P. Chunikhin, and T. I. Dzhivanyan

I. INTRODUCTION

The investigation of the role of arthropods and vertebrates in the population variability of transmissible viruses (arboviruses) is just at its beginning. Until very recently, studies in this direction were hindered by the lack of markers of arboviruses associated with vectors and vertebrates. We suggest that such markers be called ecological, since with the help of these markers ecological aspects of the population variability of viruses may be studied. Elucidation of the regularities and causes of the variability of arboviruses upon their multiplication in arthropods and vertebrates is important for the understanding of the formation of new antigenic types, as well as for the search of naturally attenuated strains. The latter is hindered by the elimination of the attenuated strains in the process of transmission, the factor of elimination being the natural virus vectors (Chunikhin, 1973).

Another cardinal problem of the ecology of arboviruses, the resolution of which is prevented by the lack of ecological markers of these viruses, is the formation of their transmissive cycles. It seems obvious that many aspects of this problem are resolved with regard to trophic specialization of virus vectors (Chunikhin, 1973). Not all the facts, however, can be confined within the framework of this resolution. The question why only some members of the groups of blood-sucking arthropods are specific vectors of viruses remains open. The

ISBN 0-12-429765-X

analysis of the relevant evidence suggests that only members of the most ancient groups of hematophagous arthropods are virus vectors, such as ticks of the suborder *Mesostigmata and* Diptera *Orthorrhapha*. No virus vectors have been found in the groups of evolutionally less ancient blood suckers such as lice, fleas, bugs, and Diptera *Cyclorrhapha*.

II. ECOLOGICAL MARKERS

Experimental investigation of the groups, at the level of the order or suborder, of specific and nonspecific vectors showed that in the majority of instances the nonspecific vectors do not get infected in the natural way, i.e., upon bloodsucking. Since generally no difficulties arise in the selection of vector-virus pairs upon parenteral inoculation, it was suggested that one of the mechanisms of this specificity lies in the intestinal barrier. However, other mechanisms of specificity have also been demonstrated. Thus, numerous experiments using cultivation of mosquito-borne viruses and viruses borne by all the other vectors (ticks, midges, sandflies) in primary and continuous mosquito tissue cultures demonstrated that the former viruses always developed in these tissues whereas the latter only as rare exclusions. No markers of this "tissue" adaptation have been found yet. The only ecological marker of tick-borne encephalitis (TBE) virus described thus far is the DS marker, which determines the stability of the plaque diameter of this virus to dextran sulfate (DS) depending on the species of the tick vector (Dzhivanyan *et al*., 1974). Studies of the DS marker in 47 strains of TBE virus isolated from *Ixodes persulcatus* and *Ixodes ricinus* ticks showed some strains to be resistant to dextran sulfate (DS$^+$) and some to be sensitive to it (DS$^-$). The former group of strains from *I. ricinus* and the latter from *I. persulca* groups of strains were characterized by large plaques (4-6 mm) under the agar overlay in the absence of dextran sulfate. On the other hand, addition of DS to the agar overlay reduced the plaques produced by the strains isolated from *I. persulcatus* to about 1-2 mm in diameter, while those produced by the strains isolated from *I. ricinus* did not differ in size from the controls. In addition to these two groups of strains (DS$^+$ and DS$^-$), a third group was found, the plaque size of which was less than 1 mm both in the presence and absence of DS. This variant of the TBE virus was designated as "pin-point" (S^{PP}). In contrast to the first other two groups of strains, this group was not associated with any definite source of isolation, species of ticks or vertebrates.

On the basis of these results two hypotheses can be postulated on the role of Ixodid ticks in the formation of *ricinus* and *persulcatus* variants of the TBE virus. The first hypothesis suggests that the Ixodid ticks play the role of the selective factor; the second assumes both the selective and mutagenic functions of Ixodids in this process.

These hypotheses were examined using various Ixodid tick species as a laboratory model. The experiments were carried out in *Hyalomma plumbeum* tick, which does not occur in natural foci of TBE virus infections and thus cannot take part in the transmissive cycles of this virus. If one assumes that during the replication of TBE virus in *I. ricinus* or in *I. persulcatus*, the DS^+ or DS^- variants gain a selective advantage, respectively, then *H. plumbeum* as a species equally far from these tick species should not play the role of the selective factor.

The first part of the experiments was carried out with two TBE virus strains isolated from *I. persulcatus* ticks collected in Estonia (EK-328) and from *I. ricinus* ticks collected in the Vologda region (VK-60). In the second part of this udy, two clones of the EK-328 virus strain (clones 718 and 76,) were used.

Prior to the experiments the strains had undergone three passages in mouse brains. The experiments were carried out in 1500 female and 400 male *H. plumbeum* ticks of one to six laboratory generations. The virus was passaged from tick to tick by parenteral inoculation into the central part of the alloscutum, and it was passaged from tick to mouse and from mouse to mouse by the intracerebral route. The material for the first passage was the supernatant of the brain suspension from TBE virus-infected white mice, and for subsequent passages a suspension of the infected ticks was used. Ticks were inoculated with a tuberculine syringe equipped with a gauge 1 needle. The volume of the inoculum was 0.001-0.008 ml with a virus dose of 1.8-2.5 log LD_{50}/0.03 ml. A plaque method as modified by Dzhivanyan and Lashkevich (1970) was used.

Both TBE virus strains produced plaques in SPEV-44 cell cultures at 4-5 days; by 9-12 days these plaques reached their maximum size of 5-6 mm. In the presence of DS, the size of the plaques produced by the EK-328 strain did not exceed 1-2 mm (DS^-), while that of the plaques produced by the VK-60 strain was 5-6 mm, i.e., did not differ from that of the control (DS^+). Both strains underwent 11 passages in *H. plumbeum* ticks, and the EK-328 strain had one additional passage in *I. ricinus* ticks. The intensity of replication of both strains was approximately similar: after 14 days of incubation in ticks at 24-25°C the virus titers increased from 1.8-2.5 to 2.9-4.8 log LD_{50}/0.03 ml (4.3-6.8 log pfu/ml). At 14-70 days after inoculation, the virus titers in ticks remained stable or changed within statistically insignificant limits. Thus,

H. plumbeum tick, which is not the natural vector of the TBE virus, has proved to be a convenient laboratory model for the study of the variability of different TBE virus strains by the DS marker.

Passages of both strains in mouse brains for 20 times resulted in no change by the DS marker. However, passages of the same TBE virus strains in Ixodid ticks revealed heterogeneity in the EK-328 strain by the DS marker. Beginning with the eighth passage, this originally homogeneous strain started showing DS-resistant plaques (DS^+) the number of which increased with successive passages. By the twelfth passage the DS-resistant portion of the strain was 26% (Table I).

Passages of the DS-mixed strain EK-328 (after 12 passages in ticks) in mouse brains resulted in the elimination of the DS^+-element foreign for this strain. In 6-16 mouse brain passages the number of DS-resistant plaques decreased to 1.4% instead of the 26% observed after 12 tick passages.

Clones 718 (DS^-) and 767 (DS^+) isolated from the EK-328 strain had 20 passages in *H. plumbeum* and mouse brain each. Passages of these clones in mouse brains resulted in no changes of these clones in their DS-characteristics. On the other hand, passages of clone 767 (*ricinus* type) in *H. plumbeum* produced phenotypic modifications in plaque size; one passage in mouse brain was sufficient to make this clone uniform by the DS-marker.

Similarly, passages of clone 718 (*persulcatus* type) in *H. plumbeum* ticks resulted in the appearance of pin-point plaques from the thirteenth passage onwards, which by the eighteenth passage completely eliminated the original variant. This pin-point variant was found to have two new characteristics differing from the original: (1) it lost its pathogenicity for mice by the peripheral routes, and (2) it lost its hemagglutinating properties. Passages of this pin-point clone (S^{PP}) in mouse brains led to the reappearance of the plaques of the original type; by the fifth passage their portion was 10%.

III. CONCLUSION

The results of the studies on the transformation of the EK-328 strain and its clones as revealed by the DS-marker indicate that passages in ticks result in the appearance of new elements in the strain or clones, whereas passages in mice lead to the elimination of these elements.

Under the appropriate experimental conditions, the DS^--strain of the TBE virus (EK-328) yielded two other variants:

TABLE I
Variability of the EK-328 TBE Virus Strain by the DS-marker in Passages in H. plumbeum *Ticks*

Passage no.	*Incubation of virus in ticks (days)*	*Plaque numbers*[a]		*Virus titers*	
		DS^-	DS^+	*log pfu/ml*	*log* LD_{50}*/0.03 ml*[b]
1	14	38/100	0/0	5.3	nd
	26	22/100	0/0	5.3	4.0
	71	63/100	0/0	5.1	3.3
2	13	56/100	0/0	4.7	4.3
	21	51/100	0/0	5.7	2.9
3	21	39/100	0/0	3.5	nd
4	23	36/100	0/0	3.9	1.5
5	14	25/100	0/0	3.3	1.8
6	14	103/100	0/0	5.0	4.0
8	14	24/89	3/11	4.4	3.7
	31	25/92	2/8	5.8	nd
9	20	35/83	7/17	6.3	nd
10	14	64/80	13/20	6.0	nd
11	14	74/79	20/21	6.8	nd
10 passages in *H. plumbeum* and 1 in *I. ricinus*	14	32/74	11/26	5.4	nd

[a]*Numerator, number of plaques; denominator, %.*
[b]*nd, Not determined.*

DS^+ (*ricinus* type) and S^{PP} (pin-point type). The fact that the transformation of the *persulcatus* variant into the *ricinus* variant did not require the participation of *I. ricinus* tick indicates a selective rather than mutagenic role of ticks. On the other hand, the appearance of the pin-point variant of the virus suggests also mutagenic function of ticks in transmissive cycles of this virus.

References

1. Chunikhin, S. P. (1973). *Trans. USSR. AMS Inst. Poliomyelitis Virus Encephalitides 21,*(1), 7-89 (in Russian).
2. Dzhivanyan, T. I., and Lashkevich, V. A. (1970). *Vopr. Virusol. 4,*395-399.
3. Dzhivanyan, T. I., Lashkevich, V. A., Chuprinskaya, M. V., Bannova, G. G., and Sarmanova, E. S. (1974). *Trans. USSR AMS Inst. Poliomyelitis Virus Encephalitides 22,* (2), 79-86 (in Russian).

Chapter 20

RAPID DIAGNOSIS OF ARBOVIRAL AND RELATED INFECTIONS

Jordi Casals

I. INTRODUCTION

The term diagnosis is used here as equivalent to specific serological diagnosis. At the present time, knowledge of physicochemical, structural, and morphological properties of a virus places it in a family or genus but does not identify the serotype.

Early and rapid diagnosis of disease is always desirable, if for no other reason because its knowledge may influence the prognosis.

With arbovirus infections early diagnosis will, in addition, determine to a considerable degree the course of action to be taken for protecting hospital staff and the other patients as

ISBN 0-12-429765-X

well as the community at large against the virus. As a general precaution, the patient will be placed in a room to which arthropod vectors have no access; with specific pathogens additional precautions will be taken. If the diagnosis is yellow fever in an area in which mosquito vectors are known to exist, vaccination of the community may be necessary; if the diagnosis is Crimean-Congo hemorrhagic fever (CCHF), special care will be taken--including isolation of the patient--in order to minimize the risk of nosocomial infections due to contact with the patient's blood; in cases of Venezuelan equine encephalitis (VEE) it may be advisable to take measures to prevent droplet spread to other individuals, as the virus can be found in throat washings; if the diagnosis is Central European tick-borne encephalitis (CETBE) and there is no evidence of a tick-bite, the possibility of transmission through virus-contaminated goat or cow milk should be looked into. As there is currently no specific treatment for any of the arbovirus infections, knowledge of the etiology will have no influence on treatment except on the type and extent of supportive measures; but it may be otherwise if and when antiviral substances are discovered.

Infections with at least one of the arenaviruses and with the Marburg-type agents constitute a situation in which rapid diagnosis is not only desirable but absolutely essential. Lassa, Marburg, and Ebola fevers, and on occasion also Bolivian hemorrhagic fever (BHF), are notorious for their tendency to spread by person-to-person contact in which urine, feces, blood, and throat secretions are variously involved. While it now appears that very close personal contact is necessary for contagion with Ebola fever, the fact remains that isolation of the patients is essential in these infections in order to prevent or minimize transmission to other patients and hospital staff, and to persons in the family and the outside community. Furthermore, international air travel has often been involved with Lassa fever patients; either an expatriate patient suspected of having the disease wishes to go home, or an unsuspected case arrives in a Western or nonendemic country and is shortly after arrival recognized as Lassa fever. In addition, early and rapid diagnosis is important in these infections because it has a direct bearing on the patient's treatment; convalescent plasma is advised and has been given to patients with Lassa and Ebola fevers. While the effect of plasma on the course of the disease is still to be fully established due to an insufficient number of observations, neverthless in view of the severe prognosis its use is strongly advocated.

The measures to be taken for the management of arbovirus infections are, under the most favorable circumstances, inconvenient to the patient and to public health personnel, to say the least. Under less favorable circumstances or in cases of Lassa, Marburg, and Ebola fevers, the measures required to save the patient and for preventing possible great loss of human life are costly in money, manpower, and convalescent plasma, which is often in short supply; furthermore, administration of plasma as a therapeutic measure may be more effective when given early in the disease. For these reasons, to implement costly measures as soon as possible when needed, but also in order not to waste resources and disrupt the life of the community unnecessarily, an early and rapid diagnosis is important.

Few of the arbovirus and arenavirus diseases of man mentioned above are likely to be seen in Arctic and sub-Arctic areas. There are exceptions, however, the main one being CETBE, which is endemic in some places in high latitudes; in addition, sporadic imported cases or, more often, suspected cases of Lassa fever can turn up nearly anywhere, as occurred in 1976 in Canada (Toronto) and Australia (Melbourne), and in 1977 in Scotland (Dundee). An important cause of human virus encephalitis in U.S. is LaCrosse virus, a member of the California antigenic group; early diagnosis in order not to apply uncalled for exploratory or therapeutic measures is advocated in this instance. It remains to be seen whether similarly urgent diagnoses will be needed with other viruses of the same antigenic group, snowshoe hare and Inkoo, which have been active in latitudes up to 70^{0}N; or with lymphocytic choriomeningitis (LCM), an arenavirus found in association with the house mouse and not infrequently with pet or experimentally used hamsters.

Another important disease to be borne in mind in the present context is hemorrhagic fever with renal syndrome (HFRS). Studies in human volunteers nearly 40 years ago showed that the disease is caused by a filtrable agent, presumably a virus. Possibly the virus may soon be maintained in the laboratory and antibodies in man against the virus may be assayed; the fact remains that for the present there is no laboratory adapted strain generally available and consequently no diagnosis, rapid or slow. The nature of the HFRS virus is, of course, unknown; on ecological, epidemiological, and to some extent clinical grounds the disease closely resembles an arenavirus infection of man. HFRS occurs, among other places, in sub-Arctic areas of the Old World and benefit would be derived from a rapid diagnostic method for better management of suspected cases and for quick recognition of foci of infection; knowledge of the latter would decide the application of measures for containment or destruction of the local rodent reservoir involved.

II. BASIS FOR DIAGNOSIS

Identification of the arbovirus or arenavirus that causes disease in a patient and demonstration of antibody development between early and late phases of the disease or convalescence are the foundations for specific diagnosis. Conventional procedures for virus identification involving isolation and propagation in a susceptible host system and subsequent typing by established serological techniques, and determination of antibodies in early and late phases, are not conducive to rapid diagnosis. (Table I).

A. Virus Identification: Conventional Methods of Diagnosis

In favorable circumstances, with viruses such as eastern and western equine encephalitis (EEE, WEE) and VEE, which have an incubation period of between 1 and 2 days in experimental hosts, it is possible by inoculation of viremic serum to newborn mice or cell cultures to have an isolate in 24 hours and an identification of its serotype in an additional few hours to one day. The time can be shortened by inoculation of different mixtures of the suspect viremic serum with each of several distinct immune reference reagents: isolation and typing would in this manner be accomplished in one operation completed in one or two days. This, however, is not the rule since most arboviruses and arenaviruses require from several to many days before their effect on inoculated animals or cell cultures is noticeable.

B. Virus Identification: Rapid Methods of Diagnosis

The fastest diagnosis is accomplished by seeing the virus and detecting its antigens directly in clinical specimens from the patient: blood, other body fluids, cells, tissues, excretions, and secretions. Visualization of a virion is not enough for establishing a specific diagnosis, which only a serological reaction can do. Some viruses have, however, such distinctive morphology that it is nearly pathognomonic, particularly when given the right epidemiological circumstances.

Should detection of the virus in clinical materials fail, isolation and identification by conventional methods, i.e., inoculation of susceptible animals or cell cultures, can be shortened by from one to several days by detecting the virus or its antigens in the inoculated animals before signs of disease develop or, preferably, in cell cultures before cytopathic effects (CPE) are noted; CPE may not even develop in the cell

TABLE I
Examples of Methods for Rapid Diagnosis of Virus Diseases of Man

Objective	*Technique*	*Diseases and viruses*
Virus, visualization in clinical specimens; or after death	Immunofluorescence (IF)	Respiratory viruses, (RSV, influenza, parainfluenza, adenoviruses), rabies, smallpox, CFT, Lassa
	Electron microscopy (EM)	Smallpox, Lassa, Ebola, gastroenteritis of children (Orbivirus-like)
	Immune electron microscopy	Hepatitis B, gastroenteritis of adults (Norwalk agent)
	Immunoperoxidase	Cytomegalovirus, herpesvirus types 1 and 2
	Radioimmunoassay (RIA)	Hepatitis B, herpesvirus types 1 and 2
Virus, visualization after isolation	IF with:	
	Insect cell cultures	Dengue, Chikungunya
	Inoculated mosquitoes	Dengue
	Vertebrate cell cultures	Lassa, Marburg, Ebola
	Foci of infection, cell cultures	Rabies, rabies-related
	EM with vertebrate cell cultures	Marburg, Ebola
Antibodies, determination	Immunofluorescence	Lassa, LCM, AHF, BHF, CCHF, CTF, and others
	Radioimmunoassay	LCM, Hepatitis B, herpes, rubella, mumps and others
	Enzyme-linked immunosorbent assay (ELISA)	Parasitic diseases; few viruses

cultures but the viral antigen may be detectable, for example, by immunofluorescence (IF) as infected cells or as foci of infection (Smith *et al.*, 1977).

C. Antibody Identification: New Methods and Improved Conventional Methods

Detection of antibody development or change is a delayed method of diagnosis; however, if virus visualization or isolation has not been attempted or has been unsuccessful, serological techniques are used for a diagnosis. The speed with which a diagnosis is reached depends on the type of antibody searched and the test used. A given test may be more sensitive than another in that it detects smaller amounts of antibody or earlier in the course of the infection, as is the case, for instance, between the neutralization and the complement-fixation (CF) tests with togaviruses, yellow fever, dengues, WEE, and others; on the other hand, the neutralization test takes longer to complete than the CF test.

The tests considered as conventional with the arboviruses and arenaviruses are the neutralization and plaque reduction tests, the hemagglutination-inhibition (HI) test, when feasible, and the CF test; it is generally observed that with most arbovirus diseases of man antibodies are first detected by neutralization test, followed within a few days by the HI test, and within a few days again by CF antibodies.

Newer methods seek to increase the sensitivity for antibody determination over conventional procedures, to decrease the length of time after infection for antibodies to be detectable, to shorten the time required for the execution of the test, and in the case of antigenically related viruses, to increase the specificity within the antigenic group. Furthermore, newer techniques in order to be of any practical use must be relatively simple, reproducible, and applicable to large numbers of sera. The relative sensitivity, specificity, serotype specificity within antigenic groups, and speed of identification by different serological tests with the arboviruses and arenaviruses infections of man are aspects that require a great deal of additional investigation.

III. METHODS FOR RAPID DIAGNOSIS: VIRUS IDENTIFICATION

A. Visualization of the Virus in Clinical Materials

1. Immunofluorescence. This method, which had long been employed for the diagnosis of smallpox by staining vesicle fluid or scrapings from dried pustules, is being used increasingly for the rapid diagnosis of viral diseases, including respiratory infections--respiratory syncythial virus, influenza, parainfluenza, adenoviruses--by staining pharyngeal secretions, throat swabs, or sputum; with measles, using urine sediments and nasopharyngeal secretions; to recognize and type herpesviruses; and in rabies, with corneal impressions, mucosal scrapings, and skin and brain biopsy materials. Among arboviruses, the technique has been successfully applied to Colorado tick fever (CTF) by Emmons and Lennette (1966) by staining smears of blood clots from patients.

Much effort has gone into attempts to improve the early diagnosis of Lassa fever; by conventional methods the virus can be isolated as early as the third day of illness from blood, urine, or throat washings (Monath and Casals, 1975). Attempts to detect the virus by IF directly in urinary sediments and the buffy coat have given inconclusive results (Wulff and Lange, 1975); on the other hand, detection of the viral antigen by IF in conjunctival scrapings from patients with prominent conjunctivitis appears to have been successful in a high proportion of recently investigated cases (K. M. Johnson and R. O. Cannon, 1977, personal communication).

2. Electron Microscopy. The diagnosis of smallpox was lately routinely made by electron microscopy (EM) of vesicle or pustule material (Cruickshank *et al.*, 1966); the morphology of the virion and the clinical circumstances together are pathognomonic.

A striking application of EM to the detection of a virus in clinical materials in the absence of actual virus isolation has been reported by Bishop *et al.* (1974). Acute nonbacterial gastroenteritis of children is a common intestinal disease of infants and young children, caused by an orbivirus-like virus, a rotavirus; the virus can be seen in specimens from the patient directly by EM of negatively stained extracts of feces, or in intestinal mucosa biopsy specimens.

EM has been used for rapid diagnosis in cases of Sudan-Zaire hemorrhagic fever (Ebola virus) employing liver fragments taken shortly after death (Johnson *et al.*, 1977); and in one case of Lassa fever by percutaneous liver biopsy also shortly after death (Winn *et al.*, 1975). The morphology of these viruses is so characteristic that their recognition as Marburg-like and arenavirus, respectively, is nearly indisputable;

the serotype, however, requires serological determination. Biopsy liver material taken early in the course of illness could probably give a rapid diagnosis by EM in these two infections; the procedure, however, is not advocated as it might result in serious threat to the patient's life through hemorrhage. It has been recommended for diagnosis of Ebola virus infection that blood samples from the patient be taken early, fixed in glutaraldehyde, and observed under the electron microscope (Johnson *et al.*, 1977); it is not clear at this time whether this procedure has been used successfully in man or has only been suggested; in experimentally infected guinea pigs it has revealed the Marburg agent, a similar virus (Peters *et al.*, 1971).

EM of other arenaviruses than Lassa or of arbovirus diseases of man has not been used as an early diagnostic procedure; even if it were successful, the serotype would have to be determined for a complete diagnosis.

3. Immune Electron Microscopy. The method permits in one operation to make visible the virus particle and to obtain the specific diagnosis as well; it has been successfully used for the rapid diagnosis of rubella and hepatitis B. More recently this technique made possible a specific diagnosis through visualization of virions associated with an acute infectious nonbacterial gastroenteritis of man, the Norwalk agent, in the absence of virus isolation in a laboratory host, (Kapikian *et al.*, 1972). The virus is detected in clarified extracts of feces incubated with a convalescent serum as source of antibody; the pelleted mixture is collected on a grid and negatively stained. In positive cases the agent appears as clusters of agglutinated virions with a characteristic halo surrounding them.

This method could conceivably be used with arenaviruses and arboviruses employing as source of antigen blood, urine, or throat washings, depending on the virus; Fauvel and associates (1977) have applied the technique to the detection and specific identification of several arboviruses, EEE, WEE, Mayaro, chikungunya, Powassan, California encephalitis, and vesicular stomatitis, grown in cell cultures.

4. Immunoperoxidase. Improvements in the technology for preparation of enzyme-labeled antibodies, according to methods and principles similar to those of IF, have made the use of enzymes as tracers in serological reactions a practical method for virus diagnosis (Kurstak and Kurstak, 1974; Kurstak *et al.*, 1977). Highly purified horseradish peroxidase can be commercially obtained; as with IF, the enzyme can be conjugated with a specific antibody for use in a direct test or with antigamma-globulin antibodies for an indirect test. An advantage of the immunoperoxidase technique is that the product resulting from the catalytic activity of the peroxidase can be seen with an ordinary light microscope and by EM as well. Viruses with which this technique has been successfully employed include cytomegalovirus and herpesvirus types 1 and 2.

5. *Radioimmunoassay.* Radioimmunoassay (RIA) is a relatively new technique that was developed and applied to the quantitation of very small amounts of biologically active substances (Yallow and Berson, 1960). In virus work, it has been extensively used in the form of solid phase assay (SPRIA) for the detection of Australian antigen, HB_sAg (Ling and Overby, 1972) and shown to have a much greater sensitivity than most other tests including CF, immunoelectroosmophoresis, and indirect hemagglutination (Peterson *et al.*, 1973). Heretofore, the technique has been used for the study of basic problems relating to togaviruses (Dalrymple *et al.*, 1972; Trent *et al.*, 1976) rather than for their rapid diagnosis in clinical cases. The technique requires reagents--antigens and antibodies--relatively pure and a dependable supply of radiolabeled compounds; a recent technical development that may promote the use of RIA is the availability of a new radiolabeled iodine compound, ^{125}I radioligand, which is more stable, easier to prepare, and easier to combine than those previously used (Bolton and Hunter, 1973).

The assay has been successfully applied to the identification of herpesvirus types 1 and 2 in specimens of human brain obtained at autopsies (Forghani *et al.*, 1974); these authors also suggested the possibility of using the technique for detecting and typing herpesvirus directly in vesicular lesion materials and leukocytes. Clinical and laboratory observations have suggested or shown that virus titers are elevated in the blood of patients in the early phase of CCHF and Lassa fever; SPRIA may, therefore, be useful for an early diagnosis of these diseases using blood as source of antigen, possibly also other clinical specimens.

B. Visualization of the Virus after Isolation from Clinical Materials

Isolation and identification of a virus either by conventional or by new techniques remains the mainstay of diagnostic virology, even though it may not be the most rapid way. The procedures described in this section consist of the application of various steps and methods, conventional or new, to the diagnostic problem, with a view to expediting the solution.

If a virus has not been detected and identified directly in clinical specimens, inoculation of these specimens to susceptible hosts is the next step; procedures have been developed in recent years with a view to increasing the sensitivity of isolation attempts and reducing the required time, some of which are listed in Table I. The new techniques used for identification of the virus are mainly IF and EM; immunoperoxidase and RIA could also be applied at this phase of the problem but they have been used little or not at all with arboviruses and arenaviruses. Due in great part to the ecological properties of the viruses and to their cell culture affinities, the procedures for isolation and identification described here have been used mostly with arboviruses, arenaviruses, and the Marburg-type agents; they may or may not be applicable to other viruses.

1. Inoculation of Insect Cell Cultures. Cultures of *Aedes albopictus* cells and to a certain extent also those from *A. aegypti* support replication of arboviruses from different taxonomic groups, including all the viruses that have mosquitoes as natural vectors; the susceptibility of these cells to arboviruses other than mosquito-borne is associated in part with the resistance or susceptibility of the virus to sodium deoxycholate (Buckley *et al.*, 1976). The virus replicates in most instances with no observable cytopathic effect (CPE) but it is revealed by subinoculation to cultures of vertebrate cells, in which the virus shows CPE, to newborn mice, or by testing the mosquito cell cultures as an antigen in the CF test (Pavri and Ghose, 1969; Singh and Paul, 1969; Ajello *et al.*, 1975). Daily staining of inoculated monolayers with a conjugated antiserum and observation by IF could possibly shorten the time needed for a specific diagnosis. *A. albopictus* cells in cultures have been used successfully for the identification of chikungunya, West Nile, and dengue viruses; on the other hand Lassa virus does not multiply in these cells (Buckley and Casals, 1970). Additional invertebrate cell lines are being successfully investigated as support for arbovirus replication (Pudney *et al.*, Chapter 16, this volume).

Mosquito cell lines have the advantage that they can be maintained at ambient temperatures, from 20 to 30°C, for up to 2 or 3 weeks with an occasional change of maintenance medium; viruses to which the cells are susceptible will replicate in cultures held at that temperature. These characteristics make the cultures useful in field conditions where access to incubators is limited or unreliable.

2. Inoculation of Mosquitoes. Certain dengue strains, particularly strains of dengue type 3, are difficult or impossible to isolate by established methods, newborn mouse inoculation, or plaquing on LLC-MK2 cell monolayers. Intrathoracic inoculation of virus-containing materials to *A. albopictus* mosquitoes has proved to be a method of far greater sensitivity than conventional ones (Rosen and Gubler, 1974; Kuberski and Rosen, 1977a); a diagnosis is made in cases when other methods fail and in all instances it reduces the time needed. Inoculated mosquitoes are killed 6 or 7 days after inoculation, their heads are squashed on a microscope slide, the slides are dried for a few minutes at room temperature, fixed with acetone, stained with a conjugated antidengue polyvalent serum, and observed in a fluorescence microscope (direct IF). Positive specimens are identified as to dengue type by CF with specific sera, using extracts in borate buffered saline of one or several whole mosquitoes (Kuberski and Rosen, 1977b). More recently, *Toxorhynchites amboinensis* mosquitoes have been used (Kuberski and Rosen, 1977a). The advantages of using these mosquitoes are that they are larger than *Aedes*, hence easier to inoculate; and they do not bite man, therefore both males and females can be used with no hazard involved should an insect escape.

3. *Immunofluorescence of Inoculated Cell Cultures.* Identification of Lassa virus has been expedited by IF staining of Vero cell cultures inoculated with suspect materials, examined before CPE was apparent. In one instance of Lassa fever the diagnosis was made 24 hours after inoculation of the cell cultures (Wulff and Lange, 1975); in most cases it could be made in 3 days. Short of recognizing the virus in the clinical specimen itself, this procedure appears at this moment to be the fastest for Lassa fever diagnosis.

Foci of infection on inoculated monolayers, detected by IF, perform the same function as plaques but with a shorter incubation (Rapp *et al.*, 1959). The assay has been successfully used with CER cell monolayers inoculated with rabies serogroup viruses, allowing a diagnosis 2 to 4 days after infection (Smith *et al.*, 1977). Detection of foci of infection by IF is particularly useful when inoculated monolayers do not develop plaques even though the virus replicates, as is the case with CCHF virus. Studies in this laboratory (G. H. Tignor, personal communication, 1977) have shown by IF foci of infection on CER monolayers with five strains of CCHF virus, in one instance two days after inoculation; since the foci can be inhibited by immune sera, the assay can be used not only for detection of virus but for that of antibodies as well.

4. *Electron Microscopy of Inoculated Cell Cultures.* Examination by EM of Vero cell cultures inoculated with specimens deriving from patients or victims of Marburg or Ebola fevers permitted a prompt diagnosis on the third or fourth days after inoculation (Gear *et al.*, 1977); the serotype diagnosis was made subsequently by IF.

IV. METHODS FOR RAPID DIAGNOSIS: ANTIBODY DETERMINATION

In the event that virus detection or isolation has failed or has not been attempted, a specific diagnosis of arbovirus or arenavirus infection can be arrived at by investigating antibody development during the course of the disease.

Among the conventional methods--neutralization, CF, and HI--neutralization, either by the intraperitoneal route of inoculation in newborn mice or as a plaque reduction assay, is generally the test with the highest sensitivity with most arboviruses. There are exceptions, one being CCHF virus, with which unless the sera are extracted with acetone and ethyl ether, nonspecific neutralization will result (Casals and Tignor, 1974; Buckley, 1974). A disadvantage of the neutralization test for rapid diagnosis is the length of time it takes before the result is known. With arenavirus diseases, particularly Lassa and LCM, the neutralization test in cell cultures is not being used much, apparently owing to inherent difficulties with the test. The HI test, with viruses that yield an antigen, is easy to perform and its result can be had in from a few to 24 hours; no agglutinating antigens are available for arenaviruses or for CTF. Complement-fixing antibodies are late in appearing in arenavirus and, although a little less, also in arbovirus diseases of man; this test is not for early diagnosis.

A. Immunofluorescence

This technique was used several years ago for serological diagnosis of CTF (Emmons *et al.*, 1969) and was shown to be more effective and faster than the CF test; more recently, it has been used with CCHF (Zgurskaya *et al.*, 1975), Bolivian hemorrhagic fever (BHF) (Peters *et al.*, 1973), Argentinian hemorrhagic fever (AHF) (Grela *et al.*, 1975), and Lassa fever (Wulff and Lange, 1975). With these diseases, IF has definitely given a more rapid diagnosis than conventional tests.

Technological developments of recent years have made the IF test a practical procedure for routine use in diagnostic problems and for seroepidemiological surveys. Incident fluorescent light microscopes are commercially available at reasonable prices. The advantage of this type of IF over the transmitted light type are several and important: more light is available, particularly for high power viewing; only the section of the specimen actually in the field of vision is excited, therefore no fading of fluorescence occurs over the entire specimen; no light of excitation wavelength is directly transmitted to the eye of the observer; and, since no immersion oil is necessary under the slide, far larger numbers of specimens can be easily and comfortably examined than was practical with transmitted light.

The development of tissue culture chamber-slides (Lab-Tek Products, Division of Miles Laboratories, Neperville, Illinois) with four or eight chambers has further facilitated the observation by IF of multiple samples by simplifying the preparation of infected monolayers and the staining. Of particular importance for seroepidemiological surveys has been the development of Teflon-coated slides (Cel-Line Associates, Inc., Minotola, New Jersey); the surface of these slides is covered with Teflon, a water-repellent substance, except on 12 circular areas 5mm in diameter. Drops of a suspension of cells infected with a virus, prepared with Vero, BHK-21, CER, LLC-MK2, or other cell cultures, are deposited on the circular areas, dried, and fixed with acetone; this type of preparation permits the simultaneous staining of 12 samples.

B. Radioimmunoassay

The assay has been used with arboviruses for detection of antigen rather than antibodies (Section III,A,5). With other viruses the technique has been employed for antibody determinations, particularly against hepatitis B antigen (Lander *et al.*, 1971); investigations about its possible diagnostic use in various diseases, rubella, mumps, measles, herpesvirus, and varicellazoster, have been reported (Forghani *et al.*, 1976). Blechschmidt and associates (1977) described the use of SPRIA for titrations of antibodies against LCM virus; the method undoubtedly offers possibilities as a tool for seroepidemiological surveys, but its general application with the arboviruses and arenaviruses needs further investigation.

C. Enzyme Immunoassays

An enzyme-labeled anti-immunoglobulin assay, similar to the immunoperoxidase assay (Section III,A,4) was developed by Engvall and Perlmann (1972) for quantitation of specific antibodies. The test, designated enzyme-linked immunosorbent assay (ELISA), has lately been widely applied for detection of antibodies in parasitic diseases such as malaria, African trypanosomiasis, Chagas disease, and toxoplasmosis (Voller *et al.*, 1976), as well as in a few virus diseases among which are measles, cytomegalovirus infection, and rubella (Voller and Bidwell, 1976). The assay can just as well be used for detection of antigen by using reference antisera. For the assay, indirect method, is used (Voller *et al.*, 1976) alkaline phosphatase conjugated with antihuman immunoglobulin; the antigen is adsorbed onto the surface of polystyrene microhemagglutination plates, and the unknown serum added followed after the required incubation by the conjugated enzyme. After another incubation, the substrate is added consisting of a solution of P-nitrophenyl phosphate; a color change follows with the end result being assessed visually or by spectrophotometry. While no reports have appeared as yet about the use of the test with arboviruses or arenaviruses, its simplicity and the fact that it does not require expensive equipment as is the case with IF, RIA, and EM, recommend the assay for immediate evaluation with these viruses.

D. Indirect Hemagglutination

Indirect methods of hemagglutination have been repeatedly described in the past. More recently, there has been a revival of interest in this technique as it applies to diseases caused by arboviruses. Glutaraldehyde-fixed sheep erythrocytes are conjugated with an antigen by means of tannic acid or bis-diazotized benzidine and on exposure to a serum containing antibodies the red cells agglutinate; the method has been recently used with Nairobi sheep disease (Davies *et al.*, 1976) and its use should be tested with other viruses. In another indirect type of hemagglutination, glutaraldehyde fixed sheep erythrocytes are sensitized or coated with immune gammaglobulin for a given virus: contact with the virus or its antigens results in agglutination. The technique has been recently applied to CCHF and CTF viruses by Gaidamovich and associates (1974). In the past, indirect hemagglutination using formalhehyde or tannic acid treated blood red cells had been faulted with a tendency to nonspecificity due to spontaneous agglutination of the cells; in view of the these newly reported results, however, the technique should be further explored with other arboviruses and arenaviruses.

V. FINAL COMMENT

Techniques are available for separation of viral structural and nonstructural proteins by physical means; selection of the subviral components as antigens and, by immunizing animals with them, as source for antibody preparation, reagents may be developed with greatly increased specificity in serological tests. Moreover, separation of IgM from other proteins particularly IgG by centrifugation in gradients, may increase the efficacy and speed of diagnosis based on the earlier appearance of the former in the course of an infection. Partially purified antigens and antibodies, as well as immunoglobulin fractions can be used in conjunction with nearly all the serological techniques described in this chapter.

Other tests may be applicable to rapid serological diagnosis in addition to those described here, such as reduction of foci of infection and immunoadherence; however, no information is available concerning their use as a rapid test with arboviruses and arenaviruses.

In appraising the techniques to use for prompt routine diagnosis of the viruses under consideration, it is important to bear in mind that the methods, in addition to being reproducible, accurate, sensitive, and relatively simple, must also be readily available in laboratories at the local level; these are the laboratories that are faced with critical and urgent diagnostic problems. It is hardly reasonable to advocate techniques that may not be available at these laboratories.

References

1. Ajello, C., Gresikova, M., Buckley, S. M., and Casals, J. (1975). *Acta Virol. 19*,441-442.
2. Bishop, R. F., Davidson, G. P., Holmes, I. H., and Ruck, B. J. (1974). *Lancet 1*,149-151.
3. Blechschmidt, M., Gerlich, W., and Thomssen, R. (1977). *Med. Microbiol. Immunol. 163*,67-76.
4. Bolton, A. E., and Hunter, W. M. (1973). *Biochem J. 133*, 529-539.
5. Bowen, E. T. W., Platt, G. S., Lloyd, G., Baskerville, A., Harris, W. J., and Vella, E. E. (1977). *Lancet I*,571-573.
6. Buckley, S. M. (1974). *Proc. Soc. Exp. Biol. Med. 146*, 594-600.
7. Buckley, S. M., and Casals, J. (1970). *Am. J. Trop. Med. Hyg. 19*,680-691.
8. Buckley, S. M., Hayes, C. G., Maloney, J. M., Lipman, M., Aitken, T. H. G., and Casals, J. (1976). *In* "Invertebrate Tissue Culture" (E. Kurstak and K. Maramorosch, eds.), pp. 3-19. Academic Press, New York.
9. Casals, J., and Tignor, G. H. (1974). *Proc. Soc. Exp. Biol. Med. 145*,960-966.
10. Cruickshank, J. G., Bedson, H. S., and Watson, D. H. (1966). *Lancet 2*,527-528.
11. Dalrymple, J. D., Teramoto, A. Y., Cardiff, R. D., and Russell, P. K. (1972). *J. Immunol. 109*,426-433.
12. Davies, F. G., Jessett, D. M., and Otieno, S. (1976). *J. Comp. Pathol. 86*, 497-502.
13. Deibel, R., Woodall, J. P., Decher, W. J., and Schryver, D. G. (1975). *J. Am. Med. Assoc. 232*,501-504.

14. Emmons, R. W., and Lennette, E. H. (1966). *J. Lab. Clin. Med. 68*,923-929.
15. Emmons, R. W., Dondero, D. V., Devlin, V., and Lennette, E. H. (1969). *Am. J. Trop. Med. Hyg. 18*,796-802.
16. Engvall, E., and Perlmann, P. (1972). *J. Immunol. 109*, 129-135.
17. Fauvel, M., Artsob, H., and Spence, L. (1977). *Am. J. Trop. Med. Hyg.26*,798-807.
18. Forghani, B., Schmidt, N. J., and Lennette, E. H. (1974). *Appl. Microbiol. 28*,661-667.
19. Forghani, B., Schmidt, N. J., and Lennette, E. H. (1976). *J. Clin. Microbiol. 4*,470-478.
20. Gaidamovich, S. Y., Klisenko, G. A., and Shanoyan, N. K. (1974). *Am. J. Trop. Med. Hyg. 23*,526-529.
21. Gear, J. H. S., Spence, I., Kirsh, M., Ryan, J., Bothwell, T., Gear, J., Gear, A., Cassell, G., and Davies, J. (1975). *M.M. Weekly Rep. 24*,89-90.
22. Grela, M. E., Garcia, C. A., Zannoli, V. H., and Barrera Oro, J. G. (1975). *Acta Bioquim. Clin. Lat. Am. 9*,141-146.
23. Johnson, K. M., Webb, P. A., Lange, J. V., and Murphy, F. A. (1977). *Lancet 1*,569-571.
24. Kapikian, A. Z., Wyatt, R. G., Dolin, R., Thornhill, T. S., Kalica, A. R., and Chanock, R. M. (1972). *J. Virol. 10*, 1075-1081.
25. Kubersky, T. T., and Rosen, L. (1977a). *Am. J. Trop. Med. Hyg. 26*,533-537.
26. Kubersky, T. T., and Rosen, L. (1977b). *Am. J. Trop. Med. Hyg. 26*,538-543.
27. Kurstak, E., and Kurstak, C. (1974). *In* "Viral Immunodiagnosis" (E. Kurstak and R. Morisset, eds.), Ch. 1, pp. 3-30. Academic Press, New York.
28. Kurstak, E., Tijssen, P. and Kurstak, C. (1977) *In* "Comparative Diagnosis of Viral Diseases. Human and Related Viruses" (E. Kurstak and C. Kurstak, eds.), Vol. II, Part B, pp. 403-448. Academic Press, New York.
29. Lander, J. J., Alter, H. J., and Purcell, R. H. (1971). *J. Immunol. 109*,834-841.
30. Ling, C. M., and Overby, L. R. (1972). *J. Immunol. 109*, 834-841.
31. Monath, T. P., and Casals, J. (1975). *Bull. Wld. Hlth. Org. 52*,707-715.
32. Pavri, K. M., and Ghose, S. N. (1969). *Bull. Wld. Hlth. Org. 40*,984-986.
33. Peters, C. J., Webb, P. A., and Johnson, K. M. (1973). *Proc. Soc. Exp. Biol. Med. 142*,526-531.
34. Peters, D., Muller, G., and Slenczka, W. (1971). *In* "Marburg Virus Disease" (G. A. Martini and R. Siegert, eds.), pp. 68-83. Springer-Verlag, New York-Heidelberg-Berlin.

35. Peterson, D. A., Froesner, G. G., and Deinhardt, F. W. (1973). *Appl. Microbiol.* *26*,376-380.
36. Rapp, F., Seligman, S. J., Jaros, L. B., and Gordon, I. (1959). *Proc. Soc. Exp. Biol. Med.* *101*,289-294.
37. Rosen, L., and Gubler, D. (1974). *Am. J. Trop. Med. Hyg.* *23*,1153-1160.
38. Singh, K. R. P., and Paul, S. D. (1969). *Bull. Wld. Hlth. Org.* *40*,982-983.
39. Smith, A. L., Tignor, G. H., Mifune, K., and Motohashi, T. (1977). *Intervirology* *8*,92-99.
40. Trent, D. W., Harvey, C. L., Qureshi, A., and LeStourgeon, D. (1976). *Infect. Immun.* *13*,1325-1333.
41. Voller, A., and Bidwell, D. E. (1976). *Br. J. Exp. Pathol.* *57*,243-247.
42. Voller, A., Bartlett, A., and Bidwell, D. E. (1976). *Trans. Roy. Soc. Trop. Med. Hyg.* *70*,98-106.
43. Winn, W. C., Jr., Monath, T. P., Murphy, F. A., and Whitfield, S. G. (1975). *Arch. Pathol.* *99*,599-604.
44. Wulff, H., and Lange, J. V. (1975). *Bull. Wld. Hlth. Org.* *52*,429-436.
45. Yallow, R. W., and Berson, S. A. (1960). *J. Clin. Invest.* *39*,1157-1175.
46. Zgurskaya, G. N., Chumakov, M. P., and Smirnova, S. E. (1975). *Vopr. Med. Virol.* (English summary), p. 293, Institute of Poliomyelitis and Viral Encephalitis.

Index

L

M

N

O

P

R

S

T